AF382863

CISM COURSES AND LECTURES

Series Editors:

The Rectors of CISM
Sandor Kaliszky - Budapest
Mahir Sayir - Zurich
Wilhelm Schneider - Wien

The Secretary General of CISM
Giovanni Bianchi - Milan

Executive Editor
Carlo Tasso - Udine

The series presents lecture notes, monographs, edited works and
proceedings in the field of Mechanics, Engineering, Computer Science
and Applied Mathematics.
Purpose of the series is to make known in the international scientific
and technical community results obtained in some of the activities
organized by CISM, the International Centre for Mechanical Sciences.

INTERNATIONAL CENTRE FOR MECHANICAL SCIENCES

COURSES AND LECTURES - No. 364

ADVANCED METHODS FOR GROUNDWATER POLLUTION CONTROL

EDITED BY

G. GAMBOLATI
UNIVERSITY OF PADUA

AND

G. VERRI
FRIULI VENEZIA GIULIA REGION

SPRINGER-VERLAG WIEN GMBH

Le spese di stampa di questo volume sono in parte coperte da
contributi del Consiglio Nazionale delle Ricerche.

This volume contains 145 illustrations

In order to make this volume available as economically and as
rapidly as possible the authors' typescripts have been
reproduced in their original forms. This method unfortunately
has its typographical limitations but it is hoped that they in no
way distract the reader.

ISBN 978-3-211-82714-7   ISBN 978-3-7091-2696-7 (eBook)
DOI 10.1007/978-3-7091-2696-7

Discovering, controlling and remediating groundwater and soil contamination is a problem of primary importance for the correct preservation, management and use of natural water resources. Pollutants introduced into the subsurface system can contaminate not only the soil (thereby damaging vegetation or making crops unfit for consumption), but also the atmosphere (through volatilization), acquifers (through percolation, leaching and recharge), and streams (through surface and subsurface runoff and seepage). Transported by groundwater, these hazardous substances may also contaminate withdrawal sites at pumping wells, and they may reappear at the surface, emerging from springs and seepage faces. The degradation of soil and water quality resulting from underground pollution can pose a serious risk to public health. Contamination can occur at point sources (e.g. isolated spills; leaking storage tanks; waste tailings from mining operations; sanitary landfills; septic tanks; radioactive waste disposal) or at nonpoint sources (e g. herbicides, pesticides, and fertilizers used in agriculture; urban runoff; sewage and waste water; atmospheric deposition including acid rain). The contaminants can be organics, trace metals, or radionuclides.

This volume addresses some of the issues concerned with groundwater pollution and constitutes the edited proceedings of the International Symposium on "Advanced Methods for Groundwater Pollution Control", held at CISM (Centre International des Sciences Mécaniques), Udine, Italy, in May 5-6, 1994. It represents a collection of invited review and research papers covering the most advanced methods for groundwater pollution analysis, prediction, control and remediation. Each contribution is authored by a leading scientist in the field and addresses a key issue in the area of modern detection, evaluation, modeling and cleanup of contaminated soils and aquifers. A total of 18 papers is provided which covers significant hydrological, geochemical, numerical and engineering aspects of the discipline and discusses new developments in both applied research and design technology to maintain sustainability of a vital resource (groundwater) which is continuously threatened by industrial, agricultural, urban and tourist contamination. It is directed to managers, professionals and researchers working in any of the areas concerned with the control, prediction and remediation of soil and groundwater contamination.

We greatly appreciate the financial support of the Friuli - Venezia Giulia Region without which both the programme of the Symposium and the publication of the Proceedings would not have been possible. We also wish to thank Dr. Paola Agnola for her helpful contribution to organizing the event in Udine and editing this book, and CISM for providing the facilities and giving the hospitality to the Symposium.

Giuseppe Gambolati
Giorgio Verri

CONTENTS

ARE WE DESCRIBING DISPERSION CORRECTLY?
SOME CONCERNS

I. Neretnieks

Royal Institute of Technology, Stockholm, Sweden

ABSTRACT

There are many observations that the dispersion length increases with the observation distance in fractured rocks. This is contrary to the common assumption on which the advection-dispersion equation is based and casts doubt on its usefulness in extrapolation to longer distances. There are several mechanisms which can cause the observed effect. Three such mechanisms are discussed. In systems with strong channeling i.e. where the flow paths are essentially independent until their waters are mixed in the collection well the dispersion of the collected waters from the different paths will depend on the velocity distribution along the path ways. The "dispersion length" evaluated for this system will be proportional to the distance. Another mechanisms which will have similar effects is the matrix diffusion in a dual porosity system. The tracer diffuses in and out of the stagnant waters in the rock matrix. The larger contact surface between the flowing water and the rock in the longer paths will increase the interaction and cause "dispersion" in addition to the hydrodynamic dispersion. A third cause can be that in a self similar system the dispersion length also will be self similar and be a constant fraction of the scale of observation. There are observations indicating that there are self similar structures in fractured rocks.

1. INTRODUCTION

It has been increasingly recognised that dispersion is not
always Fickian (Matheron and Marsily 1980, Neretnieks 1981,
Neretnieks 1983, Gelhar et al 1993.). Our investigations in
fractured crystalline rocks in the Stripa experimental mine,
together with results from other tracer tests, show that the
dispersion length increases with observation distance. Figure
1 shows a compiled results from tracer tests performed in
fractured rock, mostly in the Stripa mine and of some
laboratory tests performed on cores with axial fractures also
from Stripa.

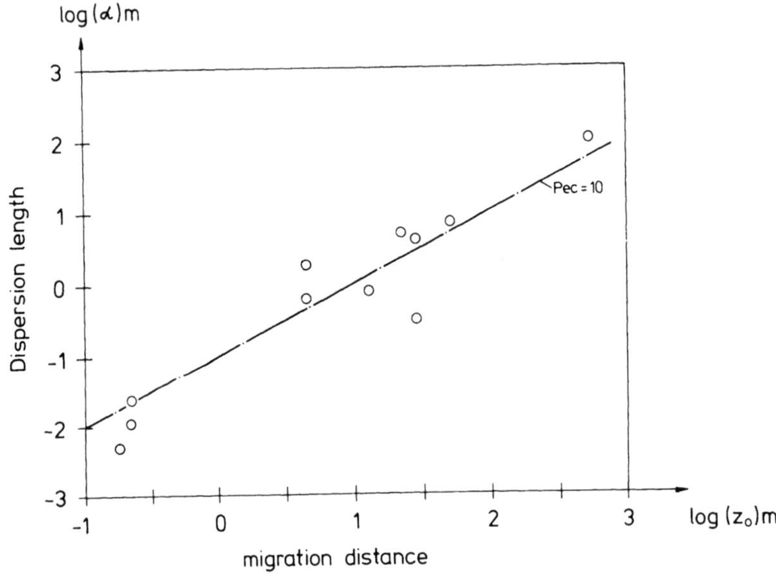

Figure 1. Dispersion length in fractured rock as function of
distance.

The Stripa tests were made over a period of 11 years in 5
different field and several laboratory investigations,
Neretnieks (1993). Gelhar (1993) has compiled and critically
reviewed a large number of tracer tests. Their compilations
also show that over a very large range of observation
distances, the dispersion length increases with distance.

It is, however, common practice to use the advection-
dispersion (AD) equation, with a constant dispersion length,
to analyse, simulate, and predict tracer transport in the

ground. For a given distance the advection-dispersion, the channeling the channel network and probably several other models can be made to adequately describe a tracer breakthrough curve. In the AD model a constant dispersion length is then typically used.

Assume that we choose a dispersion length α when simulating the tracer breakthrough curve at distance L along the flowpath. The commonly used analytical solutions have been obtained based on the assumption that α is constant all along our flowpath. With the same solution it is possible to "predict" the tracer breakthrough curve also at, say a distance L/10 along the flow path. This, however, is in violation of the observations that at distance L/10 the dispersion length is $\alpha/10$. We would thus obtain a different breakthrough curve, were we to use $\alpha/10$ to make predictions for the distance L/10. It may thus be concluded that the common analytical solution(s) based on a constant α cannot give a correct representation of the transport along the whole flowpath.

Let us then consider an alternative approach. Assume that α increases with distance from the point where the tracer is injected and downstream. This is readily incorporated in numerical schemes to solve the advection-dispersion equation. Some difficulties arise, however. To begin with the α-value(s) to use cannot be taken from the literature because these were evaluated using the equations which are based on α being independent of distance. This is probably a minor problem because data can be re-evaluated or a relation can probably be found by which the old α's can be transformed into the new, distance dependent α's.

We now proceed happily to make simulations with the distance dependent α's for the tracer, contaminant, solute etc. until we run across the case where the solute which enters at point A reacts with the solute which enters at point B, somewhere downstream. Are we to use different α's for the two solutes in the same point because solute 1 has travelled the distance Z1 and solute 2 has travelled Z2? What α should the reaction product have in a point along the flowpath? Another difficulty which has been encountered is when a solute flows through a porous medium, reacts with a stationary component to form a product P, which then is transported onward. Not only it is indeterminate what α to choose for P, and how to vary α along the path, but also if P should be allowed to move "upstream" by Fickian dispersion/diffusion. For a constant α and if the Peclet number L/α is small, meaning that dispersive transport is important compared to advective transport, the AD equation predicts that P would migrate upstream by dispersion. This is not the case. The reaction product(s) may migrate upstream by molecular diffusion but will not move up by dispersion. Also this problem could be circumvented by letting α be determined by molecular diffusion at the inlet and increase with distance. The problem, what to do with the different solutes, remains though.

The problems seem to become more pronounced when multiple solute transport and reactions are involved. Then it is very unclear what assumptions are reasonable regarding α. All programs (codes), known to me, which couple transport and chemical reaction inherently use the assumptions that α is constant, independent of distance although some codes can mimic a time or distance dependent α.

2   INDEPENDENT CHANNELS

There are known geometrical structures which have the property that the residence time distribution will give a dispersion length which increases with observation distance. Systems with independent pathways (Neretnieks 1983) have been used to model the transport of non reactive and reactive solutes.

Figure 2 shows such a system of independent pathways. Applied to flow in fractured rocks the assumption is that there is no mixing between the pathways until all their waters are collected in the sampling well. The flowrates and velocities in the different pathways are different. Neretnieks (1983) showed that in such a system the residence time distribution is such that dispersion length increases with distance. This is irrespective of the form of the velocity distribution function. For the simple case of a log normal distribution function an analytical relation can even be found between the dispersion length and the standard deviation σ of the log normal distribution.

Chesnut (1993) recently made an analysis of the dispersivity in heterogeneous porous media. He finds that the model based on the assumption of independent channels may provide more realistic results for modelling long-distance radionuclide transport than the current stochastic models which use some measure of dispersivity. He actually states that "the concept of dispersivity has outlived its usefulness and that field

data might well make more sense if they were used to determine
an effective σ rather than to calculate dispersivity."

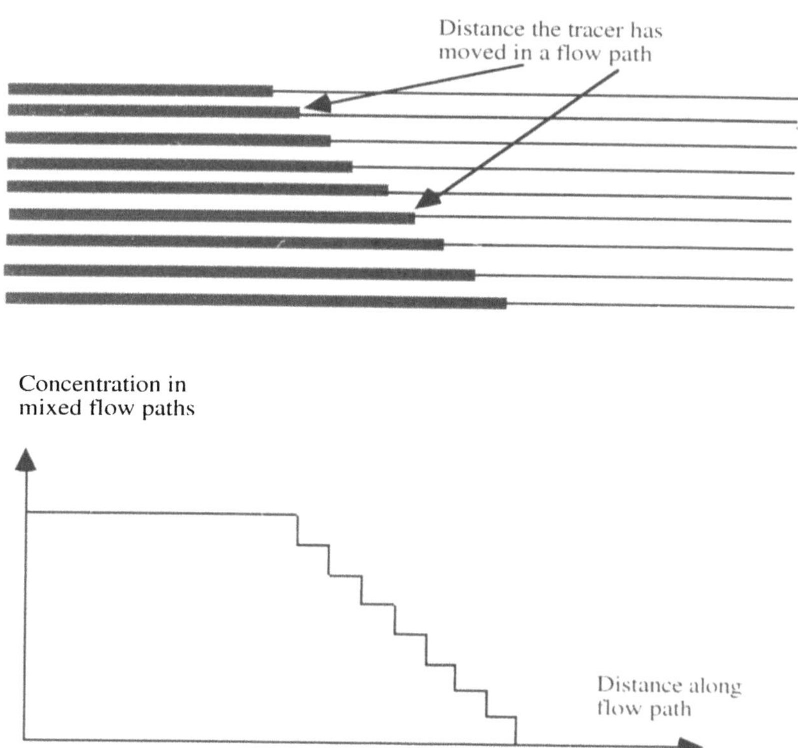

Figure 2. System of independent pathways and the resulting
concentration if the water were to be collected and
mixed at any point along the pathway.

3   SELF SIMILAR SYSTEMS

Systems with self similar structures have also been shown to
have the desired property that the dispersion length increases
with observation distance. Schweich (1993) gives an elegant
example of a simple system which has a self similar structure
as depicted in Figure 3. The basic unit consists of three
mixing cells, two in series and one in parallel to the former
two. Schweich starts by assigning each of the mixing cells the
same volume. This is called a generating pattern. Each mixing
cell can be exchanged for a set of three cells, together
having the same volume as the original cell. In this way the
total volume of the system is preserved. The process can be
repeated to obtain more and more embedded cells. He then

applies the method transfer function concept used in control
engineering or chemical engineering and the Laplace transform
to obtain a transfer function for the system of embedded
mixing cells. Already for five levels of embedded cells the
residence time distribution deviated little from that of a
system with 50 repeated embeddments. Not surprisingly it is
found that for this system the dispersion length is a constant
fraction of the scale studied.

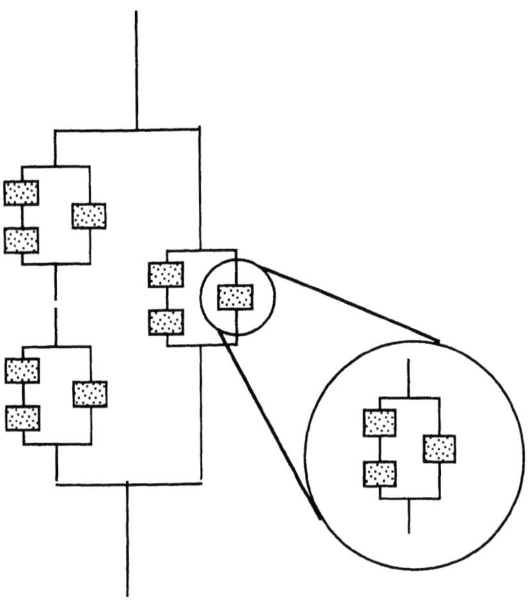

Figure 3. Example of a self similar structure

It has been noted that fractured rock systems seem to have
self similar properties over a large range of scales.
Neretnieks (1993) noted that in the Swedish site
investigations (KBS-3, 1983) there seems to be an ever
increasing scale of fractures and fracture zones of the
observer zooms out (or in) like looking at maps with ever
increasing scale. Somewhat oversimplified the impression is
that if one looks at the granite at a scale of 10-50 m one
sees the individual fractures with spacings of one or a few
per meter at most. Zooming out to look at the scale of several
hundred meters small fracture zones are seen with widths of
less than a meter and with a higher fracture density than the
"good" rock. These zones are found at spacings of 50-100
meters. Zooming out again the next order of zones are seen

with widths of a few meters or more and at spacings of half a kilometre to more than one kilometre. The actual frequencies and spacings are of course different for different rock masses but the main impression remains. In this type of system one might expect the dispersion length to increase with distance.

4   MATRIX DIFFUSION

Another mechanism that will cause a "dispersion" is what is termed matrix diffusion. In fractured rocks the water flows in fractures. The rock matrix adjacent to the fractures has a connected porosity. Although the porosity is small, typically less than one percent, the volume of water is one or more orders of magnitude larger than the mobile water in the fractures. Small solute molecules can diffuse in and out of this porosity. For long residence times a large portion of the matrix porosity may be accessed and partake in the water volume that gives the tracer a larger residence time than if only the water in the fractures can hold the solute. For shorter contact times only a fraction of the matrix porosity is accessed and the increase in residence time is smaller. The solutes diffuse into the matrix, reside for some time and may partly diffuse out again to the mobile water. This will cause a widening of the residence time distribution and seem to add to the dispersion. The spreading of a tracer pulse in the flowing water due to this mechanism differs from that for Fickian spreading. The spreading will increase with observation distance because there will be more time for the matrix diffusion. Neretnieks (1983) analysed the spreading caused by matrix diffusion. He found that this mechanism gives a very long tailing of a breakthrough curve for long residence times which considerably can contribute to "dispersion". It was also found that the dispersion if evaluated from the second moment of the breakthrough curve can have practically any values depending on the detection level for the tracer. In fact, if the tracer could be detected in arbitrarily low concentrations then the dispersion length evaluated by the moment method would approach infinity for a fractured medium with sparse fractures. This raises questions about the dispersion data reported in the literature.

5   DISCUSSION AND CONCLUSIONS

The AD model must be used with caution and afterthought - if at all. It is not known at present which of the alternative mechanisms, or some other, that has the largest influence on the dispersion. All of them will, however, give different results if used for scaling to larger distances or residence times. Alternative models which challenge the basic assumptions of the AD model and which satisfactorily handle the above issues must be developed.

6   REFERENCES

1   Matheron G., and de Marsily G. Is transport in porous media always diffusive? A counterexample, Water Resources Res. 6, p 90, 1980.

2   Neretnieks I. A note on fracture flow mechanisms in the ground, Water Resources Res. 19, p. 364-370, 1983.

3   Neretnieks I. Solute Transport in Fractured Rock - Applications to Radioactive Waste Repositories. Chapter 3. Ed Bear J., de Marsily G., Tsang C-F., Academic Press p 39-127, 1993.

4   Gelhar L.W. Stochastic Subsurface Hydrology, Prentice Hall, 1993.

5   Chesnut D. Dispersion in heterogeneous permeable media. Paper presented at the International High Level Waste Management Conference, Las Vegas, May 22-26, Proceedings, 1993.

6   Schweich D. Transport of linearly reactive solutes in porous media. Basic models and concepts. In " Migration and fate of pollutants in soils and subsoils." Ed. Petruzzelli D. and Helfferich F.G., NATO ASI series, Vol G 32, p 221-245, Springer Verlag, 1993.

7   KBS-3, Final storage of spent nuclear fuel. Report by Swedish Nuclear Fuel Supply Co, SKBF, Stockholm, Sweden, May 1983.

USING THE VOLUME AVERAGING TECHNIQUE TO PERFORM THE FIRST CHANGE OF SCALE FOR NATURAL RANDOM POROUS MEDIA

D. Bernard

L.E.P.T.-ENSAM, C.N.R.S. URA 0873, Talence, France

ABSTRACT

The volume averaging technique is one of the various theoretical methods providing a rigorous description of the change of scale procedure. Applying this method to pore scale flow through natural porous media gives rise to several theoretical and practical problems. This paper is a progress report of an ongoing effort to solve most of them in order to build physically and structurally realistic models of flow through natural porous media. We present here some recent results concerning the characterisation of the random geometry of those media, the definition of a Representative Elementary Volume (REV) appropriate for the considered change of scale, and the relevance of periodic boundary conditions to solve the closure problem originated in this process.

1. INTRODUCTION

When the flow occurring at a local scale in a porous media is slow enough, it is correctly described by the Stokes equations;

$$-\nabla p_f + \mu_f \nabla^2 \mathbf{V}_f = 0 \qquad (1)$$

$$\nabla . \mathbf{V}_f = 0 \qquad (2)$$

This equation set must be completed by boundary conditions.

Following a change of scale procedure which has been presented in details elsewhere [1, 2], we perform an averaging of the preceding equations over a Representative Elementary Volume (REV figure 1). If the following spatial constraints are verified [1];

$$l_f , l_s \langle\langle r_0 \langle\langle L \tag{3}$$

we obtain an equation describing the fluid flow at the macroscopic scale in terms of macroscopic variables. In the case considered here, those variables are the filtration velocity and the average pressure, and the macroscopic equation is DARCY's law:

$$\langle V_f \rangle = -\frac{K}{\mu_f}.(\nabla\langle p_f \rangle^f - \rho_f g) \tag{4}$$

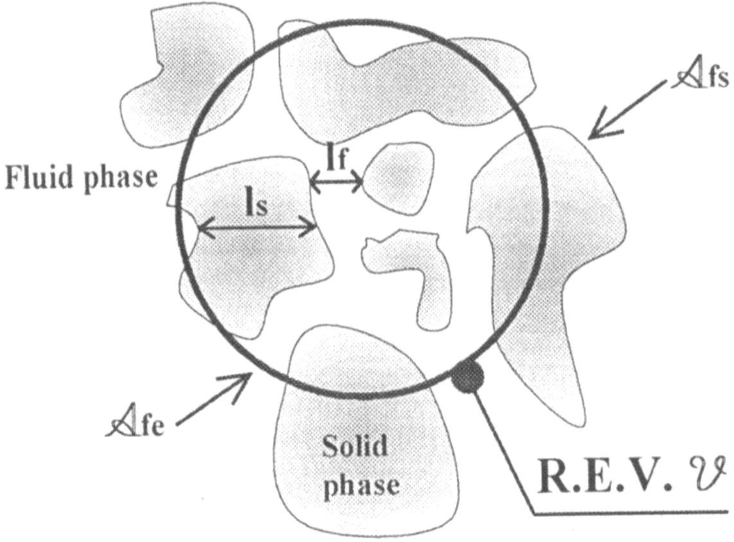

Fig. 1, Notations used at the local scale for the averaging procedure.

The permeability tensor **K** appearing in equation (4) is an intrinsic physical property of the porous media which can be calculated solving a local problem called closure problem. If we impose periodic boundary conditions at the limits of the REV, the closure problem can be written [3]:

$$-\nabla d + \nabla^2 D = I \tag{5}$$
$$\nabla . D = 0 \tag{6}$$
$$\langle d \rangle^f = 0 \tag{7}$$

$$\mathbf{D} = 0 \ \text{on} \ A_{fs} , \ \ \mathbf{D}\left(r_f + l^i\right) = \mathbf{D}\left(r_f\right) ; \ \ \mathbf{d}\left(r_f + l^i\right) = \mathbf{d}\left(r_f\right) \ , \ i = 1, N \qquad (8)$$

The permeability tensor is then given by:

$$\mathbf{K} = -\varepsilon \langle \mathbf{D} \rangle^f \qquad (9)$$

This mathematical model is valid for flow velocities satisfying the following constraint [4, 5]:

$$R_e = \frac{\rho_f V d}{\varepsilon \mu_f} < 1 \qquad (10)$$

where d is a characteristic length for the porous media at the local scale and ε the porosity. If the fluid is water at a temperature of 20°C ($\rho_f / \mu_f \sim 10^6$) and that the porosity is of about 0.3, equation (3) becomes:

$$V < \frac{3,0 \ 10^{-7}}{d} \qquad (11)$$

The maximum of the filtration velocity given by equation (11) is $3.0 \ 10^{-3}$ ms^{-1} (≈ 95 km/year) for a characteristic length of 0.1 mm. Consequently, the domain of applicability of the theoretical model presented above is clearly very broad for monophasic fluid flow in natural porous media.

One of the main consequences of the existence of the closure problem is that

the permeability tensor is exclusively a function of the local geometry of the porous media

when the constraints of equation (3) are satisfied. This leads us to the definition of a general scheme for our project (figure 2) aiming at building physically and structurally realistic models of flow through natural porous media. We first consider the last step of this scheme.

2. NUMERICAL MODEL

The complete set of numerical tests that have been performed solving the closure problem (eq. 5 to 8) using the finite element or the finite volume methods are presented in a paper by ANGUY et al. [6]. Here we just recall the principal characteristics of the finite volume version. We use an artificial compressibility algorithm [7] giving the solution of the closure problem as the steady state limit of the following transient partial differential problem:

$$\frac{\partial \mathbf{D}}{\partial t} + \nabla \mathbf{d} - \nabla^2 \mathbf{D} = -\mathbf{I} \qquad (12)$$

$$\frac{\partial \mathbf{d}}{\partial t} + c^2 \nabla . \mathbf{D} = 0 \qquad (13)$$

$$\text{at} \ t = 0, \ \ \mathbf{d} = \mathbf{d}_0 \ \text{and} \ \mathbf{D} = \mathbf{D}_0 \qquad (14)$$

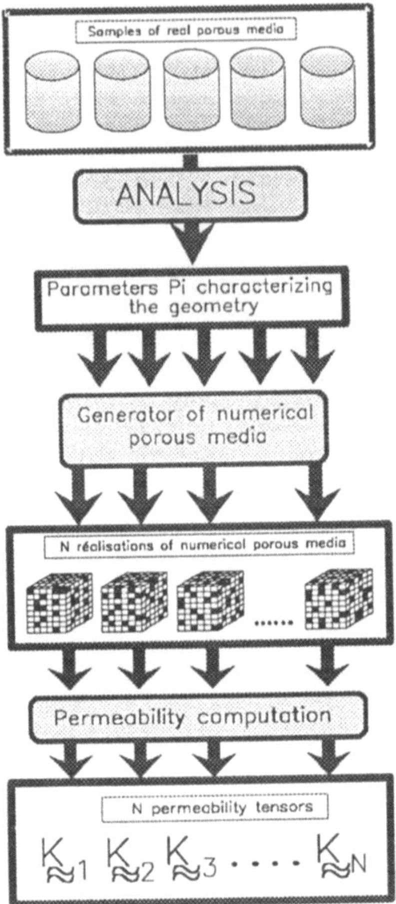

Fig. 2, General scheme adopted to study the relation between permeability tensor and micro geometry of natural porous media.

$$D = 0 \ \text{sur} \ A_{fs}, \ \ D\left(r_f + l^i\right) = D(r_f) \ ; \ d\left(r_f + l^i\right) = d(r_f) \ , \ i = 1, N \qquad (15)$$

Spatial discretisation is performed following the finite volume method [8] on a staggered grid. Time integration is done with a fully explicit scheme for which stability criteria have been developed [6].

Figure 3 shows an example of the results that can be obtained with this numerical model. For this rather simple 2D periodic media, the unit cell is defined on a regular 80 x 80 grid and the permeability tensor is equal to 0.417 and 0.125 Darcy along the principal directions of flow (see figure 3). The size of the unit cell is $10^{-4} \times 10^{-4}$ m^2.

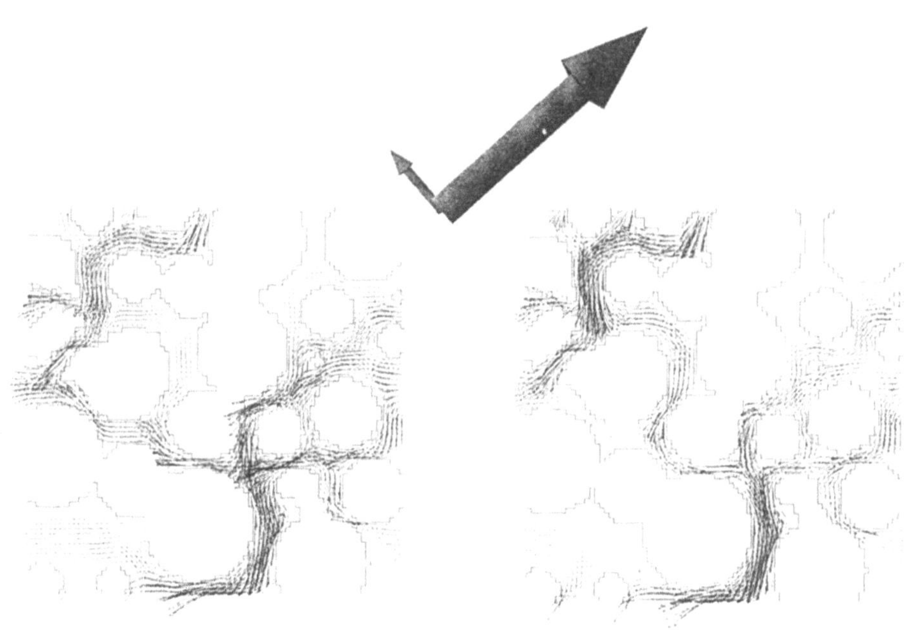

Fig. 3, Permeability tensor (above, represented along the principal directions of flow) for a 2D periodic porous media (the dashed regions are the solid phase in the two unit cells represented). Each vector field is a column of the tensor $\mathbf{D}$, solution of the closure problem (eq. 5 to 8). The permeability tensor is the average of those vector fields.

Of course natural porous media are not as simple as the porous media used in the previous example. Let's consider an other test in which some randomness is taken into account to define the micro geometry. Two different series of 2D periodic porous media have been generated. In the first one the phases distribution into the unit cells is uncorrelated (figure 4-a) and in the second one an auto correlation function has been imposed to the random phases distribution (figure 4-b) in an attempt to organise the micro porosity in a way similar to real rocks [9]. The effects of this organisation are very important:

- first for the percolation threshold (figure 5)
- second for the porosity-permeability relation (figure 6).

The percolation threshold has been calculated for large enough media (classically 1000 x 1000) in order to limit the boundary effects [10] and the permeability computations have been done using 200 x 200 grids for each unit cell.

a)   b)

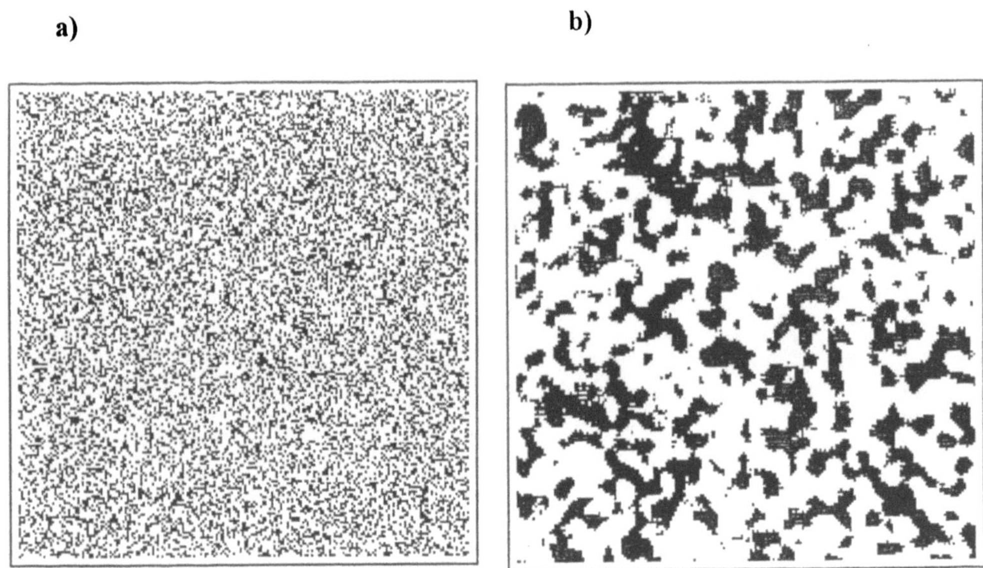

Fig. 4, Two examples of unit cells (200 pixels x 200 pixels) used to build 2D infinite periodic porous media with uncorrelated phases distribution (a) or correlated phases distribution (b). In this last series, the autocorrelation function vanishes after 8 pixels.

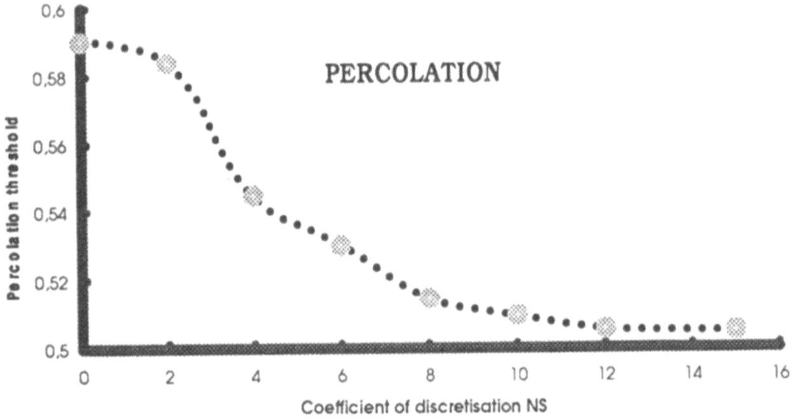

Fig. 5, Evolution of the percolation threshold with NS, the distance (expressed as a pixel number) beyond which the autocorrelation function vanishes. The larger NS, the more structured the generated medium is.

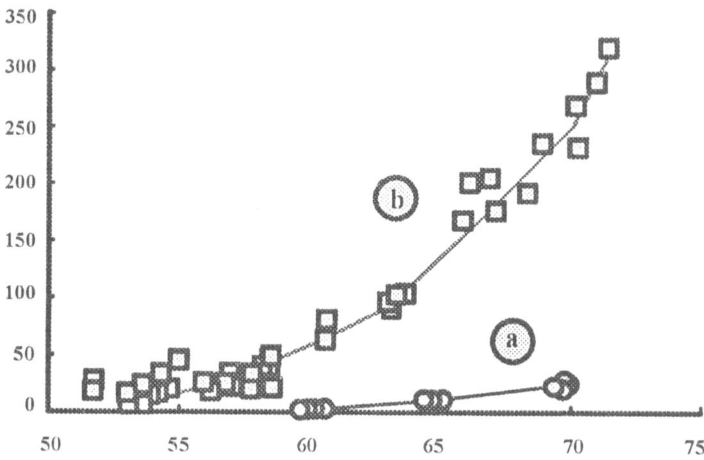

Fig. 6, Evolution of the permeability with porosity for two series of 2D periodic porous media with uncorrelated phases distribution (a) or correlated phases distribution (b).

The estimations of the percolation threshold that can be deduced from the direct computations of the permeability tensor (the media are isotropic in those examples) are in agreement with the curve of fig. 5 which has been obtained using a method entirely different. The dispersion of the values is greater in the correlated case than in the uncorrelated one indicating that the REV is larger when a structure exits than when the phases distribution is completely random.

Numerical codes being available to solve the closure problem, the question now is:

How can we characterise the micro geometry of natural porous media ?

This is the first step in the general scheme of fig. 2.

3. IMAGE ANALYSIS USING 2D FOURIER TRANSFORM

A direct three-dimensional assessment of the micro geometry of natural porous media is non-practicable in that, current means of observation are two-dimensional. Image analysis is the method we chose to obtain micro-structural information from sections through the porous medium. In the field of stereology, the two-dimensional relevance to the three-dimensional system only holds under the assumption of one or another of two conditions:
- isotropy of the micro-geometry,
- existence of a single identified direction of anisotropy, i.e. statistical cylindricity of the micro-geometry.

To a first approximation, cylindrical symmetry can be assumed in many sandstones in that permeabilities measured in direction parallel to bedding are much more similar than measurements made at high angles to bedding. This has been verified by the existence of *strong statistical correlations between geometrical characteristics measured in sections and* physical properties measured on associated samples including: permeability, formation factor and mercury porosimetry[11,12,13].

The micro-structure can be quantified from thin sections cut perpendicular to bedding (plane enclosing most of the variance of the micro-geometry) using binary images (porosity equal 1, matrix equal 0) obtained by image analysis procedures combined with Fourier transforms. Areas covered by conventional high resolution imagery (pixel edge ≤ 10 microns) being too small to evaluate the micro-structure [14], seamless mosaics of conventional views have been constructed [14,15,16], producing images combining high resolution with large area (Fig. 7, for example).

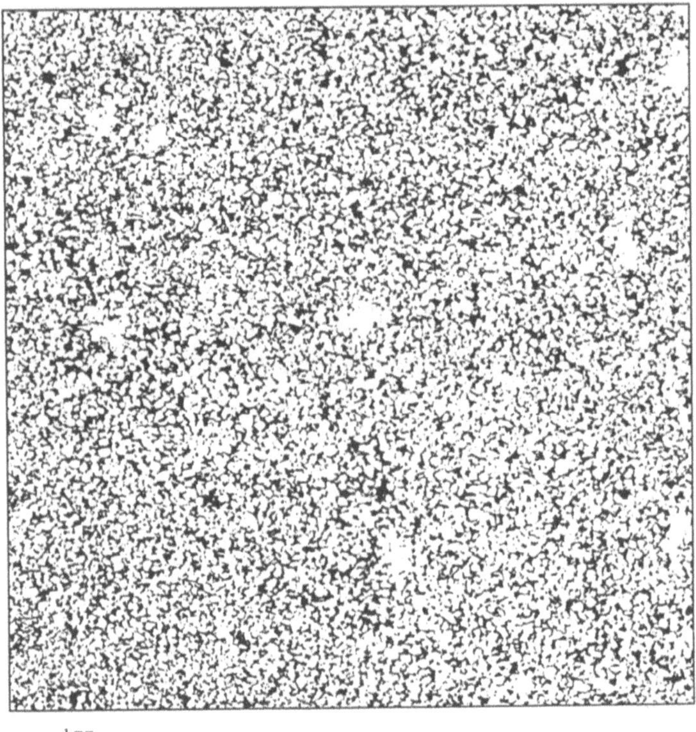

Fig. 7, 15.8 mm x 15.8 mm binary mosaic of a sandstone produced by merging 70 conventional high resolution overlapping images, each measuring 480 x 640 pixels (pixel edge : 3.861 μm). This image comprises 4096 x 4096 pixels (black represents porosity).

The Fourier transform permits quantitative assessment of the characteristic length-scales (Fig. 8) and of the nature of spatial complexity (Fig. 9). By removing all spatial wave-numbers tied to length-scales equal to or less than a specified wave-length and inverting the remainder to generate a 'filtered' image, it is possible to visualise the structure at various scale represented by peaks in the power spectrum. Such visualisation verifies the conjecture of Graton and Fraser [17] that first order structure in all sandstones consist of the juxtaposition of close-packed domains with loose-packed domains (Fig. 9a). The loose-packed domains are arrranged as circuits with large scale continuity (Fig. 9b), containing large pores and large throats, and representing a higher level of spatial organization. A local REV of geometrical type must not be sensitive to an interaction between the sample-volume and structure heterogeneities in that no significant power on the bi-dimensional power spectrum should occur at spatial wave-lengths approaching the size of the sample ('power' in Fourier parlance is equivalent to the amount to total variance tied to features expressed at any wave-length).

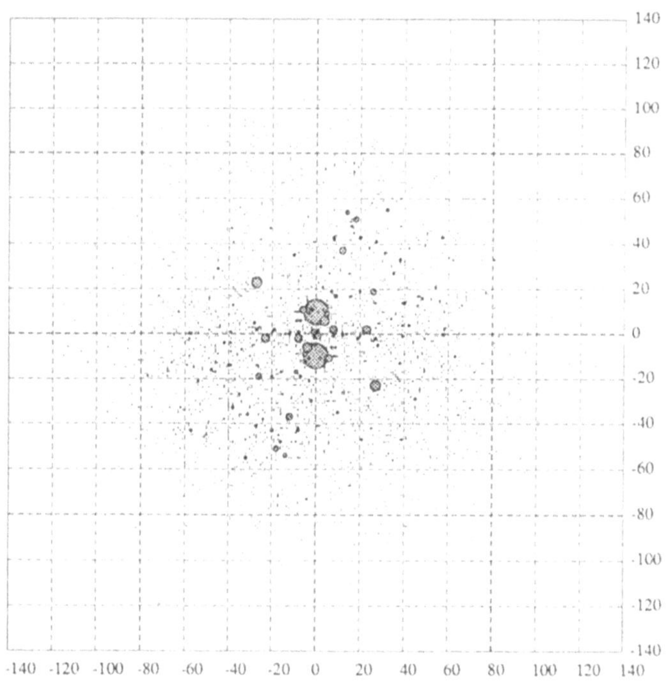

Fig. 8: 2D power spectrum of Fig. 7. Numbers on axes are the wave-numbers v_x and v_y (proportional to the inverse of wave-length). Circle size is proportional to power. The absence of power near the centre (0,0) suggests local homogeneity.

The method illustrated above permits the characterisation of the micro geometry of real rocks through the 2D Fourier transform. When local homogeneity is achieved, we have an estimate of the size of the geometrical REV. The relation between this REV and Darcy's REV (Fig. 1) is not well established, the only certainty being that the geometrical REV is included in Darcy's REV.

a)

b)

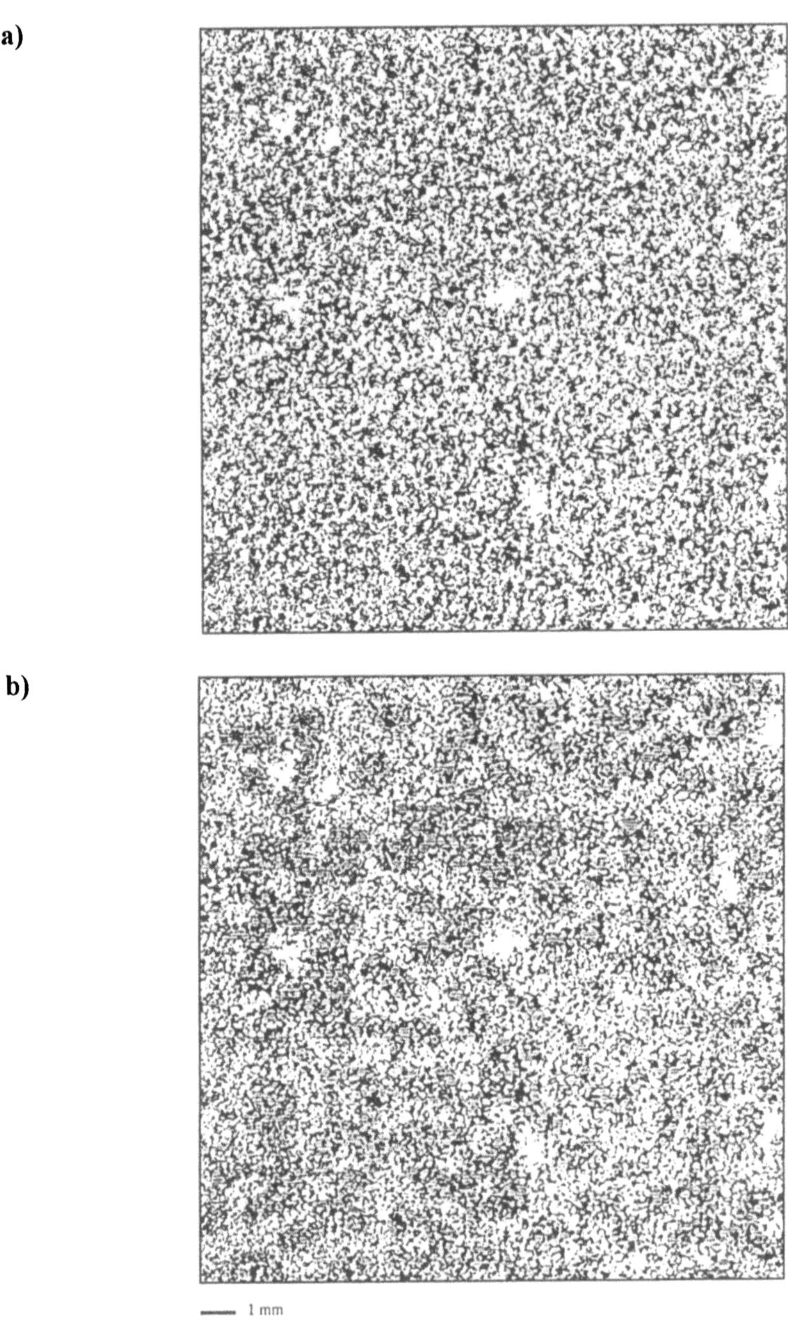

Fig. 9: Images of the sandstone presented Fig. 7 after filtering all wave-length (a) smaller than grain size (200 µm) putting into evidence the larger pores associated with packing flaws, (b) smaller than 1100 µm putting into evidence the circuits of packing flaws.

4. PERMEABILITY TENSOR OF 2D RANDOM POROUS MEDIA

Using a random generator of numerical porous media (necessary interface between the analysis of the micro geometry and the programs computing the permeability tensors, see Fig. 2) we have explored the relation between the geometrical REV and Darcy's REV for 2D structured porous media. The different porous media have been generated by convoluting a density power spectrum with a random field. This procedure is similar to the one proposed by Quiblier [9] in the sense that the autocorrelation function is the inverse Fourier transform of the power spectrum.

Fig. 10 two examples of the isotropic 2D porous media are presented. Their size is 400 pixels x 400 pixels and the geometrical REV is equal to 8 pixels x 8 pixels.

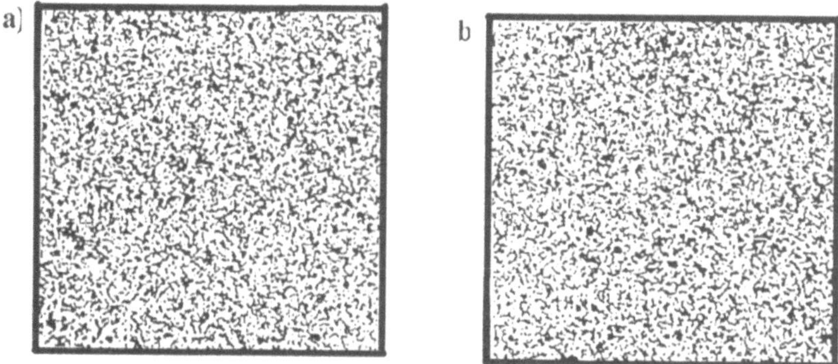

Fig. 10, a) and b): examples of artificial isotropic media. Each image includes 250 geometrical REV (50 x 50). Porosity is in white.

Unit cells of different sizes but common centres have been extracted from the 2D media and used to estimate numerically the permeability tensor **K.** Results are summarised Fig. 11 for two examples. They strongly suggest that the characteristic dimension of Darcy's REV is of about 30 times the characteristic dimension of the geometrical REV.

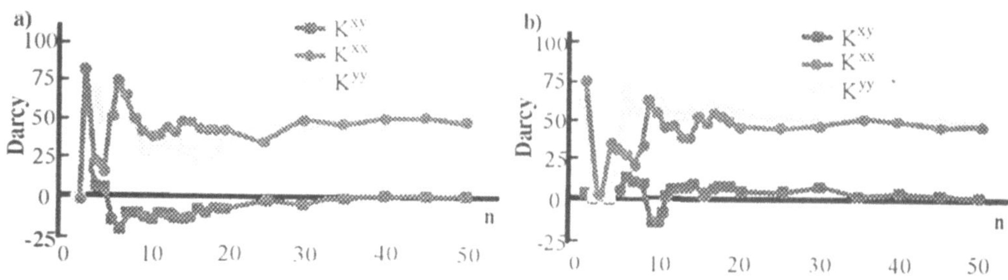

Fig. 11, a) and b): evolutions of the permeability tensors **K** with the size of the unit cells used in the computation. The number of geometrical REV included in the computation cells is equal to n^2.

For those computations we have imposed directly the periodic boundary conditions without any kind of caution placing side by side the different unit cells. When the cells are larger than 8 pixels x 8 pixels (in this particular case) there is no correlation between the phases distributions along the edges and this juxtaposition yields an artificially lowered connectivity at the limits of the computation domain (the porosity along the boundaries is not equal to 0.6 as into the cell but to 0.6 x 0.6 = 0.36 because of the lack of correlation between the edges that are put into contact). Whether it is generally admitted that the effects of the boundary conditions at the limit of the REV are much less important than the effects of the no-slip condition at the fluid-solid interface [1], there is not a lot, if any, direct test of this statement available in the literature. For one 2D porous media of the series used for the prior computations (geometry (a) Fig. 10, permeability evolution (a) Fig. 11), we have built the computation unit cell in a different way: after extraction of the n x n domain as before, we multiplied its size by a factor of four first applying a symmetry along the east edge and second applying a symmetry along the south edge. The computation cell obtained is much larger, intrinsically isotropic (the generated media are too for large enough cells, see Fig. 11), but have, by construction, totally correlated limits. The results presented Fig. 12 show that for large cells, the values obtained using both building methods are similar. Consequently, external boundary conditions can be considered as being of limited effect compared to the internal boundary conditions.

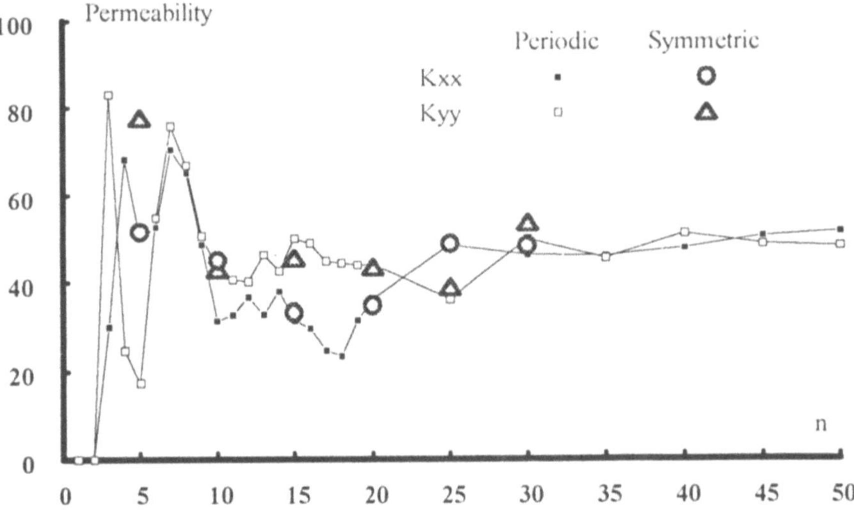

Fig. 12, Comparison of the permeability evolution with the size of the computation domain using directly the periodic boundary condition (small squares, identical to Fig. 11(a)) or artificially correlating the edges by symmetry (triangles and circles).

5. DISCUSSION

The preliminary results displayed in this paper demonstrate that the project schematised Fig. 2 is feasible using the numerical tools now available. Nevertheless several theoretical and practical problems still need important testing and development. Among them we are particularly interested in:

1. evaluating the relevance of a 2D characterisation (by image analysis) of complex 3D geometry,
2. determining the pixel size suitable for permeability computation (in chapter 3 we proposed a method to determined the larger structures present in a real porous media, we must now determined the smaller structures necessary to evaluate numerically the permeability tensor),
3. comparing our "mechanical" approach to a more statistical and empirical geological approach |11, 12, 13],
4. using 3D random geometries of large dimensions,
5. introducing the mercury injection as a validation tool (numerical and experimental) of our project.

Some rather simple 3D cases have been treated now (Fig. 13 gives an illustration of what can be easily done on a workstation, the differences between Fig. 13 b and Fig. 13 c demonstrate the importance of point number 2) but realistic computations will require to use more powerful computers [18].

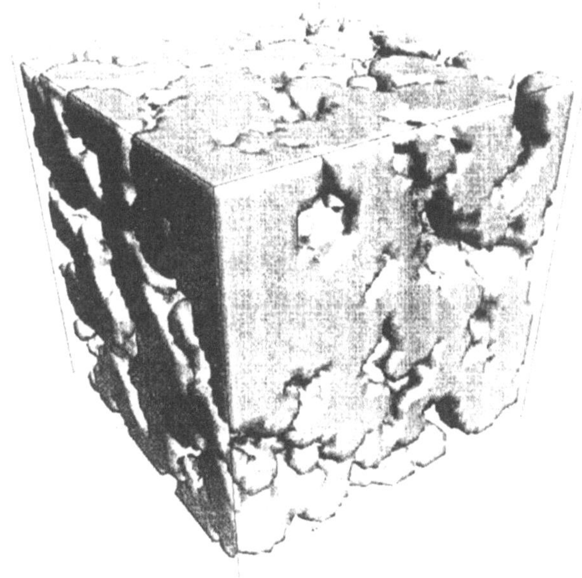

Fig. 13, a) 3D geometry of the solid phase obtained placing randomly 1200 solid spheres of random radius in a 50 x 50 x 50 domain. The porosity is equal to 0.15.

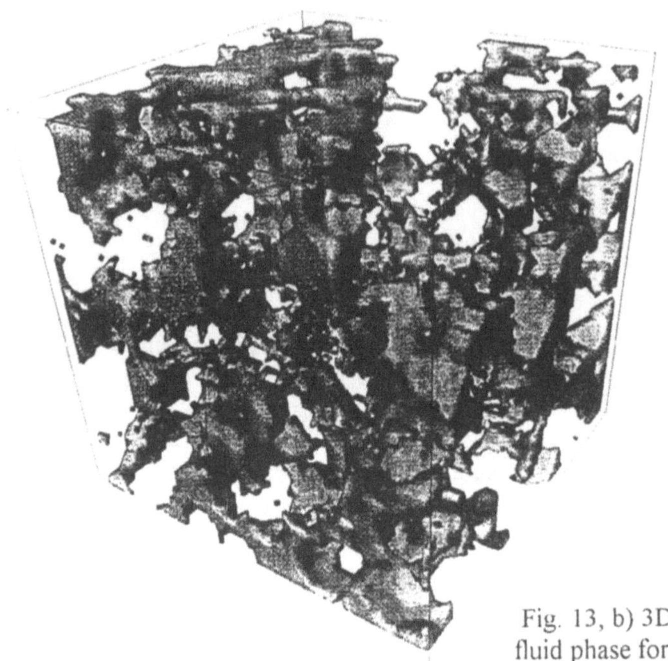

Fig. 13, b) 3D geometry of the
fluid phase for the same random
porous media as Fig. 13 a.

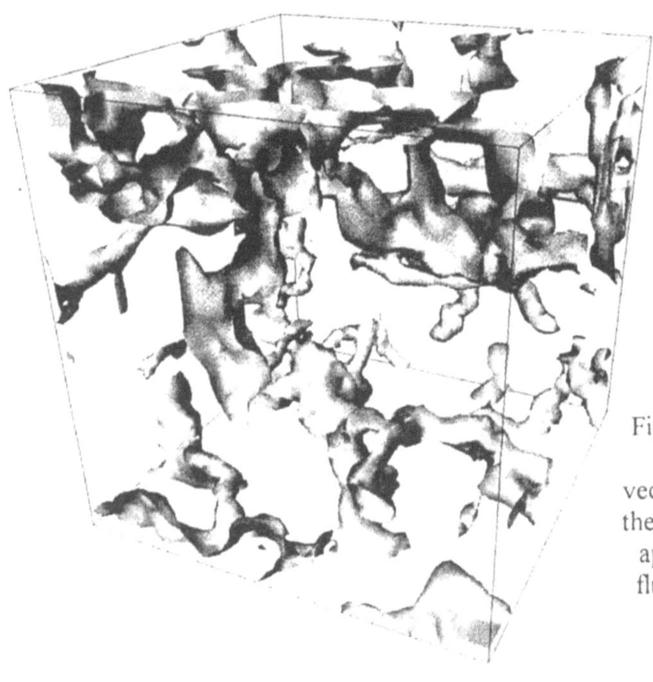

Fig. 13, c) Visualisation of the
iso-surface corresponding
vector modules equal to 0.01 of
the maximum. This surface is an
approximation of the flowing
fluid domain (to be compared
to the real fluid domain
represented Fig. 13 b).

6. CONCLUSIONS

This paper represents a progress report of an ongoing effort to build physically and structurally realistic models of flow through porous media using the local change of scale method.

Preliminary results involving the use of multiple isotropic realisations of a local geometrical R.E.V. gives insights into the scale of the Darcy's R.E.V. Results indicate that a relatively small number of local geometrical R.E.Vs. (30 or fewer) may be required. Data from natural sandstones indicate that the size of the R.E.V. of Darcy's and micro-structural types are much larger than is commonly believed. This might pose computational problems for calculating the permeability tensor.

Development of a valid methodology to characterise geometrical R.E.V. coupled with its relationship to the Darcy's R.E.V. will allow explicit calculation of the permeability tensor as an implicit function of the micro-geometry. In accordance with existence of a local closure form, *the detailed micro-structure must define the permeability tensor.* However, realistically characterising the three-dimensional micro-geometry is a complex problem since current direct observations are of porosity exposed in sections. Nonetheless, characteristics observed in section impose limits on the structural variations possible in the three-dimensional medium.

ACKNOWLEDGEMENTS

Part of this work has been funded by the research group TRABAS from the CNRS. The largest computations have been performed at the CNRS computing centre IDRIS.

REFERENCES

1. Whitaker, S.: Flow in porous media I: A theoretical derivation of DARCY's law, Transport in porous media, 1(1986), 3-25
2. Quintard, M. and Whitaker, S.: Transport processes in ordered and disordered porous media, in: Heat and Mass Transfer in Porous Media, QUINTARD, TODOROVIC Eds., Elsevier Sci. Pub., Amsterdam, 1992, 99-110
3. Barrere, J., Gipouloux, O. and Whitaker, S.: On the closure problem for DARCY's law, Transp. in Porous Media, 7(1992), 209-222
4. Dybbs, R.G. and Edwards, R.V.: A new look at porous media fluid mechanics-Darcy to turbulent, in Fundamentals of transport phenomena in porous media, Bear,J., Corapcioglu,Y. Eds, NATO ASI Series E, N° 82, Martinus Nijhoff Pub., Dordrecht, 1984, 199-256
5. Whitaker, S.: Flow in porous media III: Deformable media, Transport in porous media, 1(1986), 127-154
6. Anguy, Y., Bernard, D. and Ehrlich, R.: The local change of scale method for modelling flow in natural porous media (I): Numerical tools, Adv. Water Res., 1994, in press
7. Peyret, R. and Taylor, T.D.: Computational methods for fluid flow, Springer-Verlag, Berlin, 1983

8. Patankar, S.V.: Numerical heat transfer and fluid flow, Computational methods in mechanics and thermal sciences, Hemisphere Pub. Co., New York, 1980
9. Quiblier, J.A.: A new three-dimensional modeling technique for studying porous media, J. Coll. Interf. Sci., 98 (1984), 84-102
10. Kirpatrick, S.: Percolation and conduction, Rev. Modern Phys., 45(1973), 574-590
11. Ehrlich, R., Crabtree, S.J., Horkowitz, K.O. and Horkowitz, J.P.: Petrography and reservoir physics I : objective classification of reservoir porosity, A.A.P.G. Bulletin, 75 (1991), 1547-1562.
12. McCreesh, C.A., Ehrlich, R. and Crabtree, S.J.: Petrography and reservoir physics II : relating thin section porosity to capillary pressure, the association between pore types and throat sizes, A.A.P.G. Bulletin, 75 (1991), 1563-1578.
13. Ehrlich, R., Etris, E.L., Brumfield, D., Yan, L.P. and Crabtree, S.J.: Petrography and reservoir physics III : physical models for permeability and formation factor, A.A.P.G. Bulletin, 75 (1991), 1579-1592.
14. Anguy, Y., Ehrlich, R., Prince, C.M., Riggert, V. and Bernard, D.: The sample support problem for permeability assessment in reservoir sandstones, in A.A.P.G. Spec. Publ. : Stochastic Modeling and Geostatistics, Practical Applications and Case Histories, Eds. Yarus J. M. and Chambers R. L., 1994, in press.
15. Prince, C.M., Ehrlich, R. and Anguy, Y.: Analysis of spatial order in sandstones II : grain clusters, packing flaws, and the small-scale structure of sandstones, Jour. of Sed. Pet.,1995, in press
16. Anguy, Y.: Application de la prise de moyenne volumique à l'étude de la relation entre le tenseur de perméabilité et la micro-géométrie des milieux poreux naturels, thèse de doctorat, mécanique, Univ. Bordeaux I, 1993, 170 p.
17. Graton, L.C. and Fraser, H.J.: Systematic packing of spheres with particular relation to porosity and permeability, Jour. Geology, 43 (1935), 785-909
18. Bernard, D., Bodin, F., Goasguen, A. and Fechant, J.C.: Implementing a two-dimensional pore-scale flow model on different parallel machines, in A. Peters et al. (eds.), Computational methods in water resources X, Kluwer Acad. Pub., Dordrecht, 1992, 1507-1514

BLOCK ITERATIVE STRATEGIES FOR MULTIAQUIFER FLOW MODELS

G. Gambolati and P. Teatini

University of Padua, Padua, Italy

ABSTRACT

When groundwater flow takes place in aquifer-aquitard systems characterized by a high contrast of hydrogeological parameters and non-linear aquitard behavior, quasi - three-dimensional models of flow must be solved by a fully numerical approach. The numerical implementation with finite difference or finite element methods involves large systems whose solution requires much CPU time and computer storage. A new solution strategy of the resulting algebraic equations is suggested to overcome non-linearity in the aquitards. The global non-linear system is decoupled into a number of smaller subsystems, which are consistent with the geologic structure of the multiaquifer system and are ideally suited for updating the non-linear hydrologic parameters in the aquitards. The aquifer and the aquitard equations are solved separately with the modified conjugate gradient (MCG) and the Thomas algorithms, respectively, while the final coupled solution is obtained with an iterative procedure. The procedure, as is naturally suggested by the special block tridiagonal pattern of the coefficient matrix, can be shown to be equivalent to a block Gauss-Seidel strategy and can therefore be generalized into a block SOR (successive over-relaxation) strategy, the blocks corresponding to the aquifer and aquitard equations. The convergence properties of the new SOR scheme are initially analyzed with linear porous systems for which the optimum over-relaxation factor ω_{opt} can be theoretically computed, and the asymptotic convergence rate is studied in relation to the geometry, the hydrogeological parameters of the multiaquifer system and the size of the numerical model. The results obtained with steady state sample problems show that if flow in the various aquifers is weakly interconnected (i.e. the hydraulic exchange between the aquifer units is quite limited) ω_{opt} is close to 1, while in strongly coupled systems ω_{opt} falls into the upper range ($1.5 < \omega_{opt} < 2$) and the SOR asymptotic convergence rate can be as much as one order of magnitude larger than the Gauss-Seidel rate. Subsequently the block iterative method has been extended to non-linear porous media and early results indicate that the relaxation procedure with $\omega = \omega_{opt}$ provides a significant acceleration of convergence as well. The linear theory, however, does no more apply and ω_{opt} must be assessed empirically.

1. INTRODUCTION

The hydrologic response of multiaquifer systems to groundwater pumping is highly dependent on the difference between the physical parameter values of sandy and clayey layers. In most cases a high contrast in the hydraulic conductivity exists and groundwater flow can be assumed essentially horizontal in the aquifers and vertical in the aquitards. Quasi - three-dimensional models take computational advantage of this specific hydrogeological setting, which discards the vertical hydraulic gradient in the aquifers and the horizontal hydraulic gradient in the aquitards.

Two theoretical approaches have been developed to build computationally efficient quasi - three-dimensional models. The first approach transforms the set of partial differential diffusion equations into a set of integro-differential equations [1, 2, 3, 4, 5, 6, 7]. Only the aquifers are discretized by finite differences or finite elements and the aquifer hydraulic connection through the intervening aquitards is obtained with convolution integrals making use of an aquitard fundamental solution. The second approach relies on the full numerical integration of the original partial differential equations and proves to be also suited to non-linear problems [8, 9, 10, 11]. Discretization of the sandy and clayey units involves algebraic systems of large dimension. Special numerical techniques to efficiently solve these large problems have been consequently devised. *Chorley and Frind* [9] treat the resulting equations with an iterative technique, "solving aquifers and aquitards alternately at each time step. A favorable rate of convergence was obtained by staggering aquifer and aquitard solution by half a time step". The "adaptive explicit-implicit" technique implemented by *Neuman et al.* [10] takes advantage of the large ratio between the hydraulic aquifer-aquitard diffusivities, so that for small time step values Δt, most of the element equations related to the nodes of the aquitards can be solved explicitly with low computational cost. The remaining equations are solved implicitly with a direct scheme.

In the present paper a new strategy of solution is implemented. The procedure is shown to coincide with a block Gauss-Seidel iterative method and therefore can be mathematically generalized into a block SOR iterative procedure, in which each aquifer block is solved with the MCG (modified conjugate gradient) method and each aquitard block is solved with the direct Thomas algorithm for a tridiagonal matrix. The iterations can be accelerated with the aid of the optimum block relaxation factor ω_{opt}.

The convergence property of the new scheme is analyzed and compared in relation to the geometry and the physical properties of linear and non-linear multiaquifer systems. Two numerical linear codes are implemented. In the first code (labeled "coupled"), all the resulting linear equations are solved simultaneously with the MCG method, while in the second one (labeled "decoupled"), the block SOR iterative strategy is used. The two approaches are applied to and tested on the same linear sample problems and compared in terms of both CPU time and storage requirement. The "decoupled" code is then generalized to simulate non-linear aquifer systems. It is applied to the same multiaquifers explored with the linear analysis introducing non-linear hydrogeological aquitard parameters and is tested against the non-linearity degree of the problem.

2. MATHEMATICAL AND NUMERICAL MODEL

Consider a multiaquifer system consisting of m aquifers separated by $(m-1)$ aquitards. Neglecting point and distributed sources, the governing flow equation in the i-th aquifer confined between aquitards j and $(j+1)$ can be written as [12]:

$$\frac{\partial}{\partial x}\left(T_{xi}\frac{\partial h_i}{\partial x}\right) + \frac{\partial}{\partial y}\left(T_{yi}\frac{\partial h_i}{\partial y}\right) = S_i\frac{\partial h_i}{\partial t} + q_j - q_{j+1} \qquad i = 1,2,\ldots,m \qquad (1)$$

where T_{xi}, T_{yi}, S_i indicate the components of the anisotropic trasmissivity and the elastic storage coefficient, respectively, h_i is the hydraulic potential head, t is time, and q_j, q_{j+1} represent leakage from the overlying and underlying aquitards. In aquitard j, flow is governed by the one-dimensional vertical diffusion equation [12]:

$$K_{zj}\frac{\partial^2 h_j}{\partial z^2} = S_{sj}\frac{\partial h_j}{\partial t} \qquad j = 1,2,\ldots,m-1 \qquad (2)$$

where K_{zj} is the vertical permeability, S_{sj} is the specific elastic storage and h_j is the hydraulic potential head.

Equations (1) and (2) are solved numerically with a finite element model and coupled together by consideration of mass conservation. Applying the finite element method, each aquifer is discretized into the same triangular finite element grid of n nodes, and each vertical column within an aquitard, extending from node r in the underlying aquifer to the corresponding node in the overlying aquifer, is represented by l one-dimensional elements $((l+1)$ nodes overall, $(l-1)$ inside the aquitard plus 2 at the interfaces). Applying the Galerkin procedure to eq. (1), using linear basis functions and regarding the leakage terms as known quantities, yield the following finite element equation for each individual aquifer in matrix form:

$$H_1\mathbf{h} + P_1\frac{\partial \mathbf{h}}{\partial t} + \mathbf{q} = 0 \qquad (3)$$

where H_1 and P_1 are the n-dimensional aquifer stiffness and capacity matrices, and $\mathbf{q}$ is a n-dimensional vector accounting for leakage at the aquifer-overlying aquitard and aquifer-underlying aquitard interfaces. For the i-th aquifer confined between aquitards j and $(j+1)$, the r-th component of $\mathbf{q}$ is given by:

$$q_r = \frac{q_{j,r}}{3}\sum_e \Delta_e + \frac{q_{j+1,r}}{3}\sum_e \Delta_e \qquad (4)$$

where $q_{j,r}$ and $q_{j+1,r}$ are specific fluxes from aquitards j and $(j+1)$ at node r, Δ_e represents the area of the triangle e and the summation is taken over all elements e sharing node r.

Since the vertical elements, into which the aquitards are discretized, are not coupled, aquitard flow equation (2) is numerically integrated independently over each vertical element column, provided that Dirichlet boundary conditions are assumed. Analyzing the r-th vertical column within aquitard j, the solution of the aquitard flow equation is obtained by prescribing the flow continuity condition at the aquitard-aquifer interfaces. Applying the Galerkin procedure and using linear basis functions leads to the matrix equation:

$$H_2\mathbf{h} + P_2\frac{\partial \mathbf{h}}{\partial t} + \mathbf{q}'_{j,r} = 0 \qquad (5)$$

where H_2 and P_2 are $(l+1)$-dimensional stiffness and capacity matrices and $\mathbf{q}'_{j,r}$ is a $(l+1)$-dimensional specific leakage vector representing outflow from the aquitard column per unit horizontal area. It is $q'_{j,r} = 0$ for all column nodes except for those at the top and bottom of the column, belonging to both the aquifer and the aquitard. If we consider the node at the bottom of the column (shared by aquitard j and aquifer i), because of the flow continuity condition, $q'_{j,r} = q_{j,r}$, where $q_{j,r}$ is the leakage term used in eq. (4). The term $q_{j,r}$ appears in both eq. (3) and eq. (5), and couples the aquifer and aquitard flow together.

Writing eq. (3) for the m aquifers and eq. (5) for the n columns of the $(m-1)$ aquitards, and applying a λ-weighted scheme for the time integration (Crank-Nicolson at $\lambda = 0.5$; backward Euler at $\lambda = 1$), yields a system of equations:

$$H\mathbf{h}_{t+\Delta t} = \mathbf{b} \tag{6}$$

where H is a symmetric positive definite matrix of size $s_1 = n \cdot [m + (l-1) \cdot (m-1)]$.

3. BLOCK ITERATIVE SOLUTION PROCEDURES

Consider the sample multiaquifer system of Figure 1a. Each aquifer is discretized into 6 triangular elements (7 nodes) and each aquitard column into 4 linear elements (3 nodes inside the aquitard). With the nodal numbering as indicated in Figure 1a, the sparsity pattern of the global matrix H (eq. (6)) is shown in Figure 1b. We can observe the following:

- the non-zero structure of the aquifer matrices ($D_{1,1}$, $D_{3,3}$ and $D_{5,5}$ in Figure 1b) and that of the aquitard matrices ($D_{2,2}$ and $D_{4,4}$) are easily located within the global matrix;

- each aquifer matrix is sparse and each matrix corresponding to an aquitard nodal column is tridiagonal;

- the number of "coupling terms" is relatively small and their position ($U_{1,2} \div U_{4,5}$ and $L_{2,1} \div L_{5,4}$ in Figure 1b) is out of the square submatrices ($D_{1,1} \div D_{5,5}$) associated with each formation.

These specific sparsity properties of H naturally suggests an iterative "decoupled" solution strategy. Each unit, beginning from the upper one, is solved separately, decoupling the global system into a number of smaller subsystems. If we solve at iteration level k the subsystem of the i-th unit, $L_{i,i-1}$, $D_{i,i}$ and $U_{i,i+1}$ are the matrix blocks that are to be considered (Figure 1b). Decoupling is accomplished by the elimination of the coupling terms from the subsystem matrix: the terms of $L_{i,i-1}$ are multiplied by the corresponding nodal head values known at the current iteration $(k+1)$ because they belong to the $(i-1)$-th layer (the $(i-1)$-th layer has already been solved at the current iteration). The terms of $U_{i,i+1}$ are multiplied by the corresponding head values at the previous iteration level k. At the initial iteration, the previous time solution is used. Iterations are continued until the Euclidean norm $|\mathbf{r}^{(k+1)}|$ of the residual of the global system:

$$\mathbf{r}^{(k+1)} = \mathbf{b} - H\mathbf{h}_{t+\Delta t}^{(k+1)} \tag{7}$$

becomes less than a specified tolerance TOL. A faster rate of convergence is achieved starting the cycle from the pumped aquifer (it is not necessary that the blocks be solved in the order

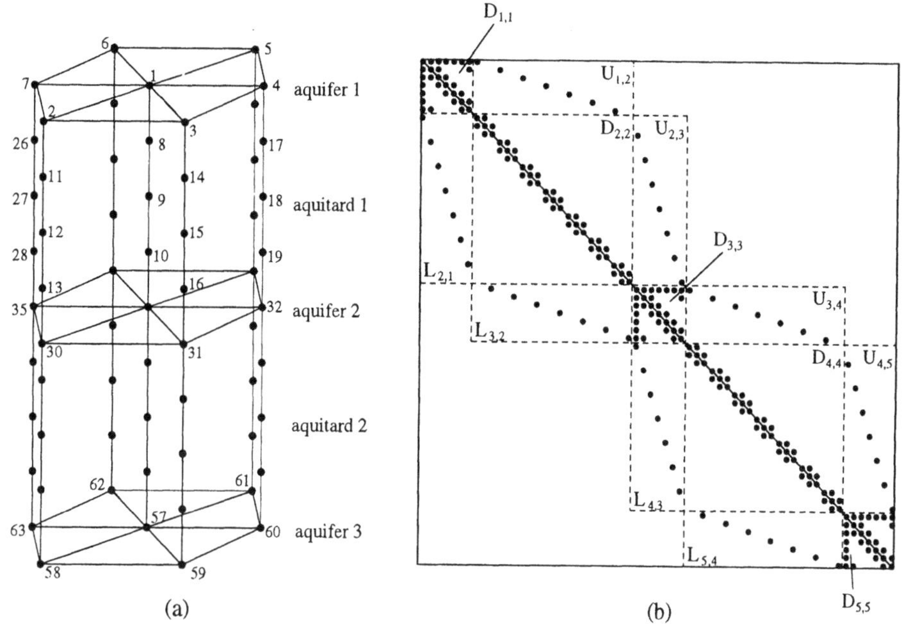

Figure 1: (a) Discretization of a sample multiaquifer system, and (b) the related matrix sparsity pattern.

$1, 2, 3, \ldots$, however the order must remain fixed [13]). If aquifers 1 and 3 of Figure 1*a* are pumped, the most convenient solution sequence is:

$$\text{aquifer 1 - aquitard 1 - aquifer 3 - aquitard 2 - aquifer 2}$$

The aquifer subsystems are solved by the MCG method [14, 15], which is best suited to large sparse positive definite set of equations [16, 17]. The direct Thomas algorithm [13] is used to solve the aquitard subsystems. Since the vertical elements into which an aquitard is discretized are not coupled, the algorithm is separately applied to each sub-block corresponding to each single vertical column.

The iterative solution scheme outlined above, which was implemented in a similar, but not identical, form by *Chorley and Frind* [9], can easily be shown to be equivalent to a block Gauss-Seidel iterative method [18]. It is well known that the block Gauss-Seidel procedure can be mathematically accelerated and generalized into a block SOR by a relaxation factor ω. If a matrix H is partitioned into $N \times N$ blocks ($N = 5$ in Figure 1) with the form:

$$H = \begin{bmatrix} H_{1,1} & H_{1,2} & \cdots & H_{1,N} \\ H_{2,1} & H_{2,2} & \cdots & H_{2,N} \\ \cdot & \cdot & & \cdot \\ \cdot & \cdot & & \cdot \\ \cdot & \cdot & & \cdot \\ H_{N,1} & H_{N,2} & \cdots & H_{N,N} \end{bmatrix} \tag{8}$$

where $H_{i,i}$, $1 \leq i \leq N$, are square and non void matrices, and we define the matrices:

$$
D = \begin{bmatrix}
H_{1,1} & 0 & \cdots & 0 \\
0 & H_{2,2} & \cdots & 0 \\
\cdot & \cdot & & \cdot \\
\cdot & \cdot & & \cdot \\
\cdot & \cdot & & \cdot \\
0 & 0 & \cdots & H_{N,N}
\end{bmatrix}
$$

$$
U = - \begin{bmatrix}
0 & H_{1,2} & \cdots & H_{1,N} \\
0 & 0 & \cdots & H_{2,N} \\
\cdot & \cdot & & \cdot \\
\cdot & \cdot & & \cdot \\
\cdot & \cdot & & \cdot \\
0 & 0 & \cdots & 0
\end{bmatrix}
\qquad
L = - \begin{bmatrix}
0 & 0 & \cdots & 0 \\
H_{2,1} & 0 & \cdots & 0 \\
\cdot & \cdot & & \cdot \\
\cdot & \cdot & & \cdot \\
\cdot & \cdot & & \cdot \\
H_{N,1} & H_{N,2} & \cdots & 0
\end{bmatrix}
$$

where L and U are respectively lower and upper block triangular with $H = D - L - U$, the associated block SOR iterative scheme is defined as [18]:

$$
(D - \omega L)\,\mathbf{h}^{(k+1)} = (\omega U + (1 - \omega)D)\,\mathbf{h}^{(k)} + \omega \mathbf{b} \qquad k \geq 0 \qquad (9)
$$

with column vectors $\mathbf{h}$ and $\mathbf{b}$ of the matrix problem $H\mathbf{h} = \mathbf{b}$ partitioned consistently with the partitioning of (8). Explicitly, if we write down the matrix problem under the form:

$$
\begin{bmatrix}
D_{1,1} & -U_{1,2} & \cdots & -U_{1,N} \\
-L_{2,1} & D_{2,2} & \cdots & -U_{2,N} \\
\cdot & \cdot & & \cdot \\
\cdot & \cdot & & \cdot \\
\cdot & \cdot & & \cdot \\
-L_{N,1} & -L_{N,2} & \cdots & D_{N,N}
\end{bmatrix}
\begin{bmatrix}
\mathbf{x}_1 \\ \mathbf{x}_2 \\ \cdot \\ \cdot \\ \cdot \\ \mathbf{x}_N
\end{bmatrix}
=
\begin{bmatrix}
\mathbf{b}_1 \\ \mathbf{b}_2 \\ \cdot \\ \cdot \\ \cdot \\ \mathbf{b}_N
\end{bmatrix}
$$

the block SOR algorithm (9) can be expressed as [18]:

$$
D_{i,i}\mathbf{x}_i^{(k+1)} = D_{i,i}\mathbf{x}_i^{(k)} + \omega \left\{ \sum_{j=1}^{i-1} L_{i,j}\mathbf{x}_j^{(k+1)} + \sum_{j=i+1}^{N} U_{i,j}\mathbf{x}_j^{(k)} + \mathbf{b}_i - D_{i,i}\mathbf{x}_i^{(k)} \right\}
$$

$$
1 \leq i \leq N \ , \quad k \geq 0 \qquad (10)
$$

If H is symmetric and positive definite, the block SOR scheme (10) is convergent for all $\mathbf{x}^{(0)}$ if and only if $0 < \omega < 2$ [18].

Let us apply the block SOR iterative procedure (10) to solve system (6) arising from the quasi three-dimensional finite element model of flow in our multiaquifer system consisting of m aquifers and $(m-1)$ aquitards. Each aquifer is discretized into a triangular grid of n nodes and each aquitard column into l vertical line elements, with $(l-1)$ nodes inside the aquitards. As Figure 1b shows, the system matrix H (with size s_1) is block tridiagonal, i.e. the only non-zero blocks lie on the main diagonal and above and below it ($L_{i,j}$ and $U_{i,j} = 0$ for $|i - j| > 1$). Algorithm (10) consequently reduces to the simplified expression:

$$
D_{i,i}\mathbf{x}_i^{(k+1)} = D_{i,i}\mathbf{x}_i^{(k)} + \omega \left\{ L_{i,i-1}\mathbf{x}_{i-1}^{(k+1)} + U_{i,i+1}\mathbf{x}_{i+1}^{(k)} + \mathbf{b}_i - D_{i,i}\mathbf{x}_i^{(k)} \right\}
$$

$$
1 \leq i \leq N \ , \quad k \geq 0 \qquad (11)
$$

where the partition number N (eq. (8)) is equal to the number of layers in the multiaquifer system, namely:

$$N = m + m - 1 = 2m - 1$$

The square diagonal block $D_{i,i}$ accounts for the non-zero structure of either the i-th aquifer matrix (i odd) or the i-th aquitard matrix (i even). The block size is $s_2 = n$ and $s_3 = n \cdot (l - 1)$, respectively. The rectangular extra-diagonal matrices $L_{i,i-1}$, $U_{i,i+1}$ (of dimension $s_2 \times s_3$, Figure 1b), contain the coupling terms between the i-th unit, and the unit just above and below, respectively. The unknowns are partitioned into N groups and each vector $\mathbf{x}_i$ rapresents the nodal values of the hydraulic head in the i-th unit.

Two different set of equations are generated from scheme (11):

1. When an aquifer block is considered (i odd), the corresponding subsystem is sparse and irreducible (i.e. the solution of the matrix equations cannot be reduced, by simultaneous row and column permutations, to the solution of at least two lower-order matrix equations). The MCG method is used to solve these equations. Writing each aquifer subsystem under the form:

$$\hat{H}\hat{\mathbf{h}} = \hat{\mathbf{b}}$$

the MCG equations read [15]:

$$
\begin{aligned}
\hat{\mathbf{h}}^{(k+1)} &= \hat{\mathbf{h}}^{(k)} + \alpha^{(k)}\mathbf{p}^{(k)} \\
\mathbf{p}^{(k+1)} &= K^{-1}\mathbf{r}^{(k+1)} + \beta^{(k)}\mathbf{p}^{(k)} \\
\mathbf{r}^{(k+1)} &= \hat{\mathbf{b}} - \hat{H}\hat{\mathbf{h}}^{(k+1)} = \mathbf{r}^{(k)} - \alpha^{(k)}\hat{H}\mathbf{p}^{(k)}
\end{aligned}
\tag{12}
$$

where:

$$
\alpha^{(k)} = \frac{(\mathbf{p}^T)^{(k)}\mathbf{r}^{(k)}}{(\mathbf{p}^T)^{(k)}\hat{H}\mathbf{p}^{(k)}} \quad ; \quad \beta^{(k)} = -\frac{(\mathbf{r}^T)^{(k+1)}K^{-1}\hat{H}\mathbf{p}^{(k)}}{(\mathbf{p}^T)^{(k)}\hat{H}\mathbf{p}^{(k)}}
$$

$$
\begin{aligned}
\hat{\mathbf{h}}^{(0)} &= K^{-1}\hat{\mathbf{b}} \\
\mathbf{p}^{(0)} &= K^{-1}\mathbf{r}^{(0)}
\end{aligned}
$$

and matrix K is set equal to:

$$K = \mathcal{L}\mathcal{L}^T$$

where $\mathcal{L}$ is the incomplete Cholesky factor of $\hat{H}$ [14] with the same sparsity pattern as $\hat{H}$.

2. When an aquitard block is solved (i even), the subsystem is tridiagonal and reducible. The system (of dimension s_3) can be reduced to n subsystems of dimension s_4, $s_4 = (l - 1)$ (Figure 1b), corresponding to each single vertical column equation. The Thomas algorithm is used to solve each single aquitard column:

$$\check{H}\check{\mathbf{h}} = \check{\mathbf{b}} \tag{13}$$

Factorizing $\check{H}$ with the product LU:

$$
\begin{bmatrix}
\check{a}_1 & \check{c}_1 & & & 0 \\
\check{b}_2 & \check{a}_2 & \check{c}_2 & & \\
& \cdot & \cdot & \cdot & \\
& & \cdot & \cdot & \cdot \\
0 & & & \check{b}_{l-1} & \check{a}_{l-1}
\end{bmatrix}
=
\begin{bmatrix}
1 & & & & 0 \\
l_2 & 1 & & & \\
& \cdot & \cdot & & \\
& & \cdot & \cdot & \\
0 & & & l_{l-1} & 1
\end{bmatrix}
\cdot
\begin{bmatrix}
u_1 & \check{c}_1 & & & 0 \\
& u_2 & \check{c}_2 & & \\
& & \cdot & \cdot & \\
& & & \cdot & \cdot \\
0 & & & & u_{l-1}
\end{bmatrix}
$$

where:

$$u_1 = \breve{a}_1$$
$$l_i = \frac{\breve{b}_i}{u_i} \quad ; \quad u_i = \breve{a}_i - l_i \breve{c}_{i-1} \qquad 2 \leq i \leq (l-1)$$

the solution of system (13) is obtained by forward and back substitution [13]:

$$g_1 = \breve{b}_1 \quad ; \quad g_i = \breve{b}_i - l_i g_{i-1} \qquad 2 \leq i \leq (l-1)$$
$$\breve{h}_{l-1} = \frac{g_{l-1}}{u_{l-1}} \quad ; \quad \breve{h}_i = \frac{g_i - \breve{c}_i \breve{h}_{i+1}}{u_i} \qquad (l-2) \leq i \leq 1$$

The iterative procedure (11) allows us to easily overcome the non-linearity in the aquitards. The coefficients of the aquitard stiffness and capacity matrices (eq. (5)) can be updated after each iteration to account for the dependence of both hydraulic conductivity and elastic storage coefficient on the effective stress which is in turn related to the hydraulic head.

4. NUMERICAL RESULTS FOR LINEAR PROBLEMS

The block iterative strategies have been applied to linear porous media where a theoretical analysis of the SOR convergence properties can be better performed.

Three sample circular multiaquifer systems are used to study the asymptotic rate of convergence of the SOR scheme and to compare the performance of the "coupled" and the new "decoupled" solving strategies. The systems are composed of 2 aquifers and 1 intervening aquitard (system A), 6 aquifers and 5 aquitards (system B), 11 aquifers and 10 aquitards (system C). The aquifers are discretized into the same grid with 324 triangles and 169 nodes and each vertical aquitard column into 6 (test case 1) and 11 (test case 2) linear elements, with 5 and 10 interior nodes, respectively. Figure 2 shows a three-dimensional sketch of system B and Table 1 gives for each system and test case the total number of nodes s_1, while the characteristic block sizes are $s_2 = 169$, $s_3 = 845$ (test case 1) or 1690 (test case 2) and $s_4 = 5$ (test case 1) or 10 (test case 2).

The boundary conditions are zero hydraulic head on the outer boundary nodes (Figure 2), and a unit pumping rate from the central nodes of both the shallowest and lowest aquifers. The capacity matrix is lumped into a diagonal matrix, i.e. a lumped formulation is used. Constant hydrogeological parameters are assumed in each formation with $T_x = T_y = T$.

A few steady-state simulations are run for various values of the dimensionless parameter $G = (K_z b_z)/T$ where b_z is the constant aquitard thickness. G is representative of the degree of aquifer hydraulic coupling. Decreasing G implies a relative decrease of the aquitard permeability, hence flow is more restricted to the pumped layers and coupling is less important. Figure 3 shows the hydraulic head changes along the radial-symmetry axis in system B. If $G = 10^{-5}$, the perturbation is confined to the units just above and just below the pumped aquifers, while with $G = 10^{-1}$, all the units are affected by a significant hydraulic head variation.

Experience shows that the steady state results are to some extent quite representative of the results obtained with transient simulations as well.

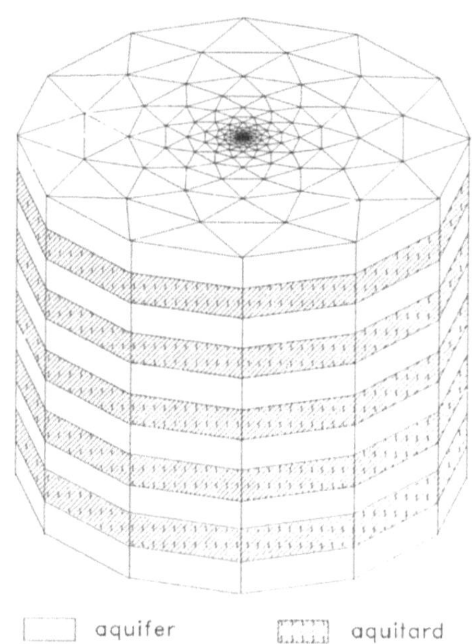

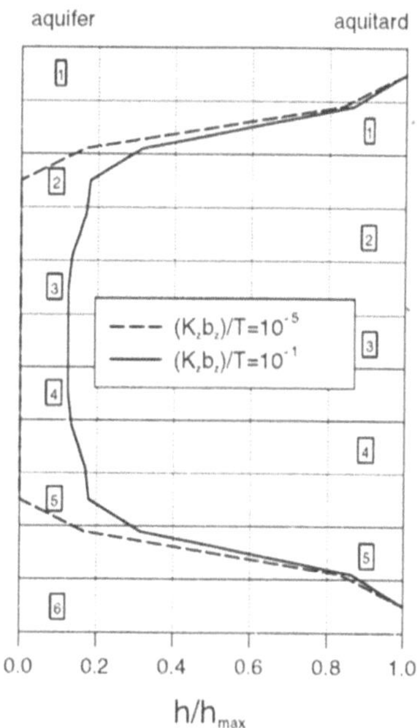

Figure 2: Three-dimensional sketch showing the finite element mesh into which the multiaquifer system is discretized (system B).

Figure 3: Dimensionless hydraulic head changes h/h_{max} along the radial-symmetry axis for weakly ($G = (K_z b_z)/T = 10^{-5}$) and pronounced ($G = (K_z b_z)/T = 10^{-1}$) aquifer hydraulic coupling (system B).

	SYSTEM A		SYSTEM B		SYSTEM C	
	Test case 1	Test case 2	Test case 1	Test case 2	Test case 1	Test case 2
s_1	1183	2028	5239	9464	10309	18759

$$s_1 = m \cdot n + (m - 1) \cdot n \cdot (l - 1)$$

$$m = 2 \text{ (Sys. A)}, \ 6 \text{ (Sys. B)}, \ 11 \text{ (Sys. C)}$$

$$n = 169$$

$$l = 5 \text{ (Test case 1)}, \ 10 \text{ (Test case 2)}$$

Table 1: Dimension of the global system of equations used in the present analysis.

4.1 Block Gauss-Seidel, Block SOR and MCG Convergence Profiles

Convergence of the block Gauss-Seidel and block SOR iterative schemes is monitored by both numerically computing the Eucledian norm $|\mathbf{r}^{(k+1)}|$ of the residual of the global system (eq. (7)), and evaluating the asymptotic rate of convergence R, which is theoretically defined as $R = -log\ \rho(E)$ where $\rho(E)$ is the spectral radius of the iteration matrix E [18]:

$$E = (D - \omega L)^{-1}(\omega U + (1 - \omega)D)$$

The iterative procedure is completed when $|\mathbf{r}^{(k+1)}|$ becomes smaller than TOL (set to 10^{-14} in the present analysis) or the specified maximum number of iterations IMAX is exceeded (IMAX=1000).

Table 2 gives the asymptotic rate of convergence R and ω_{opt} versus G for the three systems and the two test cases considered. Figures 4 and 5 display the convergence profiles of the block Gauss-Seidel and block SOR scheme with $\omega = \omega_{opt}$ for systems A and B (system C performs very much the same as system B) and test case 2. Note that R is computed as the logarithm of the ratio:

$$R = -log\frac{|\mathbf{r}^{(k+1)}|}{|\mathbf{r}^{(k)}|} \tag{14}$$

between two successive Euclidean norms of the residual for sufficiently large k (see Appendix), while ω_{opt} is evaluated by the equation [13]:

$$\omega_{opt} = \frac{2}{1 + \sqrt{1 - \rho_{GS}}} \tag{15}$$

where ρ_{GS} is the spectral radius of the block Gauss-Seidel iterative matrix, i.e. E with $\omega = 1$. The empirical computation of ω_{opt} vs the number of iterations provides values which are pretty much the same. Hence the theoretical ω_{opt} is a very reliable estimate of the actual ω_{opt} (despite the fact that H does not possess property A and is not consistently ordered [19]).

	$\frac{K_z b_z}{T}$	SYSTEM A			SYSTEM B			SYSTEM C		
		R_{GS}	R_{SOR}	ω_{opt}	R_{GS}	R_{SOR}	ω_{opt}	R_{GS}	R_{SOR}	ω_{opt}
Test case 1	10^{-5}	2.890	3.000	1.00	3.000	3.048	1.00	3.000	3.301	1.00
	10^{-4}	1.963	2.886	1.01	1.924	2.036	1.01	1.921	2.036	1.01
	10^{-3}	1.015	1.682	1.03	0.796	1.291	1.04	0.881	1.192	1.04
	10^{-2}	0.297	0.672	1.17	0.183	0.528	1.26	0.181	0.530	1.26
	10^{-1}	0.041	0.254	1.54	0.024	0.184	1.62	0.022	0.210	1.64
Test case 2	10^{-5}	2.585	2.920	1.00	2.745	3.000	1.00	2.745	3.000	1.00
	10^{-4}	1.680	2.620	1.01	1.684	1.845	1.01	1.724	1.896	1.01
	10^{-3}	0.794	1.287	1.04	0.585	1.082	1.08	0.606	1.143	1.07
	10^{-2}	0.186	0.585	1.26	0.108	0.432	1.36	0.106	0.423	1.37
	10^{-1}	0.022	0.197	1.63	0.013	0.134	1.70	0.012	0.127	1.72

Table 2: Asymptotic rate of convergence R_{GS} and R_{SOR} of block Gauss-Seidel and optimal SOR, and ω_{opt} versus $G = (K_z b_z)/T$.

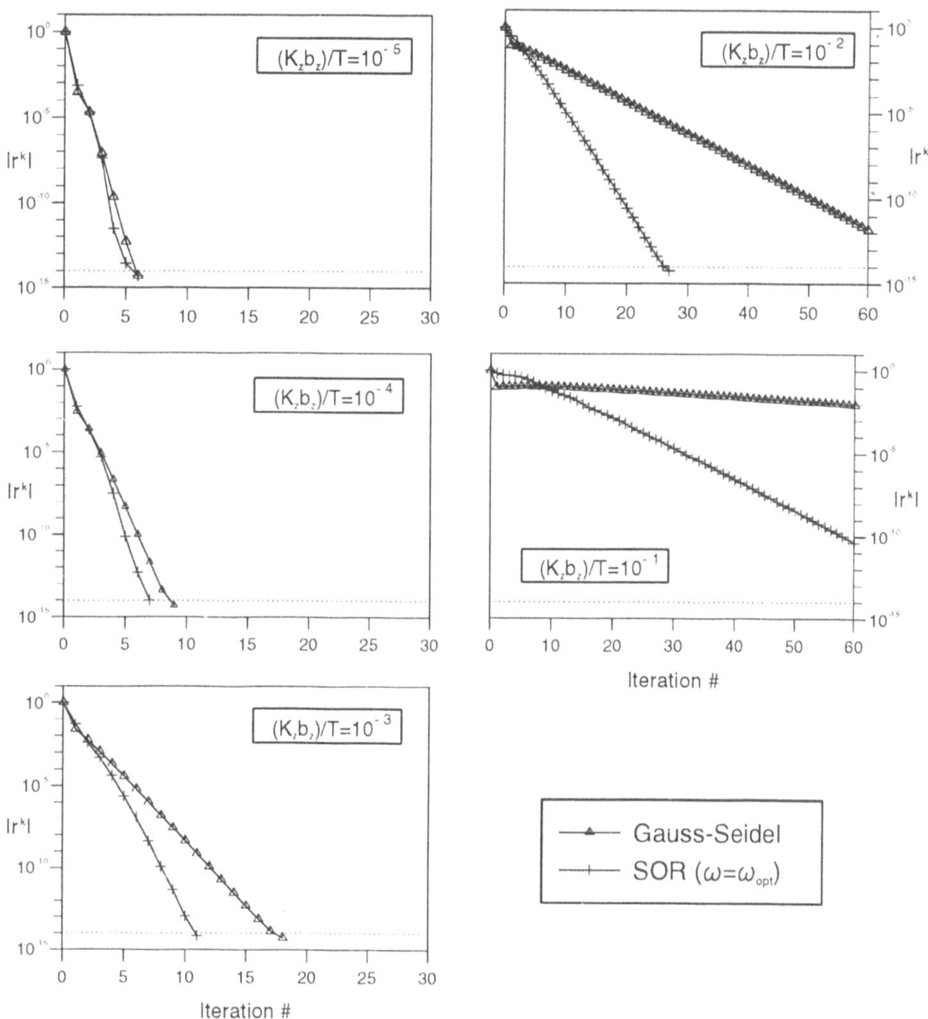

Figure 4: Convergence profiles in steady state simulations for different values of the dimensionless parameter $G = (K_z b_z)/T$ (system A and test case 2). On the vertical axis the Euclidean norm of the residual $|\mathbf{r}^{(k)}|$ is given.

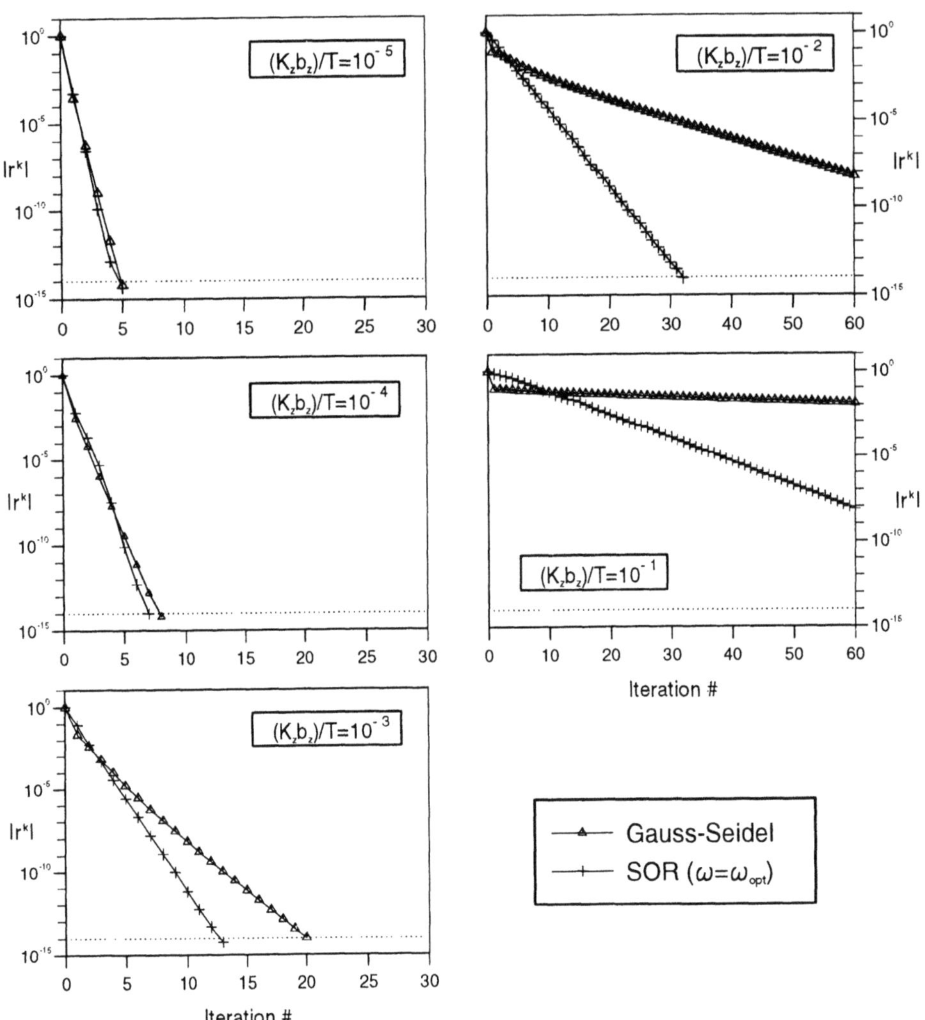

Figure 5: Convergence profiles in steady state simulations for different values of the dimensionless parameter $G = (K_zb_z)/T$ (system B and test case 2). On the vertical axis the Euclidean norm of the residual $|\mathbf{r}^{(k)}|$ is given.

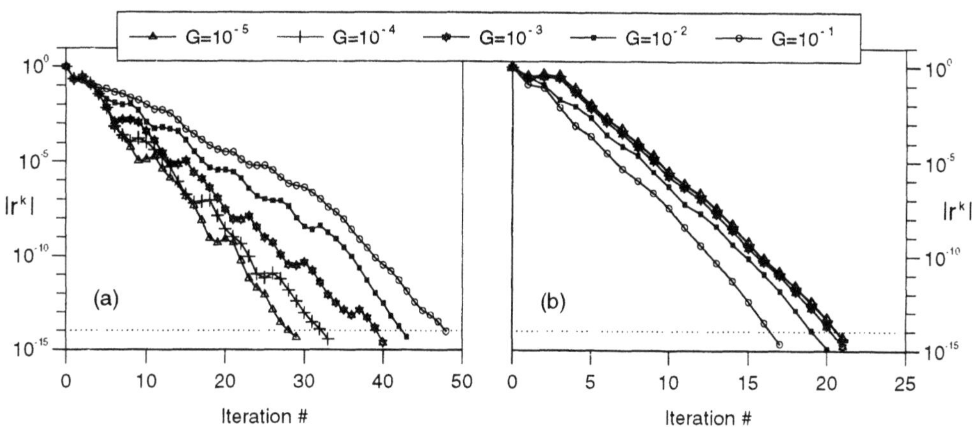

Figure 6: Convergence profiles in steady state simulations for different values of the dimensionless parameter $G = (K_z b_z)/T$ when MCG is applied to solve (a) the global system and (b) a single aquifer block (system B and test case 2).

Figures 4 and 5 and Table 2 reveal that the asymptotic rate of convergence of both the block Gauss-Seidel and optimal SOR is very much dependent on G. If coupling between the aquifers, i.e. G, decreases (the importance of the $L_{i,i-1}$ and $U_{i,i+1}$ coefficients becomes less pronounced compared to the $D_{i,i}$ coefficients), convergence improves and Gauss-Seidel approaches optimal SOR ($\omega_{opt} \to 1$). By contrast, R markedly decreases for high G values in which case ω_{opt} moves toward 2. Note that, for a strong hydraulic coupling within the system, optimal SOR behaves much better than Gauss-Seidel and R is one order of magnitude larger.

The convergence properties of MCG as applied to solve the global system (with size s_1, eqs. (12)) are also dependent on G, although to a lesser extent (Figure 6a). While the ratio between the SOR iterations at $G = 10^{-1}$ and $G = 10^{-5}$ is larger than 10, the corresponding MCG ratio is less than 2. When MCG is used in the "decoupled" strategy to solve the aquifer blocks (with size s_2), convergence occurs as is displayed in Figure 6b. It is interesting to notice that the MCG iterations needed to solve the aquifer and the global equations increase approximately from 20 to 40 (Figure 6) while the size of the corresponding systems grows from 169 to 9464 (Table 1).

4.2 Computational efficiency of the "coupled" and block "decoupled" solution schemes

The performance of the "coupled" (MCG) and "decoupled" (Gauss-Seidel and optimal SOR) strategies for solving system (6) is compared in terms of both computer storage and CPU time.

Table 3 shows the computer storage required by the corresponding codes. The iterative "decoupled" schemes are significantly less storage demanding than MCG. The advantage

| | Computer Storage | | Ratio |
	coupled (MCG)	decoupled (SOR)	$\dfrac{\text{Storage}_{\text{MCG}}}{\text{Storage}_{\text{SOR}}}$
SYSTEM A test case 1	496433	210661	2.36
SYSTEM A test case 2	723505	237741	3.04
SYSTEM B test case 1	1754154	316981	5.53
SYSTEM B test case 2	2888058	452221	6.39
SYSTEM C test case 1	3324986	490901	6.77
SYSTEM C test case 2	5589074	761461	7.34

Table 3: Computer storage (bytes) required for the simulations. Gauss-Seidel and SOR need the same memory.

grows for more complex multiaquifer systems and a large number of elements.

CPU times on an IBM Risc6000/560 vs G are provided in Figure 7 for system B. Careful inspection of Figure 7 points out that:

1. "Coupled" MCG is generally faster than the "decoupled" procedures and is much superior in strongly hydraulically coupled systems. The block iterative methods are slightly faster only if hydraulic coupling is weak (Figure 7).

2. The efficiency of the block "decoupled" strategies is higher for complex aquifer systems (Figure 7).

3. Increasing the number of aquitard elements leads to a better performance of the block iterative approach.

4.3 Discussion

The analysis of the outcome from linear multiaquifer simulations emphasizes some significant aspects:

- "Coupled" MCG is quite appropriate to linear porous media. Although the block iterative scheme needs less computer storage and is slightly faster when the system is weakly coupled and a large number of finite elements is used overall, the performance of MCG is generally better.

- An optimum over-relaxation factor ω_{opt} can be theoretically determined, which allows a higher SOR convergence rate than the Gauss-Seidel iteration. The importance of using $\omega = \omega_{opt}$ decreases as the hydraulic coupling becomes weaker. However, when coupling

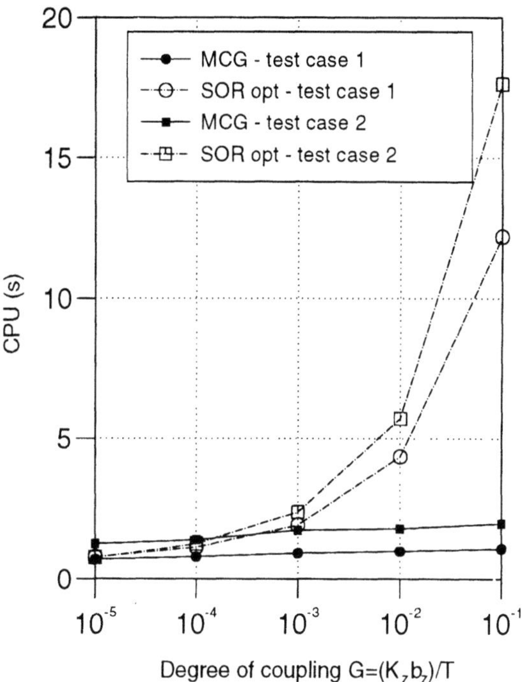

Figure 7: CPU times of the "coupled" MCG and "decoupled" optimal SOR schemes (system B).

is weak, the computational efficiency of SOR is very high anyway, the hydraulic head in each units is practically independent and, consequently, the global system solution can be obtained by simply solving each subsystem separately.

- For strongly coupled aquifers optimum block SOR converges much faster than block Gauss-Seidel. Hence a similar performance may be expected in non-linear systems where MCG is not guaranteed to converge and cannot be used.

5. PRELIMINARY RESULTS FOR NON-LINEAR PROBLEMS

When non-linear behavior of the hydrogeologic aquitard properties is introduced, the "coupled" MCG procedure may not converge and cannot be applied, and an iterative strategy of the kind developed in the present paper is to be used. Hence, the assessment of an optimum relaxation factor to accelerate convergence is of great practical interest.

Some non-linear simulations are performed with the same sample multiaquifer systems of the linear analysis, and similar initial and boundary conditions. As the simulations are steady-state, only the dependence of the vertical permeability K_z on the solution h is assumed to account for the non-linear aquitard behavior. According to *Rudolph and Frind* [20] we take:

$$dK_z = K_{z0} \left(10^{dh/m_1} - 1 \right) \tag{16}$$

where K_{z0} is the initial permeability and m_1 is a coefficient depending on the clay type. At each iteration K_z is updated in every aquitard element according to eq. (16).

Early non-linear results confirm the outcome from the linear simulations. An optimum relaxation factor ω_{opt}, providing a faster SOR convergence than Gauss-Seidel, can be numerically computed for strongly interconnected aquifer systems.

Figure 8 shows the results of the analysis for system B (test case 2) where the initial values of G is set to 10^{-1}, $m_1 = 5.5$ and the final minimum G turns to be equal to $2.1 \cdot 10^{-3}$. In Figure 8a, the number of iterations needed to solve the non-linear problem is plotted vs ω and in Figure 8b the convergence profiles for some representative underrelaxation and overrelaxation ω values are presented.

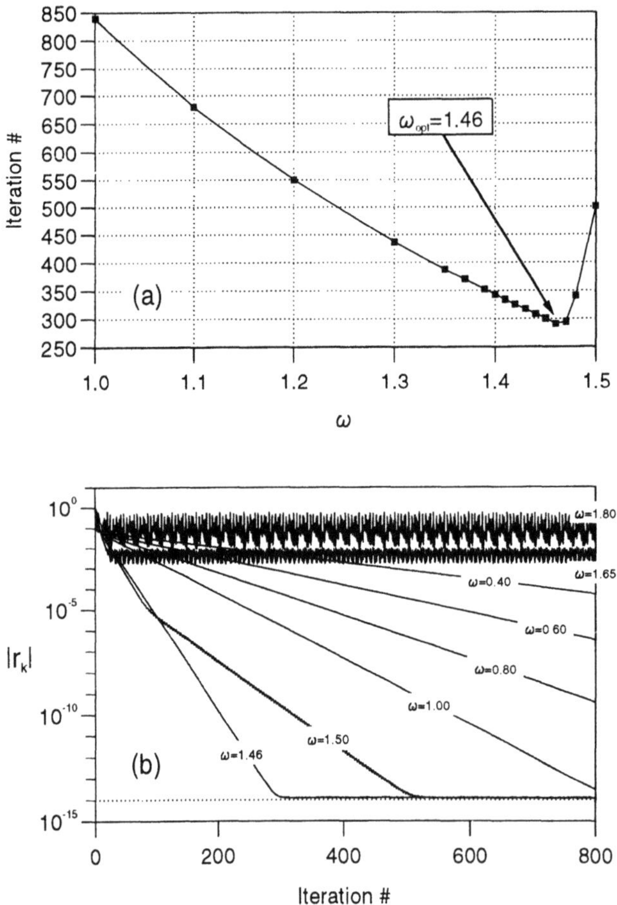

Figure 8: Non-linear steady state simulations with initial $G = (K_z b_z)/T = 10^{-1}$ and $m_1 = 5.5$ (system B and test case 2): (a) number of iterations vs ω and (b) convergence profiles for various ω values exploring both the underrelaxation and overrelaxation intervals.

Inspection of Figure 8 reveals that:

1. The theoretical calculation of ω_{opt} by (15) can no more be used. Actually the empirical computation of ω_{opt} vs the number of iterations provides a lower value (Figure 8*a* indicates an "empirical" $\omega_{opt,em}{=}1.46$ against a "theoretical" $\omega_{opt,th}{=}1.65$).

2. SOR does not converge for $\omega = \omega_{opt,th}$ (Figure 8*b*).

3. Optimal SOR with $\omega = \omega_{opt,em}$ is three times faster than Gauss-Seidel. Hence the acceleration of convergence is significant.

4. The SOR convergence behavior is very much dependent on ω:

$0 < \omega \le 1$ (underrelaxation) and $1 < \omega \le \omega_{opt,em}$	convergence is monotonic
$\omega_{opt,em} < \omega < \omega_1 < \omega_{opt,th}$	the residual starts to oscillate but the procedure is still convergent
$\omega > \omega_1$	oscillations become increasingly large and final convergence is not achieved

6. CONCLUSIONS

Block SOR iterative solution methods for a quasi - three-dimensional finite element model of flow in linear and non-linear multiaquifer systems have been developed. These methods are naturally suggested by the structure of the multiaquifer model and by the special sparsity pattern of the coefficient matrix. An optimum SOR factor ω_{opt} can be theoretically computed in linear problems using the spectral radius of the Gauss-Seidel iteration matrix (SOR with $\omega = 1$). The results from some representative sample problems show that the block SOR asymptotic rate of convergence is very fast if aquifer flow is weakly coupled. In this case, the excellent performances of optimal SOR and Gauss-Seidel tend to coincide ($\omega_{opt} \simeq 1$). If aquifer flow is strongly coupled, both Gauss-Seidel and optimal SOR converge much slowlier. However, the latter turns out to be one order of magnitude faster than the former. By distinction, MCG is computationally superior to SOR in hydraulically coupled systems, but is more computer storage demanding. The relative performance of optimal SOR tends to improve when the geometry of the multiaquifer system becomes more complex and a large number of elements are used, and outperforms MCG if hydraulic coupling is weak.

The new solution strategy has been extended to non-linear multiaquifer systems, where MCG may not converge and an iterative solution scheme of the kind developed in the present paper appears to be the most natural solver to overcome the non-linearity of the equations. Early results show that an optimal SOR factor can be numerically evaluated and the related acceleration of convergence can be very significant in hydraulically interconnected systems.

Acknowledgments. This work was developed in the CNR project "Sistema Lagunare Veneziano", Linea di Ricerca 2-7, U.O. 2.

7. APPENDIX: CALCULATION OF THE SPECTRAL RADIUS OF THE ITERATION MATRIX

We justify eq. (14) according to which the spectral radius $\rho(E)$ of the iteration matrix E may be obtained as the limiting ratio between two successive Euclidean norms of the residual.

Given the linear system $H\mathbf{h} = \mathbf{b}$, a first-degree stationary iterative procedure is commonly written in the form:

$$\mathbf{h}^{(k+1)} = E\mathbf{h}^{(k)} + \mathbf{q} \tag{17}$$

where E is known as "iteration matrix" and vector $\mathbf{q} = (I - E)H^{-1}\mathbf{b}$. Defining the k-th "residual vector", $\mathbf{r}^{(k)} = \mathbf{b} - H\mathbf{h}^{(k)}$, and the k-th "error vector", $\mathbf{e}^{(k)} = \mathbf{h} - \mathbf{h}^{(k)}$ where $\mathbf{h}$ is the solution vector, $\mathbf{r}^{(k)} = He^{(k)}$ follows.

The initial error $\mathbf{e}^{(0)} = \mathbf{h} - \mathbf{h}^{(0)}$ is an arbitrary vector that can be expressed as a linear combination of the independent eigenvectors $\mathbf{v_1}, \mathbf{v_2}, \ldots, \mathbf{v_n}$ of E associated with the ordered eigenvalues $\lambda_1, \lambda_2, \ldots, \lambda_n$ ($|\lambda_1| > |\lambda_2| > \ldots > |\lambda_n|$):

$$\mathbf{e}^{(0)} = d_1\mathbf{v_1} + d_2\mathbf{v_2} + \ldots + d_n\mathbf{v_n} \tag{18}$$

Subtracting eq. (17) from $\mathbf{h} = E\mathbf{h} + \mathbf{q}$ yields:

$$\mathbf{h}^{(k+1)} - \mathbf{h} = E(\mathbf{h}^{(k)} - \mathbf{h})$$

namely:

$$\mathbf{e}^{(k+1)} = E\mathbf{e}^{(k)} = E^{k+1}\mathbf{e}^{(0)} \tag{19}$$

Eq. (18) and eq. (19) give:

$$\begin{aligned}
\mathbf{e}^{(k+1)} &= E^{k+1}(d_1\mathbf{v_1} + d_2\mathbf{v_2} + \ldots + d_n\mathbf{v_n}) \\
&= d_1 E^{k+1}\mathbf{v_1} + d_2 E^{k+1}\mathbf{v_2} + \ldots + d_n E^{k+1}\mathbf{v_n} \\
&= \lambda_1^{k+1} d_1\mathbf{v_1} + \lambda_2^{k+1} d_2\mathbf{v_2} + \ldots + \lambda_n^{k+1} d_n\mathbf{v_n}
\end{aligned} \tag{20}$$

since the eigenvectors of E^{k+1} are tha same as those of E and the eigenvalues are equal to the $(k+1)$-th power of the eigenvalues of E.

If k is sufficiently large, from eq. (20) we have:

$$\mathbf{e}^{(k+1)} \simeq \lambda_1^{k+1} d_1\mathbf{v_1}$$

$$\mathbf{e}^{(k)} \simeq \lambda_1^{k} d_1\mathbf{v_1}$$

$$\frac{|\mathbf{e}^{(k+1)}|}{|\mathbf{e}^{(k)}|} = \frac{|H^{-1}\mathbf{r}^{(k+1)}|}{|H^{-1}\mathbf{r}^{(k)}|} \simeq \frac{|\lambda_1^{k+1} d_1 H^{-1}\mathbf{v_1}|}{|\lambda_1^{k} d_1 H^{-1}\mathbf{v_1}|} = |\lambda_1| = \rho(E)$$

REFERENCES

1. Herrera, I. and G. E. Figueroa Vega, A correspondence principle for the theory of leaky aquifers, *Water Resour. Res.* 5, 900–904, 1969.

2. Herrera, I., Theory of multiple leaky aquifers, *Water Resour. Res.* 6, 185–193, 1970.

3. Herrera, I. and R. Yates, Integrodifferential equations for systems of leaky aquifers and applications, 3, A numerical method of unlimited applicability, *Water Resour. Res.* 13(4), 725–732, 1977.

4. de Marsily, G., E. Ledoux, A. Levassor, D. Poitrinal and A. Salem, Modeling of large multiaquifer systems: Theory and application, *J. Hydrol.* 36, 1–33, 1978.

5. Hennart, J. P., R. Yates and I. Herrera, Extension of the integrodifferential approach to inhomogeneous multiaquifer systems, *Water Resour. Res.* 17, 1044–1050, 1981.

6. Premchitt, J. A., A technique in using integrodifferential equations for model simulation of multiaquifer systems, *Water Resour. Res.* 17, 162–168, 1981.

7. Gambolati, G., F. Sartoretto and F. Uliana, A conjugate gradient finite element model of flow for large multiaquifer systems, *Water Resour. Res.* 22(7), 1003–1015, 1986.

8. Fujinawa, K., Finite element analysis of groundwater flow in multiaquifer systems, 2, A quasi three dimensional flow model, *J. Hydrol.* 33, 349–362, 1977.

9. Chorley, D. W. and E. O. Frind, An iterative quasi three-dimensional finite element model for heterogeneous multiaquifer systems, *Water Resour. Res.* 14(5), 943–952, 1978.

10. Neuman, S. P., C. Preller and T. N. Narashiman, Adaptive explicit-implicit quasi three-dimensional finite element model of flow and subsidence in multiaquifer systems, *Water Resour. Res.* 18(5), 1551–1561, 1982.

11. Rivera, A., *Modèle hydrogéologique quasi-tridimensionnel non-linéaire pour simuler la subsidence dans les systèmes aquifères multicouches. Cas de Mexico.* PhD thesis, Ecole des Mines de Paris-CIG, 1990.

12. Hantush, M. S., Modification of the theory of leaky aquifers, *J. Geophys. Res.* 65, 3713–3725, 1960.

13. Westlake, J. R., *Numerical Matrix Inversion and Solution of Linear Equations.* John Wiley, New York, 1968.

14. Kershaw, D. S., The incomplete Cholesky-conjugate gradient method for the iterative solution of systems of linear equations, *J. Comp. Phys.* 26, 43–65, 1978.

15. Gambolati, G., Fast solution to finite element flow equations by Newton iteration and modified conjugate gradient method, *Int. J. Numer. Methods Eng.* 15, 661–675, 1980.

16. Gambolati, G., Perspective on a modified conjugate gradient method for the solution of linear sets of subsurface equations. In: Wang, S. Y. and et al. (eds.) *3rd Int. Conf. Finite Elements in Water Resources.* Missisipi University Press, pp 2.15–2.30, 1980.

17. Gambolati, G. and A. M. Perdon, The conjugate gradients in flow and land subsidence modeling. In: Bear, J. and Y. Corapcioglu (eds.) *Fundamentals of Transport Phenomena in Porous Media.* NATO-ASI Series, Applied Sciences 82, Martinus Nijoff B.V., The Hague, pp 953–984, 1984.

18. Varga, R. S., *Matrix Iterative Analysis*. Prentice-Hall, Englewood Cliffs, New Jersey, 1962.

19. Young, D., Iterative methods for solving partial differential equations of the elliptic type, *Trans. Amer. Math. Soc.* 76, 92–111, 1954.

20. Rudolph, D. and E. O. Frind, Hydraulic response of highly compressible aquitards during consolidation, *Water Resour. Res.* 27(1), 17–30, 1991.

MODELING VARIABLY SATURATED FLOW PROBLEMS USING NEWTON-TYPE LINEARIZATION METHODS

C. Paniconi

CRS4, Cagliari, Italy

and

M. Putti

University of Padua, Padua, Italy

ABSTRACT

Numerical procedures to solve the nonlinear equation governing flow in variably saturated porous media commonly involve Newton or Picard iteration. The former scheme is stable and quadratically convergent in a local sense, but costly and algebraically complex. The latter scheme is simple and cheap, but slower converging and not as robust. We present a common framework for comparing these two methods, and introduce other approaches that range from simplifications of the Picard scheme to approximations of Newton's method. These other approaches include explicit discretizations, first and second order accurate linearizations, and quasi-Newton schemes. Relaxation and line search algorithms to accelerate convergence of the Picard, Newton, and quasi-Newton methods will also be considered. The effectiveness of these various iterative and noniterative methods will be assessed according to criteria of efficiency and robustness.

1. INTRODUCTION

Numerical procedures for solving large scale nonlinear problems are computationally intensive and require highly efficient and robust algorithms. Efficiency ensures optimal utilization of CPU and storage resources to attain a desired level of solution accuracy, while robustness implies that a given algorithm exhibits acceptable convergence behavior across a wide spectrum of simulation scenarios. The governing equation for flow in partially saturated porous media, Richards' equation, contains nonlinearities arising from pressure head dependencies in soil moisture and hydraulic conductivity (Figure 1).

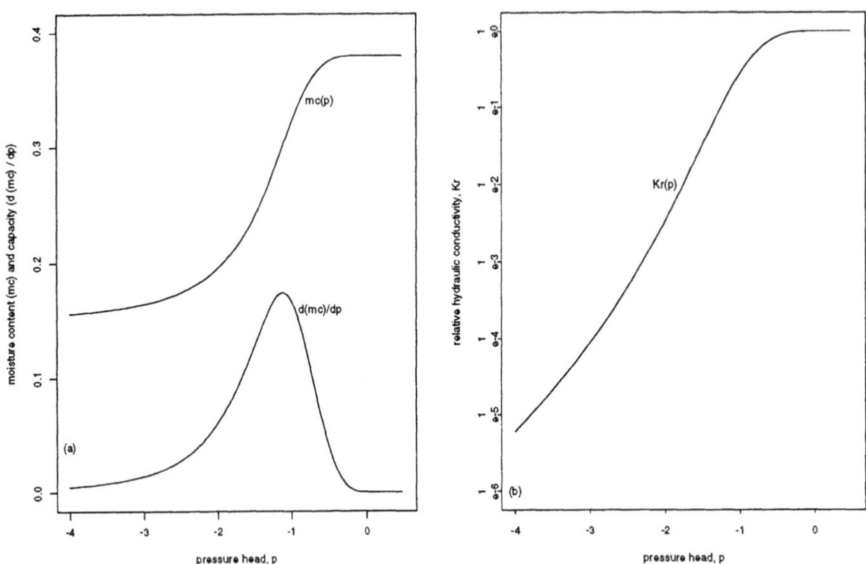

Figure 1: Typical soil curves showing the nonlinear dependencies in moisture content and hydraulic conductivity on pressure head.

In solving Richards' equation numerically, an implicit two-level time discretization is often applied, ensuring stability of the overall scheme. Picard or Newton iteration is then commonly used to linearize the resulting system of equations. The Picard method, also known as successive approximation or "simple" iteration, is a straightforward procedure which preserves symmetry of the system, is easy to implement, and is computationally inexpensive. However, the method may fail or converge very slowly under certain circumstances (such as pronounced gravity drainage zones, complex time-varying boundary conditions, strongly nonlinear characteristic equations, saturated/unsaturated interfaces, and steady state simulations). Newton iteration has better local stability and convergence properties (it is quadratically convergent), but it is more expensive per iteration (requiring evaluation and assembly of the nonsymmetric Jacobian matrix), and its global properties are not well understood (the Newton method often requires a very good initial solution estimate for convergence, and it can fail under similar conditions as Picard).

To enhance the performance of the Picard and Newton schemes, strategies such as line search and relaxation, chord slope approximation for the derivatives of the characteristic equations, parameter stepping, and mixed Picard-Newton iteration can be used. On the other hand, one can try altogether different approaches such as noniterative explicit or implicit factored schemes, or iterative quasi-Newton methods. The noniterative methods do away with the computational costs associated with repeated assembly and solution at each time step, but maintaining accuracy and stability can be tricky.

The basic idea behind quasi-Newton methods is to replace the Jacobian evaluation step of the Newton scheme with a less costly approximation to either the Jacobian or its inverse.

The methods are expressed in the form of easily calculated updates to the Jacobian or inverse Jacobian approximation. An effective quasi-Newton method will be less expensive than Newton iteration, and it will have good theoretical properties, such as superlinear convergence. Moreover, judicious choice of the initial Jacobian approximation can overcome the Newton method's sensitivity to initial solution estimates. Finally, in finite element applications, it is also important that the updates do not destroy the sparsity pattern of the system matrices. We will consider two popular quasi-Newton schemes, the Broyden and BFGS algorithms. The algorithms are coded in their inverse form, and efficient sparsity-preserving recursion formulae are used to calculate the updates. Storage and CPU requirements are further reduced by "limited memory" implementations that maintain only a specified number of iteration levels for evaluation of the updates.

Relaxation is often used to accelerate convergence of the Picard, Newton, or quasi-Newton methods. Line search algorithms systematically compute the relaxation parameter by finding the optimal step length to be taken along the search direction indicated by the iterative scheme.

We describe the implementation of these various iterative and noniterative methods, and report on some numerical tests conducted for one, two, and three-dimensional problems involving both steady state and transient flow.

2. GOVERNING EQUATIONS AND NUMERICAL PROCEDURES

2.1 Richards' Equation and Finite Element Models

Richards' equation is obtained by combining Darcy's law with the continuity equation [1]. Expressing this equation with pressure head ψ as the dependent variable, t as time, and z as the vertical coordinate (positive upward) yields

$$\eta(\psi)\frac{\partial\psi}{\partial t} = \nabla \cdot (K_s K_r(\psi) \nabla(\psi + z)) \tag{1}$$

where $\eta(\psi)$ is the general storage term or overall storage coefficient and the hydraulic conductivity tensor K is expressed as a product of the conductivity at saturation, K_s and the relative conductivity, $K_r(\psi)$. Equation (1) is highly nonlinear due to pressure head dependencies in the storage and conductivity terms.

To solve equation (1) numerically, a finite element Galerkin discretization in space with linear basis functions is used. Triangular elements are used in the two-dimensional code, and either tetrahedral or hexahedral elements in three dimensions. With tetrahedra the nonlinear coefficients in the system integrals are evaluated at the element centroids, whereas with hexahedral elements order 2 Gaussian quadrature is used to evaluate the integrals. A λ-weighted finite difference scheme is used for time discretization ($\lambda = 0.5$, Crank-Nicolson; $\lambda = 1$, backward Euler). Details of the numerical procedures can be found in standard texts, for instance [2, 3]. Discretization yields the system of nonlinear equations

$$g(\psi^{k+1}) \equiv A(\psi^{k+\lambda})\psi^{k+\lambda} + F(\psi^{k+\lambda})\frac{\psi^{k+1} - \psi^k}{\Delta t^{k+1}} + b(\psi^{k+\lambda}) - q(t^{k+\lambda}) = 0 \tag{2}$$

where g is the residual vector, $\psi^{k+\lambda} = \lambda\psi^{k+1} + (1 - \lambda)\psi^k$, ψ is the vector of nodal pressure heads, Δt is the time step size, superscript k denotes time level, A is the stiffness matrix, F is the storage or mass matrix, b contains the gravitational gradient component of equation (1), and q contains the specified Darcy flux boundary conditions.

The numerical models have the option of using either distributed or lumped mass matrices. The models can handle a variety of boundary conditions, including atmospheric inputs, seepage faces, and source/sink terms such as pumping wells. Atmospheric boundary conditions (rainfall and evaporation) are dynamic in nature, and at each iteration and time step the numerical model computes the flux and pressure head status at each soil surface node to determine whether the boundary condition at that node is to be switched from a Dirichlet (specified head) to Neumann (specified flux) condition, or vice versa [4]. Seepage faces are treated by a variant of the method described in [5, 6], where we have introduced two methods of updating the exit points, allowing us to relax the convergence requirements along seepage faces if desired [7].

2.2   Linear Solvers

One of the main drawbacks of the Newton scheme used to be the inefficiency of linear solvers for large, sparse nonsymmetric systems. This is no longer the case, as currently available conjugate gradient-type algorithms for solving nonsymmetric systems have become increasingly reliable and efficient. The solvers implemented in our models include the biconjugate gradient stabilized algorithm, BICGSTAB, the minimum residual algorithm, GRAMRB, a generalized conjugate residual method, GCRK, and the transpose-free quasi-minimal residual algorithm, TFQMR. All these schemes can be used with various preconditioners, such as incomplete Croute (LU) decomposition. Descriptions of the various algorithms can be found in [8, 9, 10, 11]. For the symmetric systems generated by Picard linearization, we use the incomplete Cholesky conjugate gradient method, ICCG [12, 13]. For small, one-dimensional simulations, a tridiagonal direct solver is available and can be used for both symmetric and nonsymmetric systems.

2.3   Characteristic Equations and Chord Slope Approximations

The nonlinear storage and conductivity terms in equation (1) can be modeled using various constitutive or characteristic relations describing the soil hydraulic properties. One of the simplest of these, useful for steady state simulations, is the exponential $K_r(\psi)$ relationship [14]

$$K_r(\psi) = \exp(\kappa\psi) \tag{3}$$

where κ is a constant.

The characteristic equations introduced in [15] are commonly used. These can be written as

$$\begin{aligned}
\theta(\psi) &= \theta_r + (\theta_s - \theta_r)[1 + \beta]^{-m} & \psi < 0 \\
\theta(\psi) &= \theta_s & \psi \geq 0
\end{aligned} \tag{4}$$

$$\begin{aligned}
K_r(\psi) &= (1 + \beta)^{-5m/2}\,[(1 + \beta)^m - \beta^m]^2 & \psi < 0 \\
K_r(\psi) &= 1 & \psi \geq 0
\end{aligned} \tag{5}$$

where θ is the volumetric moisture content, θ_r is the residual moisture content, θ_s is the saturated moisture content, $\beta \equiv (\psi/\psi_s)^n$, ψ_s is the capillary or air entry pressure head value,

n is a constant, and $m = 1 - 1/n$ for n approximately in the range $1.25 < n < 6$. The corresponding general storage term is

$$\eta = S_w S_s + \phi \frac{dS_w}{d\psi} \tag{6}$$

where $S_w \equiv \theta/\phi$ is the water saturation, ϕ $(= \theta_s)$ is the porosity, and S_s is the specific storage.

A modified version of (4) was used in [16]:

$$
\begin{aligned}
\theta(\psi) &= \theta_r + (\theta_s - \theta_r)[1 + \beta]^{-m} & \psi < \psi_o \\
\theta(\psi) &= \theta_r + (\theta_s - \theta_r)[1 + \beta_o]^{-m} + S_s(\psi - \psi_o) & \psi \geq \psi_o
\end{aligned}
\tag{7}
$$

where ψ_o is a continuity parameter and $\beta_o \equiv \beta(\psi_o) = (\psi_o/\psi_s)^n$. The general storage term corresponding to (7) is $\eta = d\theta/d\psi$.

The characteristic equations used in [17] express the water saturation S_w in terms of effective saturation S_e, in the form $S_w(\psi) = (1 - S_{wr})S_e(\psi) + S_{wr}$, where S_{wr} $(= \theta_r/\phi)$ is the residual water saturation. The characteristic relations are then written as

$$
\begin{aligned}
S_e(\psi) &= \left[1 + \kappa^\beta (\psi_a - \psi)^\beta\right]^{-\gamma} & \psi < \psi_a \\
S_e(\psi) &= 1 & \psi \geq \psi_a
\end{aligned}
\tag{8}
$$

$$K_r(\psi) = K_r(S_e(\psi)) = S_e^n \tag{9}$$

where ψ_a is the air entry pressure and κ, β, γ, and n are constants. The general storage term is given by equation (6).

The method used to evaluate the derivative term in η, and the derivatives of K_r and η needed in the Newton scheme Jacobian, may affect the convergence behavior of the iterative solvers, due to possible discontinuities, steep gradients, and points of inflection in these curves and their derivatives. Numerical differentiation is often used to prevent floating point overflow near singularities, or to avoid oscillations around points of inflection. Analytical and various numerical techniques for differentiation of the characteristic equations can be implemented in the numerical models, including local or global chord slope and tangent slope formulae [7, 17].

2.4 Dynamic Time Stepping

Time step sizes during a transient simulation can be dynamically adjusted according to the convergence behavior of an iterative technique such as Newton or Picard. The time step size can be increased if convergence at the current time level is achieved in very few iterations, and decreased if the solution converged slowly. Note that a small time step improves the initial solution estimate used in the iterative procedure. If convergence is not attained, that is, if some specified maximum number of iterations is exceeded, the solution at the current time level can be recomputed using a smaller time step size.

3. LINEARIZATION AND RELAXATION METHODS

3.1 Newton's Method

Newton's method for the solution of equation (2) can be written as

$$J(\psi_m^{k+1})s_m = -g(\psi_m^{k+1}) \tag{10}$$

where m is the iteration index, $s_m \equiv \psi_{m+1}^{k+1} - \psi_m^{k+1}$, and

$$J_{ij} = \lambda A_{ij} + \frac{1}{\Delta t^{k+1}} F_{ij} + \sum_s \frac{\partial A_{is}}{\partial \psi_j^{k+1}} \psi_s^{k+\lambda} + \frac{1}{\Delta t^{k+1}} \sum_s \frac{\partial F_{is}}{\partial \psi_j^{k+1}} (\psi_s^{k+1} - \psi_s^k) + \frac{\partial b_i}{\partial \psi_j^{k+1}} \quad (11)$$

is the ij-th component of the Jacobian $J(\psi^{k+1})$. The nodal pressure heads are calculated at the new iteration as $\psi_{m+1}^{k+1} = \psi_m^{k+1} + s_m$.

Newton's method can be derived by taking a Taylor series expansion of equation (2) about its fixed point. The method can also be thought of as a parallel-chord method with updating, that is, we use the "tangent" J as the iteration matrix, and update it at each m [18]. Substituting $J(\psi_0^{k+1})$ for $J(\psi_m^{k+1})$ in (10), i.e., no updating, gives the "simplified" Newton scheme. If updating is performed selectively rather than at every iteration, we get the "modified" Newton scheme. The secant method is yet another variant of Newton's method, with the Jacobian or iteration matrix approximated by a finite difference formula.

The initial solution estimate can have a big effect on the behavior of the Newton (and other) iterative schemes. Unlike iterative techniques for linear systems, where theorems exist that guarantee global convergence under specified conditions, for nonlinear equations only local convergence results are available, that is, provided the initial estimate is "close enough" to the solution.

3.2 Picard Iteration

The Picard method is usually arrived at by evaluating all nonlinear terms in equation (2) at the previous iteration level, m, and the linear terms at $m + 1$. This yields

$$\left(\lambda A_m^{k+\lambda} + \frac{1}{\Delta t} F_m^{k+\lambda} \right) \psi_{m+1}^{k+1} = \left(\frac{1}{\Delta t} F_m^{k+\lambda} - (1 - \lambda) A_m^{k+\lambda} \right) \psi^k + q(t^{k+\lambda}) - b_m^{k+\lambda} \quad (12)$$

By simple algebraic manipulation, the above equation can be rearranged to the form

$$\left(\lambda A_m^{k+\lambda} + \frac{1}{\Delta t} F_m^{k+\lambda} \right) s_m = -g(\psi_m^{k+1}) \quad (13)$$

Comparing (10) and (13), it is apparent that the Picard scheme can be viewed as an approximate Newton method. It can be shown that under suitable conditions the Newton scheme is quadratically convergent [19], while Picard converges only linearly. Another important difference between the two schemes is that Newton linearization generates a nonsymmetric system matrix, whereas Picard preserves the symmetry of the original discretization. This factor is important in assessing the relative efficiency of the two schemes, since different storage and linear solver algorithms can be used to exploit these structural differences. A final observation to make is that calculation of the three derivative terms in the Jacobian makes the Newton scheme more costly and algebraically complex than Picard. In our numerical tests, the per iteration CPU cost of the Newton method was found to be approximately twice that of the Picard method, independent of the dimensionality of the problem.

3.3 Mixed Picard-Newton Approach

In many of our test simulations we observed the Newton scheme to be quite sensitive to the initial solution estimate. With a poor initial estimate the Newton scheme can diverge, whereas when the estimate is good Newton converges very rapidly. Our observations also suggest that the Picard method does not generally diverge — poor Picard performance is more often manifested by small or zero average convergence rate. In order to exploit the best features of both methods, a mixed Picard-Newton approach is introduced, based on the idea of using Picard iteration to improve the "initial" solution estimate for Newton's method. In this approach the Picard scheme is used for the first few iterations, until the solution begins to converge steadily, and then the Newton scheme is used for the remaining iterations. The switch from Picard to Newton can be made after a specified reduction in convergence error has been achieved.

3.4 Noniterative Methods

The Newton and Picard schemes preserve the order of accuracy of the time differenced equation (2). For instance, with $\lambda = 0.5$ in (2), the Picard or Newton linearized equation is second order accurate. If we linearize equation (2) to $O(\Delta t)$ by Taylor series expansion, or, equivalently, if we solve Richards' equation by taking only one Newton or Picard iteration, we obtain the so-called linearized Newton and linearized Picard methods, which are first order accurate and noniterative [16]:

$$J(\psi^k)(\psi^{k+1} - \psi^k) = -g(\psi^k) \tag{14}$$

where

$$J_{ij}^k = \frac{1}{2}A_{ij}^k + \frac{1}{\Delta t}F_{ij}^k + \frac{1}{2}\sum_s \frac{\partial A_{is}^k}{\partial \psi_j}\psi_s^k + \frac{1}{2}\frac{\partial b_i^k}{\partial \psi_j} \tag{15}$$

and

$$\left(\frac{1}{2}A^k + \frac{1}{\Delta t}F^k\right)(\psi^{k+1} - \psi^k) = -g(\psi^k) \tag{16}$$

To develop an $O(\Delta t^2)$ noniterative method which does not require storing three time levels of information and computing second order derivatives of the mass matrix F, we express Richards' equation in the form

$$\frac{\partial \psi}{\partial t} = \frac{1}{\eta(\psi)}\nabla \cdot (K_s K_r(\psi)\nabla(\psi + z)) \tag{17}$$

Applying a Crank-Nicolson time discretization and linearizing to $O(\Delta t^2)$ by Taylor series expansion, we obtain the implicit factored method [16]

$$\left(\frac{1}{2}[B^k + B'^k + c'^k - r'^k] + \frac{1}{\Delta t}G\right)(\psi^{k+1} - \psi^k) = r^k - c^k - B^k\psi^k \tag{18}$$

where B, G, r, and c are matrices and vectors which arise from the finite element discretization of (17), and B', c', and r' contain the derivative components of B, c, and r.

The final noniterative scheme we consider is the linear second order accurate three-level Lees [20] discretization

$$\frac{1}{3}A^k(\psi^{k+1} + \psi^k + \psi^{k-1}) + F^k\frac{\psi^{k+1} - \psi^{k-1}}{2\Delta t} = q(t^k) - b^k \tag{19}$$

3.5 Quasi-Newton Methods

The quasi-Newton family is defined by substituting in the Newton equation (10) an approximation K to the Jacobian J, yielding $K(\psi_m^{k+1})s_m = -g(\psi_m^{k+1})$. Rather than calculate the Jacobian at each iteration, as is done in the Newton method, matrix K is updated using recursion formulae that require fewer operations than a full evaluation of J. The update formulae can be derived as follows [21, 22, 23]. Let $y_m = g_{m+1} - g_m$ be the residual difference. Then a Taylor series expansion of the residual vector g gives

$$g_{m+1} = g(\psi_m^{k+1} + s_m) = g_m + J_m s_m + \cdots \tag{20}$$

that is, $y_m = J_m s_m + \ldots$. The approximation K to the Jacobian J is chosen so that it satisfies the previous equation exactly up the the first order terms. We obtain therefore the equation $K_{m+1} s_m = y_m$, which is known as the quasi-Newton or secant condition. This expression enables the direct calculation of K_{m+1} or K_{m+1}^{-1}. In the limit as $m \to \infty$, the matrix K_{m+1} thus defined tends to J. In implementations of the quasi-Newton method, updates for K_{m+1}^{-1} are often used instead of K_{m+1} updates, as these inverse updates avoid the need to solve a linear system at each iteration.

Several update formulae have been proposed in the literature for the case where the Jacobian is symmetric. We will describe only the Broyden and BFGS updates, which are considered to be amongst the most efficient quasi-Newton algorithms. In our initial trials with quasi-Newton methods, we have applied these update formulae without modification to the nonsymmetric Jacobian arising from Richards' equation. The expression for the Broyden inverse update is

$$K_m^{-1} = K_{m-1}^{-1} + \frac{\left(s_m - K_{m-1}^{-1} y_m\right) s_m^T K_{m-1}^{-1}}{s_m^T K_{m-1}^{-1} y_m} \tag{21}$$

while the BFGS inverse update can be written as

$$K_m^{-1} = \left(I - \omega_m s_m y_m^T\right) K_{m-1}^{-1} \left(I - \omega_m y_m s_m^T\right) + \omega_m s_m s_m^T \tag{22}$$

where $\omega_m = (s_m^T y_m)^{-1}$. In our implementation of the Broyden and BFGS algorithms, we use the exact Jacobian J_0 for the initial update K_0. The inverse is never calculated explicitly. Instead, the LU factorization of K is used, and sparsity-preserving recursion expressions that calculate directly the solution vector s_{m+1} are developed [24].

The main difficulty in the implementation of these recursion formulae is that two vectors have to be saved at each iteration. Thus storage requirements become excessive when a large number of iterations is required for convergence. This problem may be circumvented by the use of so-called limited memory quasi-Newton (LMQN) implementations, which are based on different restart strategies after a given number of iterations M has been completed.

We employed three distinct LMQN strategies. In implementation 1 all the previous saved vectors are discarded and the iteration restarts using the vectors at level M, and K_0 as the initial Jacobian approximation. Implementation 2 also uses K_0 as the initial Jacobian approximation, but always saves M vectors, with the oldest vectors replaced by the most recently calculated ones. Implementation 3 is a full restart of the scheme obtained by discarding all the previous vectors and recalculating the Jacobian, that is, $K_{M+1} = J_{M+1}$. A template describing the general procedure for the BFGS and Broyden quasi-Newton methods is given in Figure 2.

<table>
<tr><td>

Start with ψ_0^{k+1} and K_0; $p = 0$

for i=0,1,...

  1. $p = p + 1$; if $p \geq M$ do a restart .

  2. $g_i = g(\psi_i^{k+1})$

  3. evaluate $K_i^{-1} = f(K_{i-1}^{-1})$ based on (21) (Broyden) or (22) (BFGS)

  4. $s_{i+1} = -K_i^{-1} g_i$

  5. $\psi_{i+1}^{k+1} = \psi_i^{k+1} + s_i$

  6. if $\left\| \psi_{i+1}^{k+1} - \psi_i^{k+1} \right\| \leq \epsilon$ stop;

end for;

</td><td>

1. start with ITER $= 1$, $\alpha_1 = 0$ and $\alpha_2 = 1$

2. $\alpha_3 = (\alpha_1 + \alpha_2)/2$

3. if $(\alpha_1 - \alpha_2) <$ TOLL or ITER $>$ ITMAX exit with $\alpha = \alpha_3$

4. if $\Phi'(\alpha_2)\Phi'(\alpha_3) < 0$ then $\alpha_1 = \alpha_3$, else $\alpha_2 = \alpha_3$

5. if $\Phi(\alpha_3) > \Phi(\alpha_2)$ exit with $\alpha = \alpha_2$

6. set ITER $=$ ITER$+1$ and go to step 2

</td></tr>
</table>

Figure 2. Quasi-Newton algorithm.   *Figure 3.* Line search algorithm.

3.6 Relaxation and Line Search Methods

Convergence of an iterative scheme can be enhanced by introducing a relaxation (or damping) parameter α to weight the solution vector from the current and previous iterations, i.e.:

$$\psi_{m+1}^{k+1} = \psi_m^{k+1} + \alpha(\psi_{m+1}^{k+1} - \psi_m^{k+1}) = \psi_m^{k+1} + \alpha s_m \tag{23}$$

In applications of Picard and Newton iteration to Richards' equation, ad hoc relaxation methods, such as constant α, have been used [5, 6, 7]. Line search algorithms, originally developed for optimization problems, provide a systematic procedure for determining α. In the context of unconstrained optimization, the line search strategy applied to Newton's method may be described as follows. In equation (10), the vector s_m can be viewed as a search direction in the N-dimensional vector space along which the new approximate solution will be sought. The relaxation parameter α, as employed in (23), can be interpreted as the length of the step that will be taken along s_m. The optimal value for α can be chosen as the point that minimizes the objective function along the search direction. In our implementation, this minimum is found by solving the one-dimensional nonlinear problem [25, 26, 27]

$$s_m^T g(\psi_m^{k+1} + \alpha s_m) = 0 \tag{24}$$

is important to note that only if the Jacobian (or the Hessian of the objective function) is symmetric are we guaranteed that a minimum of Φ exists. Otherwise equation (24) may have no or more than one solution within the admissible interval for α. Furthermore, for nonsymmetric systems, s_m may not be a strict descent direction, or the objective function may not decrease monotonically along that direction.

To be effective for Richards' equation, the line search algorithm must therefore be able to detect and compensate for the special circumstances that may arise in the nonsymmetric case. The algorithm employed here, described in Figure 3, is a simple line search procedure

Table 1: Number of Broyden and BFGS Iterations to Convergence for the 1-D Steady State Test Problem

	Broyden Method			BFGS Method		
Iteration Levels	LMQN Implementation			LMQN Implementation		
Saved (M)	1	2	3	1	2	3
4	failed	497	failed	221	failed	failed
20	256	527	31	393	failed	56
50	263	968	57	252	failed	71

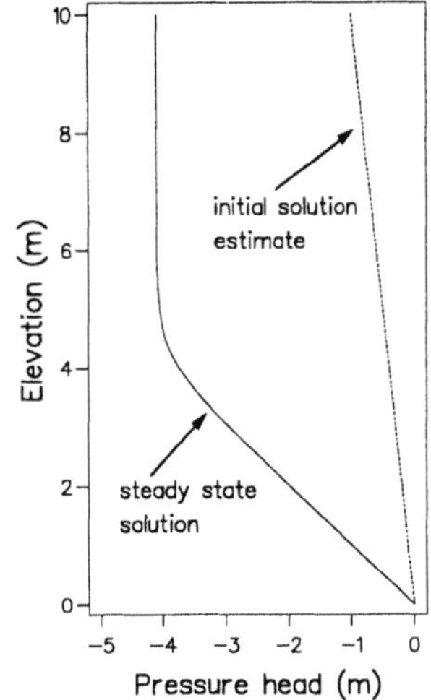

Figure 4: Initial estimate and final solution for the 1-D steady state test problem.

based on a bisection iteration (steps 2, 3, and 4), complemented by a check for a decreasing objective function; if the objective function does not decrease, the procedure terminates with the value of α corresponding to the lower objective function (step 5). This algorithm requires the evaluation of the residual vector g_m at each iteration. This is a costly operation, and should be performed as seldom as possible. As suggested by [28] and others, it is not critical for the line search procedure to be highly accurate when the current approximate solution is still far from the real solution. For this reason the value of TOLL in Figure 2 can be set relatively high, and ITMAX can be kept small.

4. NUMERICAL TESTS

4.1 Steady One-Dimensional Infiltration

The first example tests the various implementations of the Broyden, BFGS, and line search methods, and compares their convergence behavior against the Picard and Newton schemes. The test case, whose solution is shown in Figure 4, involves steady state flow in a

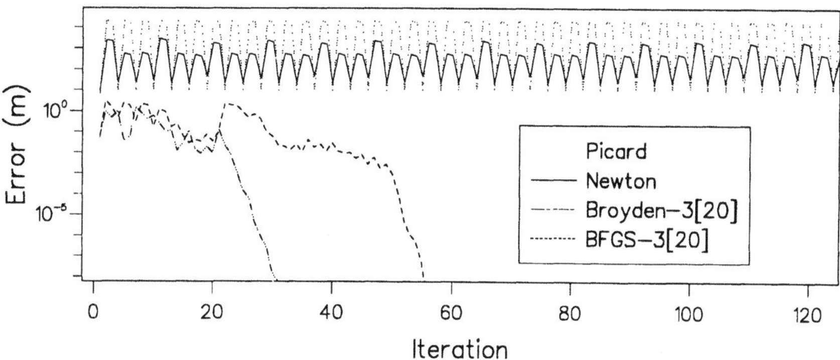

Figure 5: Convergence profiles for the 1-D steady state test problem run without relaxation. The notation "Broyden-3[20]" refers to limited memory implementation 3 of the Broyden method, with 20 iteration levels saved.

one-dimensional soil column. For this test case we used a uniform grid spacing Δz of 0.1 m, $\psi = 0$ at $z = 0$, $K(\psi)((\partial\psi/\partial z) + 1) = 0.0001$ at $z = 10$, $K_s = 0.01$ m/h, and characteristic equation (5) with $n = 5$ and $\psi'_s = -3$.

The results for the test problem run without relaxation are summarized in Table 1, and the convergence profiles for some of these runs are plotted in Figure 5. For this test case both the Picard and Newton methods produced oscillations and failed to converge. Several of the quasi-Newton methods also failed or converged very slowly, with particularly poor performance observed for limited memory implementations 1 and 2. This result is not surprising given the poor performance of the Newton scheme, since LMQN 1 and 2 restart with the initial update K_0, which in our implementation is the exact initial Newton Jacobian J_0. The best results were obtained for LMQN 3 with 20 or 50 iteration levels saved, which converged in 31–71 iterations. The fastest of these was the 20-level Broyden-3 method.

The performance of the line search algorithm is shown in Figure 6. We see that the Newton scheme, which failed for this test case, converged in as few as 34 iterations with a fixed relaxation parameter α, and when line search was applied only 12 iterations were required for convergence. The top left plot in Figure 6 is a good illustration of the usefulness of line search to calculate the relaxation parameter, as it shows how widely the "optimal" value of α can vary from one iteration to the next, particularly in the early stages of an iterative solution procedure.

4.2 Transient One-Dimensional Infiltration

The computational efficiency of competing numerical schemes is best evaluated on the basis of CPU time expended to achieve a given level of solution accuracy. Since closed form analytical solutions are nonexistent except for simplified forms of Richards' equation, a numerical solution obtained using a very fine grid and time discretization can be used as a surrogate "exact" solution with respect to which accuracy can be measured [16]. This procedure is used in the following test problem to evaluate the Newton, Picard, linearized Newton, linearized Picard, implicit factored, and three-level Lees schemes. The problem is

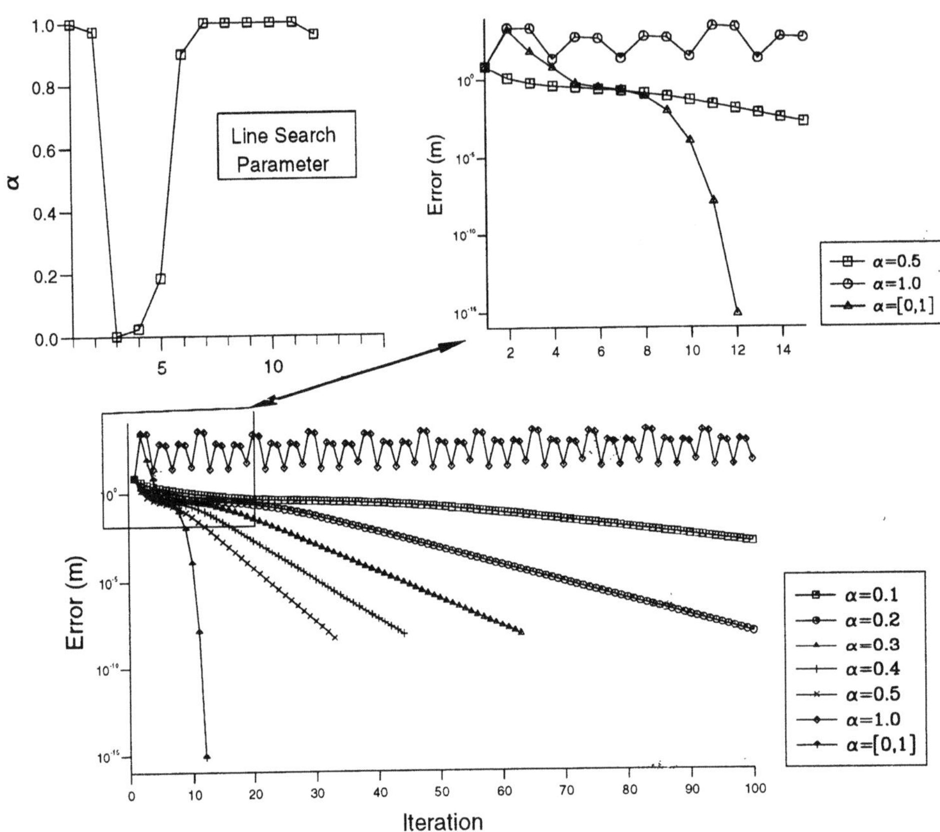

Figure 6: Convergence profiles for the 1-D steady state test problem: Newton scheme with constant relaxation parameter ($\alpha = 1$ is unrelaxed Newton) and with the relaxation parameter computed by line search ($\alpha = [0, 1]$). Top right plot is a zoom on the first 15 iterations, and top left plot shows the line search-computed α value at each iteration.

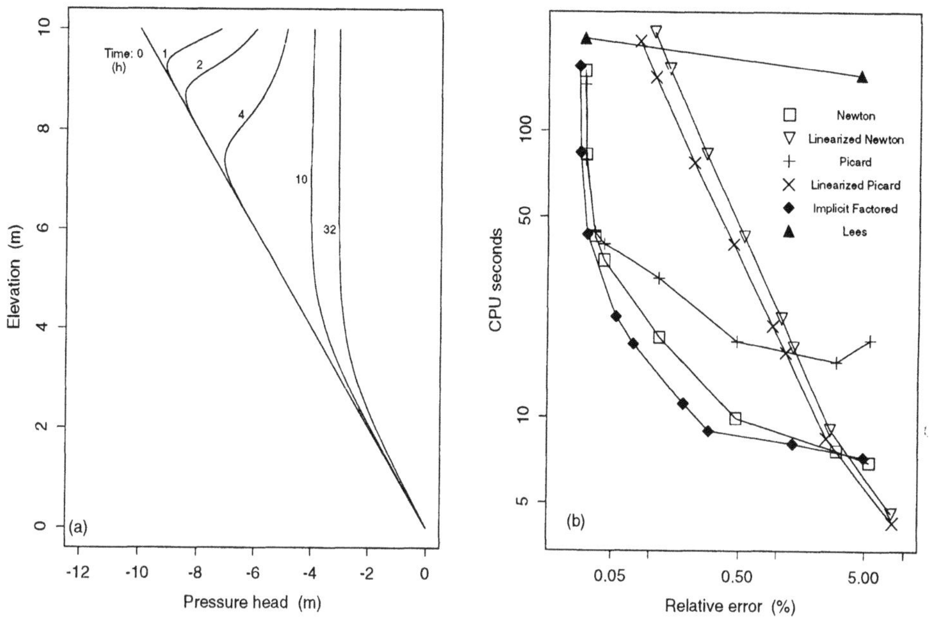

Figure 7: (a) Solution for the 1-D transient test problem at various times; (b) computational efficiency of six linearization methods.

one of infiltration into a soil column initially at hydrostatic equilibrium, with the infiltration flux increasing linearly with time. The resulting pressure head profiles at various times are shown in Figure 7a. For this test case we used a distributed mass matrix, $\lambda = 0.5$, a uniform grid spacing Δz of 0.1 m, $\psi = 0$ at $z = 0$, $K(\psi)((\partial\psi/\partial z) + 1) = t/64$ at $z = 10$, $K_s = 5$ m/h, and characteristic equations (7) and (5) with $\theta_r = 0.08$, $\theta_s = 0.45$, $S_s = 0.001$, $\psi_o = -0.191055$, $n = 3$, and $\psi_s = -3$.

The computational efficiency of the six linearization methods is shown in Figure 7b. For this test problem the three-level Lees scheme required a very small time step size to overcome severe oscillations. At time step sizes greater than 0.01 h, this oscillatory behavior produced highly inaccurate solutions. It therefore appears that the Lees scheme is only conditionally stable for nonlinear problems. From the efficiency plot we observe that the first order accurate linearized Picard and Newton schemes are only competitive with the implicit factored and iterative schemes at low accuracies. As the accuracy requirements become more stringent, the first order schemes become increasingly inefficient relative to the second order accurate schemes. The implicit factored scheme is very competitive with the iterative methods. The superior performance of the Newton scheme over Picard for this test case is likely due to the continuous forcing provided by the time-varying boundary condition, which results in sustained slow convergence for the Picard scheme. Another factor contributing to the relatively poor performance of the Picard scheme is the gravity drainage zone which develops in the solution at later times [7]. In this test case the Picard scheme required 2–3 times more iterations overall than Newton, and this is roughly the threshold at which the

Table 2: Summary of Results for the 2-D Steady State Test Problem With $K_r(\psi)$ Relationship (9) (Parameters β and n)

			Number of Nonlinear Iterations			
β	n	$\Delta x = \Delta z$	Picard	Newton	Mixed Picard-Newton	Relaxed Picard
4	4	10	21	failed	10 + 15 †	35
4	4	2	1138	failed	52 + 64	53
2	2	10	34	44	14 + 19	
2	2	2	78	failed	10 + 27	
1	2	10	89	35	10 + 28	
1	2	2	87	34	6 + 25	
1	4	10	failed	58	9 + 31	44
1	4	2	failed	failed	7 + 30	87

† Picard iterations + Newton iterations.

Newton method becomes more efficient than Picard, regardless of the dimensionality of the problem.

4.3 Steady Two-Dimensional Flow Through a Square Embankment

This test problem is taken from [5] and involves steady state flow through a square embankment. It will be used here to illustrate the effects of grid discretization, soil characteristics, chord slope approximations, mixed Picard-Newton iteration, and relaxation on Picard and Newton convergence. For this test case we used a distributed mass matrix, uniform grid spacings of $\Delta x = \Delta z = 10$ m or $\Delta x = \Delta z = 2$ m, an initial solution estimate of $\psi(x, z) = 100 - z$, $K_s = 0.01$ m/h, and characteristic equation (9) with $\kappa = \sqrt[4]{0.1}$, $\psi_a = 0$, $\gamma = 1$, $\beta = 1, 2$, or 4, and $n = 2$ or 4, or characteristic equation (5) with $n = 1.5, 3$, or 5 and $\psi_s = -3$ or -20. The boundary conditions are: $\psi = 100 - z$ at $x = 0$, $0 \le z \le 100$, $\psi = 20 - z$ at $x = 100$, $0 \le z \le 20$, seepage face at $x = 100$, $20 < z \le 100$, and no-flow conditions on all other segments of the boundary. A representative solution using one of the grid and soil parameter combinations is shown in Figure 8, along with a plot of the conductivity curves for the range of parameter values used in the two $K_r(\psi)$ relationships.

Tables 2 and 3 summarize the convergence results obtained using two $K_r(\psi)$ relationships, with several parameter combinations, and two grid discretizations. The performance of the Picard and Newton schemes deteriorated appreciably for the finer grid in almost all runs, in some cases passing from rapid convergence with the 10 m grid to nonconvergence with the 2 m grid. The mixed Picard-Newton scheme was very effective in overcoming convergence problems at both grid discretizations, in some cases providing rapid convergence where the Picard and/or Newton schemes failed to converge. The switching from Picard to Newton iteration in the mixed approach was done after the convergence error was reduced by one order of magnitude, measured relative to the second Picard iteration.

From the conductivity plots in Figure 8 and the results in Tables 2 and 3, it can be seen that the $K_r(\psi)$ curves that are most strongly nonlinear — those spanning many orders

Table 3: Summary of Results for the 2-D Steady State Test Problem With $K_r(\psi)$ Relationship (5) (Parameters n and ψ_s)

n	ψ_s	$\Delta x = \Delta z$	Number of Nonlinear Iterations		
			Picard	Newton	Mixed Picard-Newton
5	-20	10	16	24	$6 + 14$ †
5	-20	2	20	29	$5 + 19$
5	-3	10	17	176	$8 + 15$
5	-3	2	74	failed	$10 + 33$
3	-3	10	24	50	$8 + 21$
3	-3	2	74	failed	$8 + 30$
1.5	-3	10	failed	44	$9 + 29$
1.5	-3	2	failed	failed	$6 + 29$

† Picard iterations + Newton iterations.

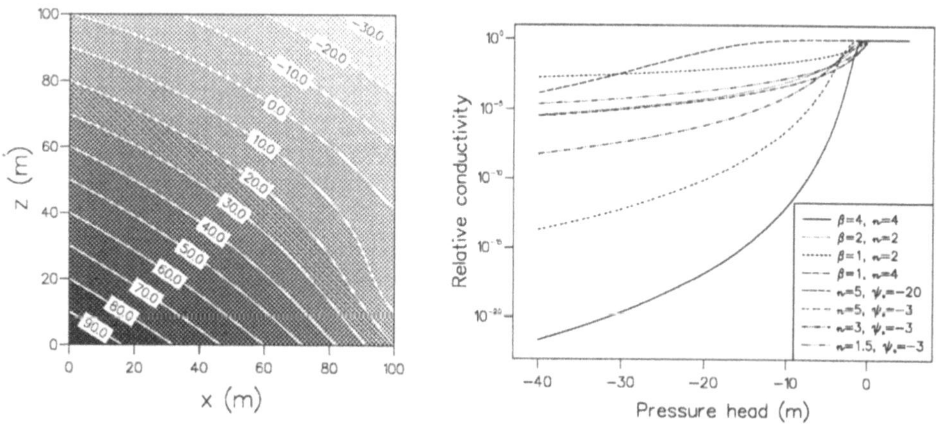

Figure 8: Pressure head solution contours for the 2-D steady state test problem with $\Delta x = \Delta z = 2$ m and equation (9) parameters $\beta = 4$ and $n = 4$, and relative hydraulic conductivity profiles from equations (9) (parameters β and n) and (5) (parameters n and ψ_s).

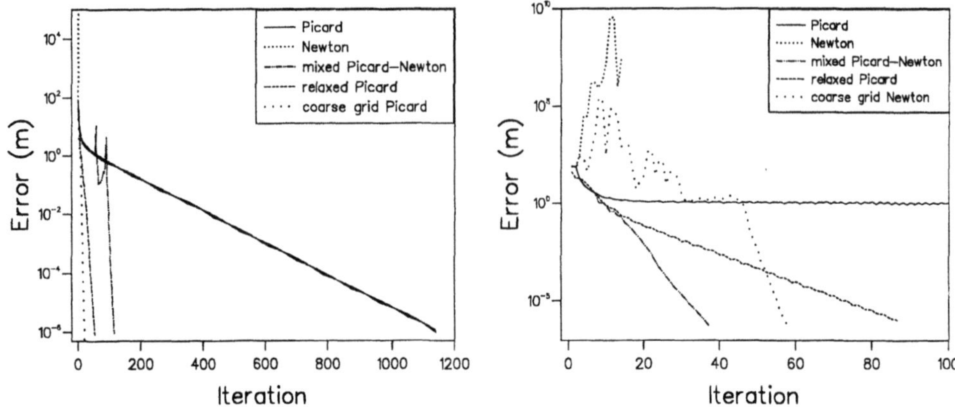

Figure 9: Convergence profiles for the 2-D steady state test problem with equation (9) parameters $\beta = 4$, $n = 4$ (left) and $\beta = 1$, $n = 4$ (right).

of magnitude in K_r or having very steep or near-discontinuous gradients around $\psi = 0$ — caused the greatest convergence difficulties for the Picard and Newton schemes. As before, the mixed Picard-Newton scheme was very effective in improving convergence behavior.

Chord slope approximations and ad hoc relaxation (without line search) were tried on the four most difficult cases represented in Table 2 — the coarse and fine grid runs for $\beta = 4$, $n = 4$ and $\beta = 1$, $n = 4$. The relaxed Picard results reported in the table are for constant relaxation parameter $\alpha = 0.5$. $\alpha = 1.2$, 0.8, and 0.2, and relaxation with iteration-dependent α, all gave worse results than $\alpha = 0.5$. Both fixed and variable-α relaxation were unsuccessful for the Newton scheme. The convergence profiles for some of the $\beta = 4$, $n = 4$ and $\beta = 1$, $n = 4$ runs are plotted in Figure 9, showing the dramatic effect that grid discretization, relaxation, and the mixed Picard-Newton scheme can have on the performance of the Picard and Newton methods. The various chord slope approximations that were tried gave disappointing results [7].

4.4 Transient Three-Dimensional Flow in a Pumped Unconfined Aquifer

This test problem involves three-dimensional flow in a pumped unconfined aquifer, with prevailing flow in the aquifer along the x direction (Figure 10). The domain is discretized into 2079 nodes and 9600 tetrahedra. The principal components of the saturated conductivity tensor are $K_{sx} = 5$ m/d, $K_{sy} = 2$ m/d, and $K_{sz} = 0.5$ m/d, the porosity is $\phi = 0.5$, and the specific storage is $S_s = 0.01$ m^{-1}. We used a distributed mass matrix and $\lambda = 0.5$.

Variations of this test problem have been used to examine the effects of mass lumping, time weighting, time stepping, grid discretization, soil characteristics, chord slope approximations, seepage faces, and linear solvers on Picard and Newton convergence [7, 29]. We will summarize here the results of two sets of runs. In the first set of runs, denoted by "h", we used characteristic equations (8) and (9) with $\kappa = 0.5$, $\beta = 2$, $\gamma = 1$, $n = 2$, $\psi_a = 0$, and $S_{wr} = 0.05$. A 5 time step simulation of 7.4 days was run using variable time stepping. The time step sizes were $\Delta t = 0.5$, 0.9, 1.38, 1.96, and 2.65 d. In the second set of runs,

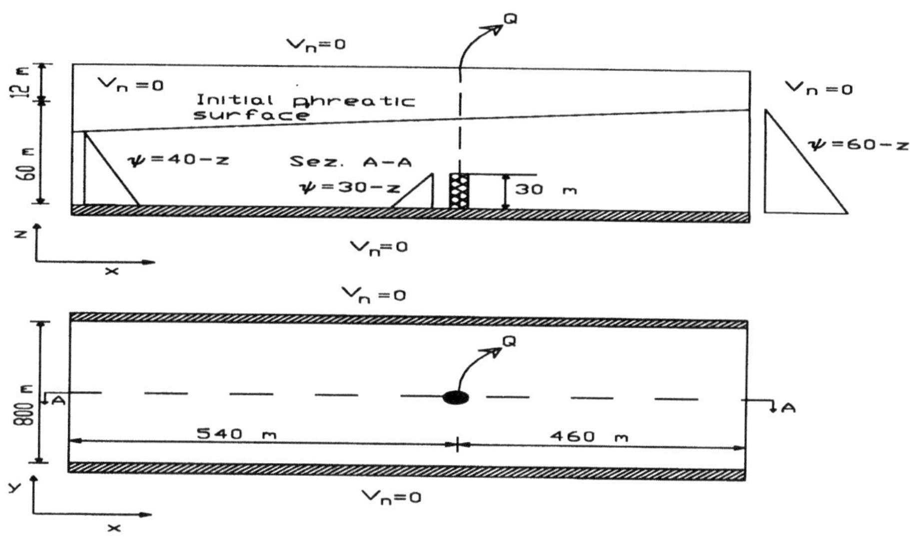

Figure 10: Description of the 3-D transient test problem, with boundary and initial conditions indicated.

denoted by "v", we used characteristic equations (4) and (5) with $n = 3.35$, $\psi_s = -0.3$, and $\theta_r = 0.025$. A fixed time step simulation of 8 days was run, using $\Delta t = 0.2$ d.

Various methods for evaluating the derivative terms in the characteristic equations were considered. For the Newton scheme we used either analytical evaluation of all the derivatives terms in the Jacobian (we label this method "Newton 1"), or analytical differentiation for first derivatives and a chord slope formula for second derivatives ("Newton 2"). For the Picard scheme we used either analytical evaluation of the derivative terms ("Picard 1"), or chord slope differentiation with either a large tolerance of 0.05 m ("Picard 2") or a small tolerance of 10^{-4} m ("Picard 3").

The graph on the left in Figure 11 shows the convergence profiles for the third time step of the first set of runs. Newton's method converged very rapidly, while the Picard scheme converged more slowly and showed a marked sensitivity to the method used to differentiate the characteristic equations. Picard with chord slope differentiation converged faster than Picard with analytical differentiation, although the convergence profiles for the chord slope case are quite irregular. This can be seen in the "Picard 2h" results of Figure 11, where the chord slope formula was used if the norm of the pressure head difference s_m was larger than 0.05 m (the first 8 iterations), and analytical differentiation was used otherwise. In "Picard 3h" the tolerance was decreased from 0.05 to 10^{-4} m, resulting in chord slope differentiation for the entire simulation. The result was faster but even more irregular convergence than "Picard 2h" or "Picard 1h". The chord slope approximation had a very minor effect on convergence rate for the Newton runs, although the irregular convergence pattern is evident here too.

The graph on the right in Figure 11 shows the convergence profiles for the sixth time step of the second set of runs. For these runs "Picard 1h" failed to converge, while "Picard 2h" showed very slow and oscillatory convergence behavior. The Newton scheme converged very rapidly with both analytical and chord slope differentiation.

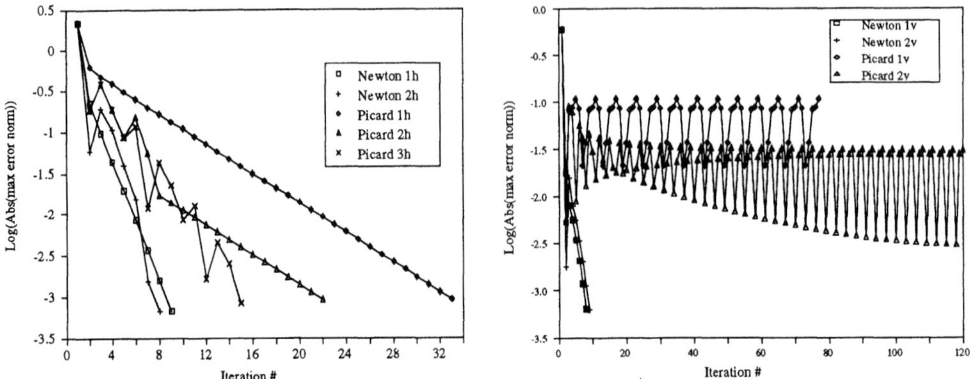

Figure 11: 3-D transient test problem convergence results for the third time step of the first set of runs (left) and for the sixth time step of the second set of runs (right).

5. CONCLUSIONS

The two most commonly used methods for solving Richards' equation, Picard and Newton iteration, have been presented and evaluated, along with other approaches including noniterative, hybrid, and quasi-Newton methods. Relaxation, line search, and chord slope procedures to accelerate convergence were also considered.

When it works, the Picard scheme is clearly the simplest and most efficient method for linearizing Richards' equation. However, there are cases where the Picard scheme fails or converges very slowly. We note in particular the difficulties encountered with gravity drainage zones, complex time-varying boundary conditions, strongly nonlinear characteristic equations, and saturated/unsaturated interfaces. Relaxation is sometimes successful in overcoming these convergence troubles. The drawback of ad hoc relaxation methods (constant or empirically calculated relaxation parameter) can be overcome with line search algorithms, which systematically compute this parameter. Only limited success was obtained with the various chord slope approximations for the characteristic equations.

The Newton scheme is generally more robust and faster converging than Picard, although it too can fail to converge, in some cases for the same reasons as Picard (strongly nonlinear characteristic equations and saturated/unsaturated interfaces), and in other cases due to poor initial solution estimates. The hybrid Picard-Newton approach can effectively overcome the Newton scheme's sensitivity to initial solution estimates, but it has the disadvantage of requiring both symmetric and nonsymmetric storage modes (unless the Picard scheme is stored and solved as a nonsymmetric system).

Preliminary tests of the Broyden and BFGS quasi-Newton methods showed good performance against Picard and Newton iteration. Work on quasi-Newton and line search procedures that are better suited to nonsymmetric systems is ongoing. Of the noniterative strategies, we found the first order accurate linearized Newton and Picard methods to be inefficient compared to second order accurate schemes. Of the two $O(\Delta t^2)$ accurate noniterative methods, the three-level Lees scheme may be only conditionally stable, and the implicit factored scheme can be quite compotetive with the Newton and Picard methods

Acknowledgments. This work has been supported by the Italian CNR (Gruppo Nazionale per la Difesa dalle Catastrofi Idrogeologiche, linea di Ricerca n. 4), by Fondi Ministeriali 40%, and by the Sardinia Regional Authorities.

REFERENCES

1. Philip, J. R., Theory of infiltration, *Adv. Hydrosci.* 5, 215–296, 1969.

2. Ames, W. F., *Numerical Methods for Partial Differential Equations.* Academic Press, San Diego, CA second edition, 1977.

3. Huyakorn, P. S. and G. F. Pinder, *Computational Methods in Subsurface Flow.* Academic Press, New York, NY, 1983.

4. Paniconi, C. and E. F. Wood, A detailed model for simulation of catchment scale subsurface hydrologic processes, *Water Resour. Res.* 29(6), 1601–1620, 1993.

5. Cooley, R. L., Some new procedures for numerical solution of variably saturated flow problems, *Water Resour. Res.* 19(5), 1271–1285, 1983.

6. Huyakorn, P. S., E. P. Springer, V. Guvanasen and T. D. Wadsworth, A three-dimensional finite-element model for simulating water flow in variably saturated porous media, *Water Resour. Res.* 22(13), 1790–1808, 1986.

7. Paniconi, C. and M. Putti, A comparison of Picard and Newton iteration in the numerical solution of multidimensional variably saturated flow problems, *Water Resour. Res.* 30(12), 3357–3374, 1994.

8. Axelsson, O., Conjugate gradient type methods for unsymmetric and inconsistent systems of linear equations, *Linear Algebra Appl.* 29, 1–16, 1980.

9. Pini, G., G. Gambolati and G. Galeati, 3-D finite element transport models by upwind preconditioned conjugate gradients, *Adv. Water Resour.* 12, 54–58, 1989.

10. van der Vorst, H., Bi-CGSTAB: A fast and smoothly converging variant of BI-CG for the solution of nonsymmetric linear systems, *SIAM J. Sci. Stat. Comput.* 13, 631–644, 1992.

11. Freund, R. W., A transpose-free quasi-minimal residual algorithm for non-Hermitian linear systems, *SIAM J. Sci. Comput.* 14(2), 470–482, 1993.

12. Kershaw, D. S., The incomplete Cholesky-conjugate gradient method for the iterative solution of systems of linear equations, *J. Comput. Phys.* 26, 43–65, 1978.

13. Gambolati, G. and A. Perdon, Conjugate gradients in subsurface flow and land subsidence modeling. In: Bear, J. and M. Y. Corapcioglu (eds.) *Fundamentals of Transport Phenomena in Porous Media.* Martinus Nijhoff, Dordrecht, Holland, pp 953–984, 1984.

14. Pullan, A. J., The quasilinear approximation for unsaturated porous media flow, *Water Resour. Res.* 26(6), 1219–1234, 1990.

15. van Genuchten, M. T. and D. R. Nielsen, On describing and predicting the hydraulic properties of unsaturated soils, *Ann. Geophys.* 3(5), 615–628, 1985.

16. Paniconi, C., A. A. Aldama and E. F. Wood, Numerical evaluation of iterative and noniterative methods for the solution of the nonlinear Richards equation, *Water Resour. Res.* 27(6), 1147–1163, 1991.

17. Huyakorn, P. S., S. D. Thomas and B. M. Thompson, Techniques for making finite elements competitive in modeling flow in variably saturated porous media, *Water Resour. Res.* 20(8), 1099–1115, 1984.

18. Ortega, J. M. and W. C. Rheinboldt, *Iterative Solution of Nonlinear Equations in Several Variables.* Academic Press, New York, NY, 1970.

19. Stoer, J. and R. Bulirsch, *Introduction to Numerical Analysis.* Springer-Verlag, New York, NY, 1980.

20. Lees, M., A linear three-level difference scheme for quasilinear parabolic equations, *Math. Comp.* 20, 516–522, 1966.

21. Dennis, J. E. and J. J. Moré, Quasi-Newton methods, motivation and theory, *SIAM Review* 19(1), 46–89, 1977.

22. Matthies, H. and G. Strang, The solution of nonlinear finite element equations, *Int. J. Numer. Meth. Eng.* 14, 1613–1626, 1979.

23. Geradin, M., S. Idelsohn and M. Hogge, Computational strategies for the solution of large nonlinear problems via quasi-Newton methods, *Computers & Structures* 13, 73–81, 1981.

24. Putti, M. and C. Paniconi, Quasi-Newton methods for Richards' equation. In: Peters, A., G. Wittum, B. Herrling, U. Meissner, C. A. Brebbia, W. G. Gray and G. F. Pinder (eds.) *Computational Methods in Water Resources X, Volume 1.* Kluwer Academic, Dordrecht, Holland, pp 99–106, 1994.

25. Paniconi, C. and M. Putti, Quasi-Newton and line search methods for the finite element solution of unsaturated flow problems. In: Wang, S. S. Y. (ed.) *Second International Conference on Hydro-Science and Engineering*, Beijing, China, 1995.

26. Papadrakakis, M., Solving large-scale nonlinear problems in solid and structural mechanics. In: Papadrakakis, M. (ed.) *Solving Large-Scale Problems in Mechanics* 183–223, John Wiley & Sons, New York, NY, 1993.

27. Dennis, J. E. and R. B. Schnabel, *Numerical Methods for Unconstrained Optimization and Nonlinear Equations.* Prentice-Hall, Englewood Cliffs, NJ, 1983.

28. Fletcher, R., *Practical Methods of Optimization, Vol. 1: Unconstrained Optimization.* John Wiley & Sons, New York, NY, 1980.

29. Putti, M. and C. Paniconi, Evaluation of the Picard and Newton iteration schemes for three-dimensional unsaturated flow. In: Russell, T. F., R. E. Ewing, C. A. Brebbia, W. G. Gray and G. F. Pinder (eds.) *Proceedings of the IX International Conference on Computational Methods in Water Resources, Vol. 1, Numerical Methods in Water Resources.* Computational Mechanics Publications, Southampton, UK, pp 529–536, 1992.

FINITE ELEMENT MODELING OF SALTWATER INTRUSION PROBLEMS WITH AN APPLICATION TO AN ITALIAN AQUIFER

M. Putti

University of Padua, Padua, Italy

and

C. Paniconi

CRS4, Cagliari, Italy

ABSTRACT

Density dependent transport in groundwater may be described by a coupled model of flow and solute transport. The coupling is nonlinear because salt concentration affects the water motion, which in turn controls the fate of the contaminant. For the highest concentrations commonly encountered in the sea (between 25 and 40 kg/m^3) the influence of coupling is weak for the flow equation, but stronger for the transport equation. The solution of the algebraic system of nonlinear equations arising from the finite element discretization of the coupled problem may be efficiently obtained by decoupling flow and transport and iterating between the two equations. Using this strategy, at each iteration a linear flow equation and a nonlinear transport equation have to be solved. While for the flow equation the usual finite element approach can be used, linearization techniques have to be employed for the transport equation. The most common solution method uses a Picard linearization applied to the transport equation. However, a more efficient scheme, named the partial Newton method, can be obtained by substituting the Picard with a Newton type linearization. This approach leads to a scheme with better convergence properties but approximately the same computational cost, on a per iteration basis.

Results from two and three-dimensional test simulations show that the partial Newton scheme gives improved convergence and robustness compared to Picard linearization, especially for highly advective problems or large density ratios. The fully three-dimensional finite element model is used to simulate the saltwater contamination of a coastal aquifer in Southern Italy.

1. INTRODUCTION

The mathematical model of seawater intrusion into aquifers can be formulated as a coupled system of two partial differential equations, one describing mass conservation for the water-salt solution (the flow equation), and the other mass conservation for the salt contaminant (the transport equation) [1, 2, 3]. From a physical point of view, the salt contained in the groundwater affects the solution density and induces changes in the flow field, which becomes dependent not only on the hydrogeological parameters of the aquifer, but also on the salt concentration. This interdependence produces a coupling between the flow and transport equations.

The numerical solution of the salt intrusion problem, or, more generally, of density dependent flow and transport problems, poses two main difficulties. One is related to coupling, which necessitates the simultaneous solution of two mathematically different partial differential equations. The other difficulty is connected to the nonlinearity of the flow and transport equations, indirectly caused by coupling. The degree of nonlinearity depends mainly on the salt concentration and on the ratio between the freshwater and saltwater densities [4].

The usual solution procedure is to decouple the problem by first solving the flow equation, then calculating the velocity field, and finally solving the transport equation. This three-step sequence is repeated until convergence is achieved. The procedure halves the dimensionality of the problem by sequentially solving two systems of size $N \times N$ instead of a single $2N \times 2N$ system, where N is the number of nodes in the domain discretization. Linearization is performed by freezing any nonlinear coefficient at the previous iteration level. This approach, known as the Picard scheme, has been successfully employed in a number of cases. However, when the density ratio is large, or, in transient simulations, when the time step size is large, convergence may be problematic [4, 5, 6]. Relaxation schemes similar to those developed for the solution of the unsaturated flow equation [7] have achieved limited success for saltwater intrusion problems, in particular for the steady state case [5].

In this paper we propose a solution approach similar to the Picard scheme, but based on a Newton linearization of the transport equation. This procedure can be derived by neglecting the off-diagonal blocks of the $2N \times 2N$ Jacobian matrix of the fully coupled system. For appropriate values of the equation coefficients, dropping these off-diagonal terms does not appreciably decrease the convergence rate relative to the full Newton approach. Moreover, this "partial" Newton scheme produces the same computational advantage as the Picard approach, namely that of avoiding the need to solve a $2N \times 2N$ system. The system is decoupled and the flow and transport equations are solved sequentially, in this case with Newton rather than Picard linearization applied to the transport equation. In both the Picard and partial Newton approaches, discretization of the flow equation yields an $N \times N$ symmetric system, while linearization of the transport equation yields an $N \times N$ nonsymmetric system of equations. On a per iteration basis, therefore, the CPU cost of the partial Newton scheme is not significantly greater than that of the Picard method.

We will first analyze the governing equations, examining the nature and effects of the various nonlinear terms. The three-dimensional finite element model is then developed, and three linearization strategies — the Picard method, the Newton scheme applied to the full system, and the proposed partial Newton technique — are presented. The performance of the Picard, relaxed Picard, and partial Newton methods is compared for two test problems, where we examine in particular convergence sensitivity to the density ratio and dispersivity coefficients.

2. MATHEMATICAL MODEL

The mathematical model of saltwater intrusion into aquifers can be stated using two equations of mass conservation for the fluid mixture and for the contaminant, respectively, supplemented by a constitutive relationship relating the variation of the mixture density with concentration. The continuity equation for the mixture can be expressed in terms of an equivalent freshwater head, defined as [5, 8]:

$$h = \frac{p}{\gamma_0} + z$$

where γ_0 is the freshwater specific weight and z is the elevation above a datum. The density of the saltwater solution can be written in terms of a reference density ρ_0 and the normalized solute concentration c:

$$\rho = \rho_0(1 + \epsilon c)$$

where ρ_0 is the freshwater density, $\epsilon = (\rho_s - \rho_0)/\rho_0$ is the density ratio, typically $\ll 1$, and ρ_s is the maximum solution density (density of seawater). Assuming that the salt is completely dissolved in water, that the temperature is constant, that viscosity does not affect the permeability of the porous medium, and that both Fick's and Darcy's laws hold [6], the equations governing the movement of the salty plume in an aquifer can be written as [5, 9]:

$$\frac{\partial}{\partial x_i}\left[k_{ij}\left(\frac{\partial h}{\partial x_j} + \epsilon c \eta_j\right)\right] = S_s \frac{\partial h}{\partial t} + n \frac{\rho_0}{\rho}\epsilon \frac{\partial c}{\partial t} - q \tag{1a}$$

$$v_i = -k_{ij}\left(\frac{\partial h}{\partial x_j} + \epsilon c \eta_j\right) \qquad i = 1, 2, 3 \tag{1b}$$

$$\frac{\partial}{\partial x_i}\left(D_{ij}\frac{\partial c}{\partial x_j}\right) - v_i \frac{\partial c}{\partial x_i} = n \frac{\partial c}{\partial t} + q(c - c^*) - f \tag{1c}$$

where x_i $(i = 1, 2, 3)$ is the spatial coordinate system ($x_3 = z$ is vertical and positive upward), k_{ij} is the saturated hydraulic conductivity tensor, p is the pressure, g is the gravitational constant, η_j is the unit vector ($\eta_1 = \eta_2 = 0$; $\eta_3 = 1$), S_s is the elastic storage coefficient, t is time, n is the porosity of the medium, q is the injected/extracted volumetric flow rate, v_i is the Darcy velocity in the i-th direction, c^* is the normalized concentration of salt in the injected/extracted fluid, f is the volumetric rate of injected/extracted solute that does not affect the velocity field, and D_{ij} is the usual dispersion tensor as defined in [3], a function of the longitudinal and transverse dispersivities, α_L and α_T, the molecular diffusion coefficient, D_0, and the Darcy velocity. The further hypothesis that $\epsilon c/(1 + \epsilon c) \ll 1$ is assumed to hold. Initial and Dirichlet, Neumann, or Cauchy boundary conditions are added to complete the mathematical formulation of the flow and transport problems expressed in (1) [10].

Coupling in the seawater intrusion model arises from the concentration terms in the flow equation (1a) ($\epsilon c \eta_j$ and $n(\rho_0/\rho)\epsilon(\partial c/\partial t)$) and the head terms that appear in the transport equation (1c) via the Darcy velocities (1b). The model is nonlinear because of the

$n(\rho_0/\rho)\epsilon(\partial c/\partial t) = n(\epsilon/(1 + \epsilon c))(\partial c/\partial t)$ term in (1a) and the $\epsilon c n_j$ term in the advective and dispersive components of (1c) when (1b) is substituted for velocities.

To examine the coupling and nonlinearity effects, it is convenient to consider the simpler case of one-dimensional vertical flow with $S_s = 0$, $D_o = 0$, and constant coefficients. Assuming $v_3 > 0$, substituting (1b) into the transport equation, and collecting the ϵ-terms, we obtain

$$\frac{\partial}{\partial x_3}\left(k_{33}\frac{\partial h}{\partial x_3}\right) + q = \epsilon\left(-k_{33}\frac{\partial c}{\partial x_3} + \frac{n}{1 + \epsilon c}\frac{\partial c}{\partial t}\right) \tag{2a}$$

$$-\frac{\partial}{\partial x_3}\left[\alpha_L\left(k_{33}\frac{\partial h}{\partial x_3}\right)\frac{\partial c}{\partial x_3}\right] + k_{33}\frac{\partial h}{\partial x_3}\frac{\partial c}{\partial x_3} - n\frac{\partial c}{\partial t} + q(c - c^*) + f$$

$$= \epsilon\left\{\frac{\partial}{\partial x_3}\left[(\alpha_L k_{33}c)\frac{\partial c}{\partial x_3}\right] - k_{33}c\frac{\partial c}{\partial x_3}\right\} \tag{2b}$$

For $\epsilon = 0$, equations (2) can be decoupled by first solving the flow equation (2a) for potential head, then using equation (1b) to calculate the Darcy velocities, and finally solving the transport equation (2b) for concentration. In this case both equations are linear.

If $\epsilon \neq 0$, the flow and transport equations are coupled and nonlinear, and must be solved simultaneously for h and c. The coupling term is now given by

$$\epsilon\left(-k_{33}\frac{\partial c}{\partial x_3} + \frac{n}{1 + \epsilon c}\frac{\partial c}{\partial t}\right) \tag{3}$$

If this term is zero the system is uncoupled, and the two equations can be solved separately. In this case the flow equation is linear, while the transport equation is nonlinear. If term (3) is nonzero, the nonlinearity in the flow equation is weak because, since $\epsilon \ll 1$ and $c < 1$, $\epsilon c \ll 1$, and therefore $[n/(1 + \epsilon c)]\partial c/\partial t \approx n\partial c/\partial t$. The transport equation contains two nonlinear components. The first, on the left hand side, is

$$-\frac{\partial}{\partial x_3}\left[\alpha_L\left(k_{33}\frac{\partial h}{\partial x_3}\right)\frac{\partial c}{\partial x_3}\right] + \left(k_{33}\frac{\partial h}{\partial x_3}\right)\frac{\partial c}{\partial x_3} \tag{4}$$

The degree of nonlinearity of this component is controlled by the dependence of $\partial h/\partial x_3$ on term (3) via equation (2a). When concentration gradients in space and time are small the coupling is weak, and consequently (4) is weakly nonlinear. The second nonlinear component, on the right hand side of (2b), is

$$\epsilon\left\{\frac{\partial}{\partial x_3}\left[(\alpha_L k_{33}c)\frac{\partial c}{\partial x_3}\right] - k_{33}c\frac{\partial c}{\partial x_3}\right\} \tag{5}$$

Note that this component does not depend on h, but only on c, ϵ, and the spatial concentration gradients. When transport is dispersion dominated, spatial and temporal concentration gradients are generally small. In this case the influence of the coupling and nonlinear terms (3) and (4), and to a lesser extent (5), is weak.

In summary, the flow equation is weakly nonlinear, and we expect the importance of coupling and the degree of nonlinearity in the transport equation to decrease as ϵ decreases or as dispersion becomes dominant.

3. NUMERICAL MODEL

The numerical model for the solution of the system of equations (1) is a standard Galerkin scheme, with tetrahedral elements, and linear basis functions, complemented by weighted finite differences for the discretization of the time derivatives. The discretized problem can be written as [5, 9]

$$\left(\theta_f H + \tfrac{1}{\Delta t}P\right) h^{k+1} + \left[(1-\theta_f)H - \tfrac{1}{\Delta t}P\right] h^k + q_f^{k+\theta_f} - b_f^{k+\theta_f} + g_3^{k+\theta_f} = 0 \quad (6a)$$

$$v_i^{k+1} = V_i h^{k+1} + g_i^{k+1} \quad i = 1,2,3 \quad (6b)$$

$$\left[\theta_c(A+B+D)^{k+\theta_c} + \tfrac{1}{\Delta t}C\right] c^{k+1} + \left[(1-\theta_c)(A+B+D)^{k+\theta_c} - \tfrac{1}{\Delta t}C\right] c^k$$
$$+ q_c^{k+\theta_c} - b_c^{k+\theta_c} = 0 \quad (6c)$$

where θ_f and θ_c are the time discretization weights for the flow and transport equations, respectively ($\theta_f, \theta_c \in [0.5, 1]$), Δt is the time step size, k is the time step index, H is the flow stiffness matrix, P is the flow mass matrix, h is the vector of nodal equivalent freshwater head values, q_f is the mass flux vector for the flow equation, b_f is the vector containing the boundary conditions and the discretized $[n\epsilon/(1+\epsilon c)]\partial c/\partial t$ term of the flow equation, g_i is the vector containing the gravity term $k_{ij}\epsilon c\eta_j$, v_i is the velocity vector in the ith direction, $i = 1,2,3$, V_i is the velocity matrix, A, B, and D are the transport stiffness matrices, containing the dispersive component, the advective component, and the Cauchy boundary component, respectively, C is the transport mass matrix, c is the vector of nodal concentration values, q_c is the mass flux vector for the transport equation, and b_c is the vector of boundary conditions for the transport equation. In equation (6a), g_3 and b_f are dependent on concentration c, while in (6c), matrices A, B, and D depend on both h and c. The algebraic system (6) is therefore coupled and nonlinear in the dependent variables h and c, and the flow and transport equations must be solved simultaneously. Note that system (6) has dimension $2N \times 2N$, where N is the number of nodes of the computational grid.

4. LINEARIZATION STRATEGIES

4.1 Picard Iteration

The most commonly used algorithm for the numerical solution of the nonlinearly coupled system (6) can be described as follows. Denoting with m the iteration counter, at the $(m+1)$-st iteration the flow equation is solved for $h^{k+1,(m+1)}$ using, as initial guess, the values of head, $h^{k+1,(m)}$, and concentration, $c^{k+1,(m)}$, from the previous iteration. The velocity field (6b) is then calculated using $h^{k+1,(m+1)}$ and $c^{k+1,(m)}$. Using these updated values of velocity and head, the transport equation becomes linear and is solved for $c^{k+1,(m+1)}$. This three-step procedure is repeated until convergence is achieved. We refer to this iterative procedure as the Picard or sequential solution method.

The algorithm can be expressed mathematically as

$$\left(\theta_f H + \frac{1}{\Delta t} P\right) h^{k+1,(m+1)} = -\left[(1-\theta_f)H - \frac{1}{\Delta t}P\right]h^k$$

$$- q_f^{k+\theta_f} + b_f^{k+\theta_f,(m)} - g_3^{k+\theta_f,(m)} \tag{7a}$$

$$v_i^{k+1,(m)} = V_i h^{k+1,(m+1)} + g_i^{k+1,(m)} \quad i = 1,2,3 \tag{7b}$$

$$\left[\theta_c (A+B+D)^{k+\theta_c,(m)} + \frac{1}{\Delta t}C\right] c^{k+1,(m+1)} = \tag{7c}$$

$$- \left[(1-\theta_c)(A+B+D)^{k+\theta_c,(m)} - \frac{1}{\Delta t}C\right]c^k - q_c^{k+\theta_c} + b_c^{k+\theta_c} \tag{7d}$$

Note that in (7b) we write $v_i^{k+1,(m)}$ instead of $v_i^{k+1,(m+1)}$ because, while the velocity field is calculated using the new estimate $h^{k+1,(m+1)}$ for the head, the previous estimate $c^{k+1,(m)}$ is used for concentration. Analogously, the transport matrices are expressed at iteration m since they are evaluated using $v_i^{k+1,(m)}$. In the Picard method two $N \times N$ linear systems are solved sequentially at each iteration, the symmetric flow system (7a) and the nonsymmetric transport system (7d).

4.2 Newton Scheme

An alternative procedure for solving (6) is to use the Newton technique. Let f be the vector-valued function

$$f(u) = \left[\begin{array}{c} f_1(u) \\ f_2(u) \end{array}\right] \tag{8}$$

where $u = \left(h^{k+1}, c^{k+1}\right)^T$ and

$$f_1(u) = \left(\theta_f H + \frac{1}{\Delta t}P\right)h^{k+1} + \left[(1-\theta_f)H - \frac{1}{\Delta t}P\right]h^k$$

$$+ q_f^{k+\theta_f} - b_f^{k+\theta_f} + g_3^{k+\theta_f} \tag{9}$$

$$f_2(u) = \left[\theta_c(A+B+D)^{k+\theta_c} + \frac{1}{\Delta t}C\right]c^{k+1} + \left[(1-\theta_c)(A+B+D)^{k+\theta_c} - \frac{1}{\Delta t}C\right]c^k$$

$$+ q_c^{k+\theta_c} - b_c^{k+\theta_c} \tag{10}$$

We want to find a correction

$$\Delta u = \left(\begin{array}{c} \Delta h \\ \Delta c \end{array}\right)$$

such that $f(u + \Delta u) = 0$. Using a Taylor expansion truncated after the first derivative terms, we obtain $f(u) + f'(u)\Delta u \simeq 0$ or $f'(u)\Delta u = -f$, where f' is the Jacobian matrix of f, defined as

$$f' = \left[\begin{array}{cc} \dfrac{\partial f_1}{\partial h^{k+1}} & \dfrac{\partial f_1}{\partial c^{k+1}} \\[2ex] \dfrac{\partial f_2}{\partial h^{k+1}} & \dfrac{\partial f_2}{\partial c^{k+1}} \end{array}\right] \tag{11}$$

The Newton scheme for system (6) can thus be written as

$$f'\left(u^{(m)}\right)\Delta u \;=\; -f\left(u^{(m)}\right) \tag{12}$$

$$u^{(m+1)} \;=\; u^{(m)} + \Delta u \tag{13}$$

Note that the Jacobian matrix is a $2N \times 2N$ nonsymmetric matrix subdivided into four matrix blocks. The calculation of the four terms which make up the Jacobian matrix is given in the appendix.

Because of the cost of assembling and solving a $2N \times 2N$ system, and the complexity of the off-diagonal Jacobian terms $\partial f_1/\partial c$ and $\partial f_2/\partial h$, we did not implement the Newton method. Instead we developed a partial Newton scheme which takes into consideration the most important of the nonlinear derivative components of the Jacobian (11). This scheme, described below, produces a decoupled system, in the manner of the Picard scheme or other similar approaches used in petroleum reservoir simulation [11], while retaining some of the important advantages of the Newton technique.

4.3　Partial Newton Scheme

If we rewrite equations (7) with the velocities expressed using $c^{k+1,(m+1)}$ rather than $c^{k+1,(m)}$, we obtain

$$\left(\theta_f H + \frac{1}{\Delta t}P\right) h^{k+1,(m+1)} = -\left[(1-\theta_f)H - \frac{1}{\Delta t}P\right]h^k$$
$$-\,q_f^{k+\theta_f} + b_f^{k+\theta_f,(m)} - g_3^{k+\theta_f,(m)} \tag{14a}$$

$$\left[\theta_c\,(A+B+D)^{k+\theta_c,(m+1)} + \frac{1}{\Delta t}C\right]c^{k+1,(m+1)} = -\left[(1-\theta_c)\,(A+B)^{k+\theta_c,(m+1)} - \frac{1}{\Delta t}C\right]c^k$$
$$-\,q_c^{k+\theta_c} + b_c^{k+\theta_c} \tag{14b}$$

where

$$v_i^{k+1,(m+1)} = V_i h^{k+1,(m+1)} + g_i^{k+1,(m+1)}$$

The transport equation is now nonlinear and can be linearized using the Newton scheme.

The $N \times N$ Jacobian matrix for equation (14b), i.e., the derivative of the transport equation with respect to concentration, yields:

$$J\left(h^{k+1},c^{k+1}\right) = \theta_c\,(A+B+D)^{k+\theta_c} + \left[\frac{\partial}{\partial c^{k+1}}(A+B+D)^{k+\theta_c}\right]c^{k+\theta_c} + \frac{1}{\Delta t}C \tag{15}$$

This term is identical to the fourth block of the Jacobian matrix for the full Newton scheme, that is, $\partial f_2/\partial c$ of equation (11). Evaluation of the different components of the derivative term in (15) is reported in the appendix. The Newton equation for (14b) is therefore written as

$$J\left(h^{k+1,(m)},c^{k+1,(m)}\right)\Delta c = -f_2\left(h^{k+1,(m)},c^{k+1,(m)}\right) \tag{16}$$

where $\Delta c = c^{k+1,(m+1)} - c^{k+1,(m)}$ and

$$f_2\left(h^{k+1,(m)}, c^{k+1,(m)}\right) = \left[\theta_c\left(A + B + D\right)^{k+\theta_c,(m)} + \frac{1}{\Delta t}C\right]c^{k+1,(m)}$$

$$+ \left[(1 - \theta_c)\left(A + B + D\right)^{k+\theta_c,(m)} - \frac{1}{\Delta t}C\right]c^k + q_c^{k+\theta_c} - b_c^{k+\theta_c}$$

The solution procedure for one iteration of the partial Newton method is a sequence similar to the Picard scheme, and consists of the following two steps:

1. calculate $h^{k+1,(m+1)}$ by solving the flow equation using $c^{k+1,(m)}$ and $h^{k+1,(m)}$ as initial values;

2. solve the Newton-linearized transport equation (16) to evaluate the new values of concentration $c^{k+1,(m+1)}$.

Comparing equation (15) with the left hand side of (7d), we see that the additional cost of the partial Newton scheme relative to the Picard scheme amounts to the two matrix-vector products represented in (18) and (19) in the appendix, and the assembly of the corresponding elements. The remaining operations are formally the same, and the final system matrix is nonsymmetric for both schemes.

The partial Newton scheme can also be derived directly from the full Newton procedure. Since the flow equation is weakly nonlinear, we can neglect the $\partial f_1/\partial c$ term of (11), obtaining the head difference solution

$$\Delta h = -\left[\frac{\partial f_1}{\partial h^{k+1}}\right]^{-1} f_1\left(h^{k+1,(m)}, c^{k+1,(m)}\right)$$

The concentration difference can now be calculated solving the system

$$\left[\frac{\partial f_2}{\partial c^{k+1}}\left(h^{k+1,(m)}, c^{k+1,(m)}\right)\right]\Delta c = -f_2\left(h^{k+1,(m)}, c^{k+1,(m)}\right) - \frac{\partial f_2}{\partial h^{k+1}}\left(h^{k+1,(m)}, c^{k+1,(m)}\right)\Delta h$$

$$(17)$$

If the $(\partial f_2/\partial h)\Delta h$ term is neglected, equation (17) reduces to equation (16) of the partial Newton scheme. The importance of this term on the convergence behavior of the iterative scheme will be evaluated in future work.

Because of the application of a quadratically convergent Newton linearization to the transport equation, we expect faster convergence for the partial Newton scheme compared to the Picard technique. As noted previously, the degree of nonlinearity of the transport equation varies with the magnitude of the density ratio and dispersion coefficient. We therefore expect the partial Newton scheme to converge successfully where Picard linearization may fail, that is, when ϵ is large or dispersion is small. For problems where ϵ is small or dispersion is dominant, on the other hand, the Picard scheme should give adequate performance.

5. NUMERICAL EXAMPLES

The convergence behavior of the Picard and partial Newton schemes are analyzed by means of a steady state two-dimensional flow and saltwater transport problem in a rectangular domain. The problem has been adapted from [5]. The relaxation technique suggested by [5] is also tested. A second sample test considers the three-dimensional simulation of the saltwater intrusion in a Southern Italian aquifer [12]

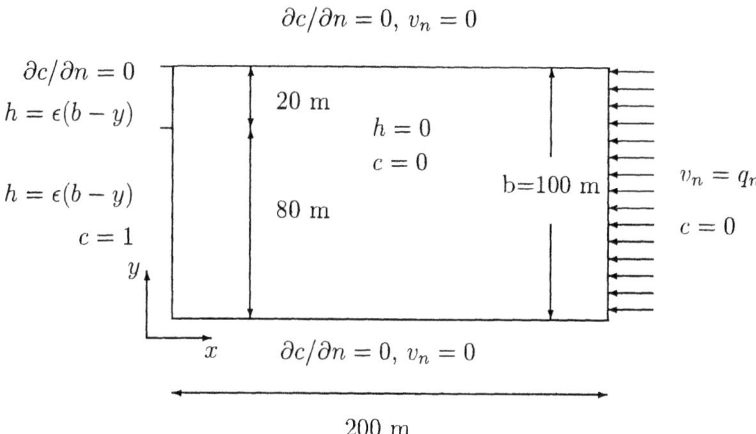

Figure 1: Two-dimensional domain with boundary and initial conditions for Example 1. In this figure n is the normal to the boundary.

5.1 Two-Dimensional Homogeneous Flow and Transport in a Rectangular Domain

The geometry and initial and boundary conditions of the first problem are shown schematically in Figure 1. The incoming flux of freshwater on the right boundary is $q_n = 6.6 \times 10^{-5}$ m/s. The two-dimensional domain is discretized using a three-dimensional grid of unit width. The discretized domain contains 693 nodes and 2400 tetrahedra.

The following parameter values are used for all runs: $k_{11} = k_{22} = 0.01$ m/s, $n = 0.35$, $D_o = 0$ m^2/s. The values for the dispersivities α_L and α_T range from 0.0035 m to 0.1 m, and the value of the density ratio ϵ range from 0.0245 to 0.1. Ten runs were made, and the convergence results are summarized in Table 1. The solution to this problem is given in Figure 2, where the steady state head, concentration, and velocity fields are shown for run 1. The effects of the density variation can be seen from the velocity field, where a flux inversion occurs near the left boundary.

Convergence plots for four of the ten runs are shown in Figure 3. The convergence error plotted on the ordinate is the infinity norm of the concentration vector, $\|c^{(m+1)} - c^{(m)}\|_\infty$. For all test runs the convergence behavior for the coupled problem is dictated by the concentration errors.

From Table 1 and Figure 3, we note that, for this example, the partial Newton method is generally more robust than the both the relaxed and unrelaxed Picard schemes. Note that in runs 2 and 4 the partial Newton scheme displays an overshoot at the beginning of the convergence process. This phenomenon is drastically amplified in run 5, where partial Newton diverges after few iterations. On the other hand, the Picard method shows a smoother initial convergence behavior, followed by slower convergence, or even nonconvergence in some cases. This situation suggests that an effective linearization strategy for strongly nonlinear problems may be to use the Picard method for the first few iterations and the partial Newton method for subsequent iterations. In this way the Picard scheme is used to provide an improved initial

Table 1: Parameter Values and Number of Iterations to Convergence for the Test Runs of Example 1

Run #	α_L (m)	α_T (m)	ϵ	Partial Newton	Partial Newton with Relaxation	Picard	Picard with Relaxation
1	0.0350	0.0350	0.0245	9	10	n.c.	26
2	0.0350	0.0350	0.0500	21	32	n.c.	n.c.
3	0.0500	0.0500	0.0245	8	9	n.c.	11
4	0.0500	0.0500	0.0500	19	29	n.c.	n.c.
5	0.0350	0.0350	0.1000	n.c.	n.c.	n.c.	n.c.
6	0.1000	0.1000	0.0245	7	7	11	7
7	0.1000	0.1000	0.1000	n.c.	n.c.	n.c.	n.c.
8	0.0035	0.0035	0.0245	n.c.	n.c.	n.c.	n.c.
9	0.0100	0.0100	0.0245	17	17	n.c.	n.c.
10	0.0350	0.0035	0.0245	18	19	n.c.	n.c.
n.c.: convergence not achieved after 50 iterations							

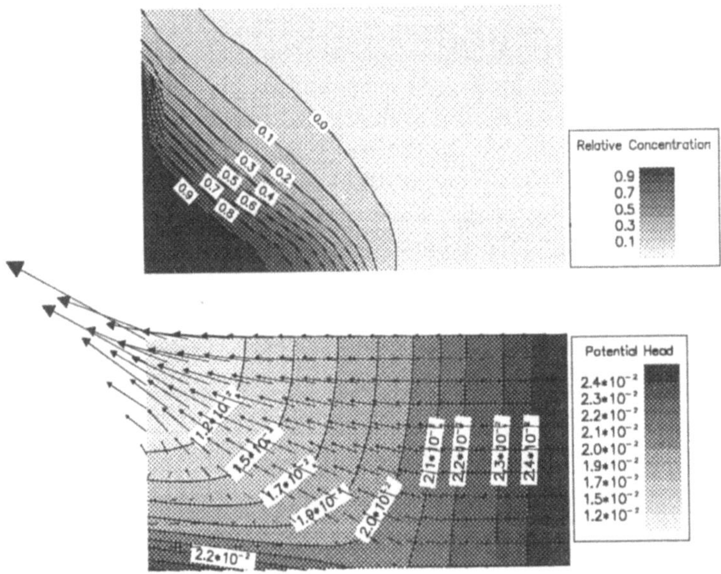

Figure 2: Steady state head, velocity, and concentration fields for run 1 of Example 1.

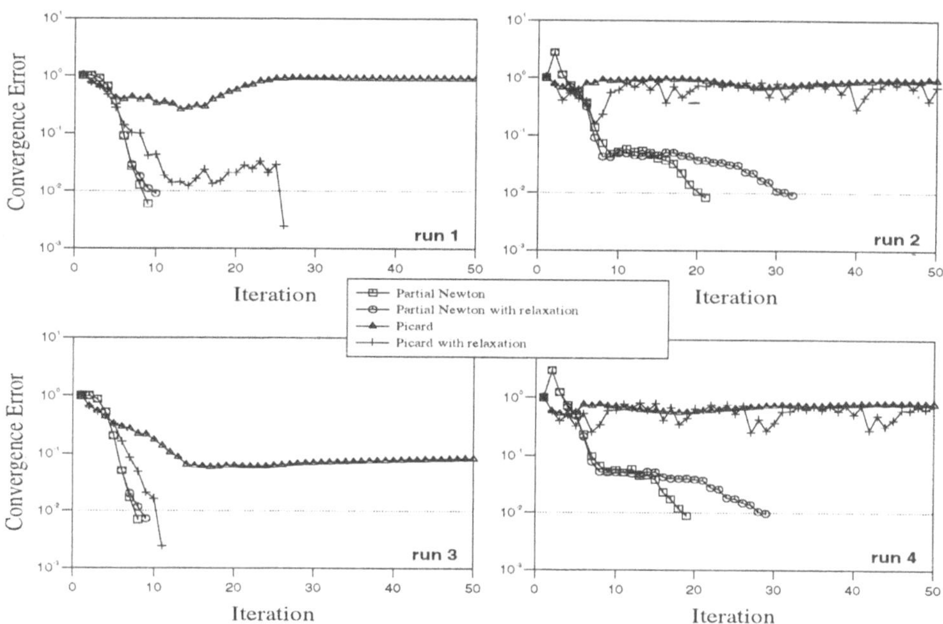

Figure 3: Convergence plots for runs 1, 2, 3, and 4 of Example 1.

solution estimate for Newton. This "mixed" Picard-Newton approach has been successfully used to solve the unsaturated flow equation [13].

The behavior of the Picard and partial Newton schemes with respect to the density ratio and dispersivity values confirms the observations made in previous sections. The partial Newton scheme is more effective than the Picard method at high density ratios and lower dispersivity values, and divergence is noticed only for runs 5 and 7, characterized by a large ϵ ($\epsilon = 0.1$), and for run 8, characterized by small dispersive fluxes ($\alpha_L = 0.35$ m and $\alpha_T = 0.0035$ m).

Convergence difficulties at low dispersivity values are due to instabilities in the finite element discretization that are typical of advection dominated problems. The results in Table 1 indicate that the partial Newton scheme is less sensitive to these instabilities than the Picard method.

5.2 Three-Dimensional Intrusion in a Southern Italian Coastal Aquifer

This test case deals with the simulation of saltwater intrusion into a coastal aquifer in Southern Italy [12]. The 36 km^2 study site is located on the Tyrrhenian Sea, and is bounded by rivers on its northern and southern borders, by the sea on the west, and by a high plane on the east (Figure 4).

The construction of a channel system, used to discharge water into the sea, requires the lowering of the water table of the underlying aquifer for a period of 6 months. The extensive pumping needed for the complete dewatering of the construction site may affect the natural

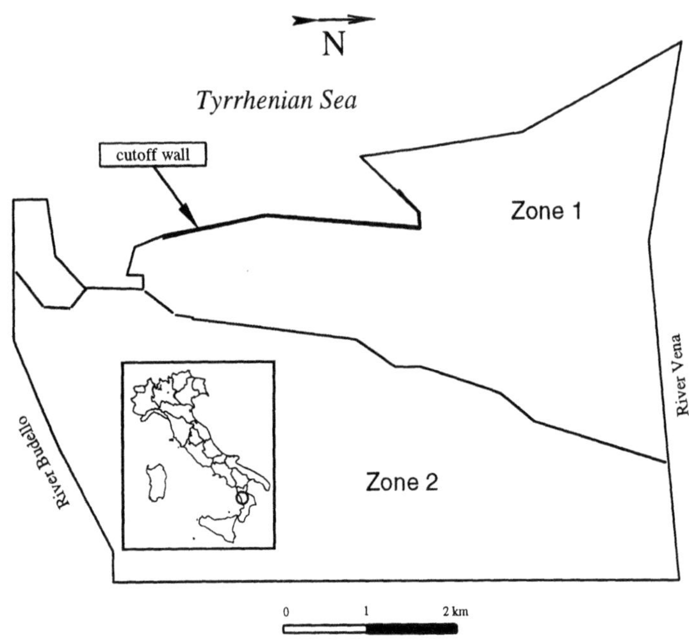

Figure 4: Location and geometry of the coastal aquifer for Example 2.

flow field and alter the behavior of the saltwater front. The consequent contamination of the aquifer may pose a threat to the crops of the surrounding farmland. A three-dimensional model of the aquifer system is therefore used to evaluate the impact of dewatering on the aquifer contamination by seawater and the effectiveness of a cutoff wall which was constructed with the aim to contain the seawater intrusion.

The boundary conditions of the system are as follows. At the top and bottom of the aquifer and on the north and south boundaries we impose a zero flux condition for both the flow and transport equations. Dirichlet conditions are imposed on the east boundary, with zero concentration prescribed for the transport equation and a range of head values for the flow equation, from 66.2 m for the nodes at the north and south corners to 96.7 m for the middle node. The seaside boundary contains three zones. The central zone is characterized by a 26.3 m deep cutoff wall, while the two lateral zones are in direct contact with the sea. In these lateral zones, hydrostatic head is prescribed, along with zero concentration flux for the top 10.9 m and unitary (seawater) concentration in the lower part. In the central zone, zero flux is prescribed for the flow and transport equations for the top 26.3 m, while for the lower part hydrostatic head and unitary concentration are imposed.

The surface mesh contains 2829 triangles and 1459 nodes, and is duplicated vertically to form 10 layers of depths 3.7, 3.7, 1.5, 2.0, 7.9, 7.5, 7.5, 7.5, 7.5, and 7.5 m from top to bottom, for a total aquifer depth of 56.3 m. The three-dimensional grid contains 84870 tetrahedra and 16049 nodes.

The aquifer contains two distinct hydrogeological zones (Figure 4). Zone 1 is stratified,

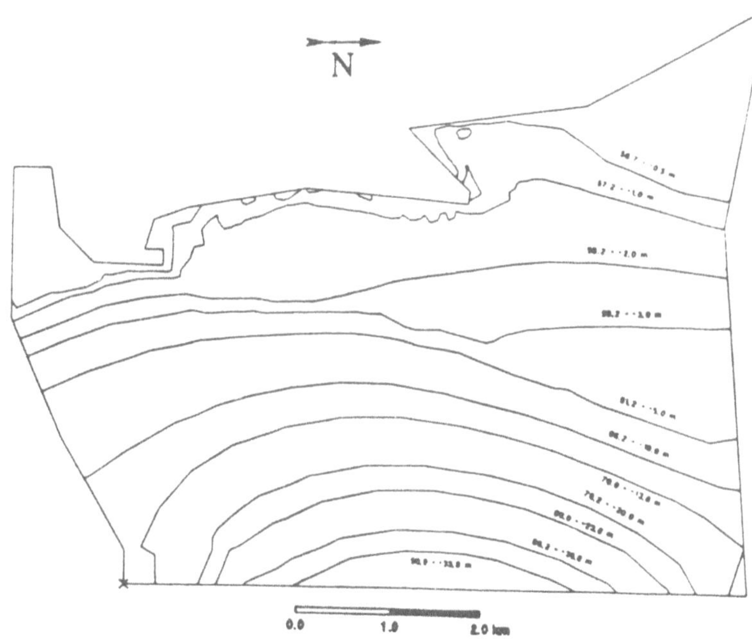

Figure 5: Example 2: steady state equipotential lines at the bottom of the aquifer (after [12]).

with hydraulic conductivity values of $k_{11} = k_{22} = 1.8 \times 10^{-3}$ m/s and $k_{33} = 3.0 \times 10^{-4}$ m/s in the top 7.4 m, and $k_{11} = k_{22} = k_{33} = 6.0 \times 10^{-4}$ m/s in the bottom 48.9 m. In zone 2 we have $k_{11} = k_{22} = k_{33} = 8.0 \times 10^{-5}$ m/s. In the entire aquifer, we have $D_o = 0.0$ and $n = 0.22$, $\epsilon = 0.03$, $\alpha_L = 10$ m, and $\alpha_T = 1$ m.

Figures 5 and 6 show the steady state head contour lines and the steady state equiconcentration lines at the bottom of the aquifer before pumping starts. Note that this steady state solution is not obtained directly, as neither the Picard nor the partial Newton schemes are able to converge within the maximum number of iterations allowed (500). Rather, the solution is achieved by running a transient simulation until variations in head and concentration are negligible. The time at which steady state is considered achieved is 1342 days. Using time step sizes varying from 1 to 25 days, less than 12 Picard iterations per time step are needed.

Starting from the steady state conditions, pumping at a rate of 2.189 m^3/day is applied to the aquifer. A period of 185 days (which is the expected duration of the channel construction phase) is simulated. Figure 7 shows the head drawdown at the end of the pumping period, while Figure 8 shows the saltwater equiconcentration lines for the same period. Both figures are referred to a horizontal section at the bottom of the aquifer. Note that, in Figure 8, the pumping location is clearly visible as all the concentration contour lines tend to converge together with the flow field towards the extraction wells. The results of the simulations show that the water withdrawal is responsible for only a limited increase in the total saltwater intrusion.

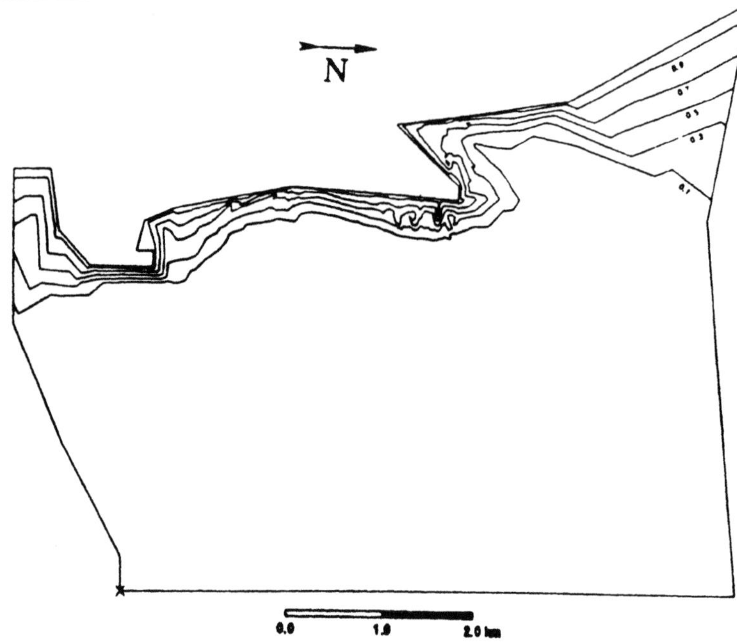

Figure 6: Example 2: steady state equiconcentration lines at the bottom of the aquifer (after [12]).

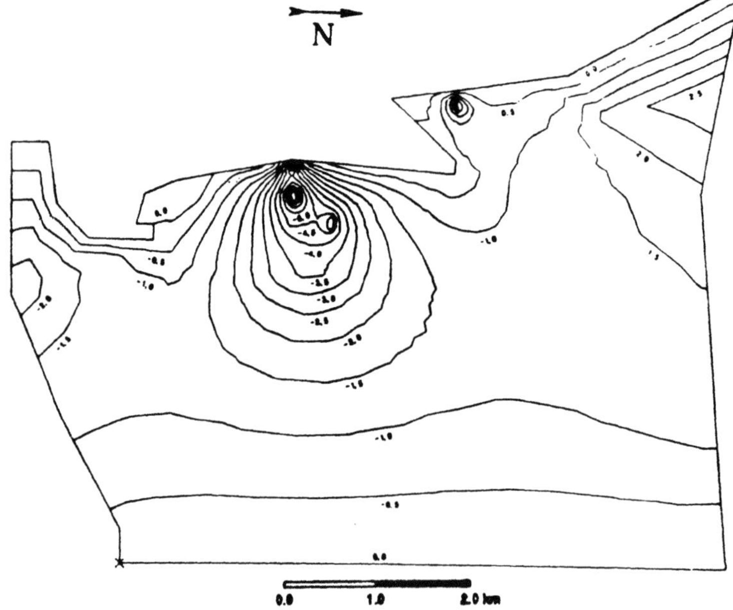

Figure 7: Example 2: drawdown contour lines (m) at the bottom of the Southern Italian aquifer after 185 days of pumping (after [12]).

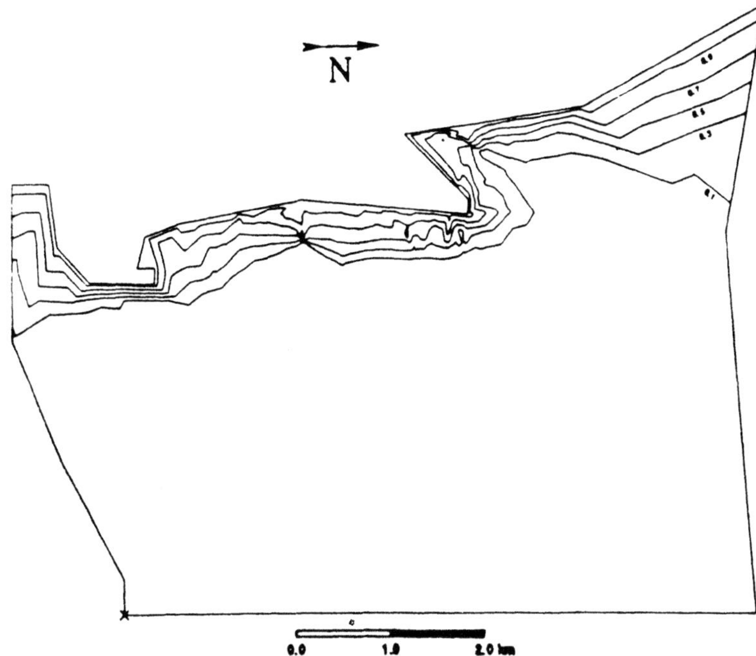

Figure 8: Example 2: equiconcentration lines at the bottom of the aquifer (after [12]).

6. CONCLUSIONS

Newton and Picard type linearization strategies for the numerical solution of the coupled flow and saltwater transport equations in aquifers have been presented and evaluated. First it was shown that the Picard and the partial Newton methods can be obtained by applying to the transport equation a Picard or Newton-type linearization, respectively. Better convergence properties of the partial Newton technique over Picard are due to the more accurate linearization applied to the most significant nonlinear terms present in the coupled system.

The convergence behavior of the Picard and partial Newton schemes is exemplified by means of a two-dimensional sample problem, where the effects of relaxation are also examined. An application of a fully three-dimensional model to a Southern Italian aquifer system is also described.

The results of the simulations confirm that the partial Newton scheme is generally more robust and efficient than the Picard method, with and without relaxation. The presence of oscillations in the solution affects the convergence behavior of the iterative schemes, in some cases causing divergence. However, with small oscillations, the partial Newton method proves to be more robust than the Picard scheme.

For very large density ratios, both the partial Newton and Picard methods fail to converge, suggesting that schemes even more robust than the partial Newton method are needed for these cases, like for example, a full Newton scheme. Other alternatives worth investigating include quasi-Newton methods, applied either to the fully coupled system or to the transport equation alone, and parameter stepping algorithms [4, 14].

Acknowledgments. This work has been partially supported by the Italian CNR, Gruppo Nazionale per la Difesa dalle Catastrofi Idrogeologiche, linea di Ricerca n. 4, and by the Sardinia Regional Authorities.

APPENDIX: CALCULATION OF THE JACOBIAN MATRIX

We develop the four terms which make up the Newton scheme Jacobian given by equation (11).

First term

$$
\begin{aligned}
\frac{\partial f_1}{\partial h^{k+1}} &= \frac{\partial}{\partial h^{k+1}} \left\{ \left(\theta_f H + \frac{1}{\Delta t} P \right) h^{k+1} \right. \\
&\quad \left. + \left[(1 - \theta_f) H - \frac{1}{\Delta t} P \right] h^k + q_f^{k+\theta_f} - b_f^{k+\theta_f} + g_3^{k+\theta_f} \right\} \\
&= \left(\theta_f H + \frac{1}{\Delta t} P \right)
\end{aligned}
$$

Second term

$$
\begin{aligned}
\frac{\partial f_1}{\partial c^{k+1}} &= \frac{\partial}{\partial c^{k+1}} \left\{ \left(\theta_f H + \frac{1}{\Delta t} P \right) h^{k+1} \right. \\
&\quad \left. + \left[(1 - \theta_f) H - \frac{1}{\Delta t} P \right] h^k + q_f^{k+\theta_f} - b_f^{k+\theta_f} + g_3^{k+\theta_f} \right\} \\
&= \frac{\partial}{\partial c^{k+1}} \left(g_3^{k+\theta_f} - b_f^{k+\theta_f} \right)
\end{aligned}
$$

For a general tetrahedral element with nodes i, l, m, n and basis functions ς, ζ, and ξ,

$$
\begin{aligned}
g_{3_i}^{k+\theta_f} - b_{f_i}^{k+\theta_f} &= -\operatorname{sign}(V^e) k_{33} \frac{\epsilon}{4} \xi_i \left[\theta_f \left(c_i^{k+1} + c_l^{k+1} + c_m^{k+1} + c_n^{k+1} \right) \right. \\
&\quad \left. + (1 - \theta_f) \left(c_i^k + c_l^k + c_m^k + c_n^k \right) \right] \\
&\quad - n \left\{ \frac{\theta_f}{\left[1 + \epsilon \left(c_i^{k+1} + c_l^{k+1} + c_m^{k+1} + c_n^{k+1} \right)/4 \right]} \right. \\
&\quad \left. + \frac{(1 - \theta_f)}{\left[1 + \epsilon \left(c_i^k + c_l^k + c_m^k + c_n^k \right)/4 \right]} \right\} \frac{\epsilon}{\Delta t} \frac{|V^e|}{10} \cdot \\
&\quad \left[(c_i^{k+1} - c_i^k) + \tfrac{1}{2}(c_l^{k+1} - c_l^k) + \tfrac{1}{2}(c_m^{k+1} - c_m^k) + \tfrac{1}{2}(c_n^{k+1} - c_n^k) \right]
\end{aligned}
$$

where V^e is the elemental volume and k_{33} is the vertical component of the conductivity tensor (assuming the reference frame is aligned with the principal directions of anisotropy).

Differentiating with respect to $\mathbf{c}^{k+1}$, we obtain

$$
\left(\frac{\partial f_1}{\partial \mathbf{c}^{k+1}}\right)_{ij} = -\operatorname{sign}(V^e)k_{33}\xi_i\frac{\epsilon}{4}\theta_f
$$

$$
- n\left[\frac{\theta_f}{1+\epsilon\bar{c}^{k+1}} + \frac{1-\theta_f}{1+\epsilon\bar{c}^k}\right]\frac{|V^e|}{10}\frac{\epsilon}{\Delta t}\left[\frac{1}{2}(\delta_{ij}+1)\right]
$$

$$
+ \frac{n}{4}\frac{\epsilon^2}{\Delta t}\frac{|V^e|}{10}\frac{\theta_f}{[1+\epsilon\bar{c}^{k+1}]^2}\left[(c_i^{k+1}-c_i^k)+\frac{1}{2}(c_l^{k+1}-c_l^k)\right.
$$

$$
\left.+ \frac{1}{2}(c_m^{k+1}-c_m^k)+\frac{1}{2}(c_n^{k+1}-c_n^k)\right]
$$

$$
= \epsilon\left\{-\operatorname{sign}(V^e)k_{33}\frac{\theta_f}{4} - \frac{n|V^e|}{10\Delta t}\left(\frac{\theta_f}{1+\epsilon\bar{c}^{k+1}}+\frac{1-\theta_f}{1+\epsilon\bar{c}^k}+\right)\left[\frac{1}{2}(\delta_{ij}+1)\right]\right\}
$$

$$
+ \epsilon^2\frac{n|V^e|}{40\Delta t}\frac{\theta_f}{(1+\epsilon\bar{c}^{k+1})^2}\left[(c_i^{k+1}-c_i^k)+\frac{1}{2}(c_l^{k+1}-c_l^k)\right.
$$

$$
\left.+ \frac{1}{2}(c_m^{k+1}-c_m^k)+\frac{1}{2}(c_n^{k+1}-c_n^k)\right]
$$

where δ_{ij} is the Kronecker delta and $\bar{c} = (c_i + c_l + c_m + c_n)/4$.

Third term

$$
\frac{\partial f_2}{\partial \mathbf{h}^{k+1}} = \frac{\partial}{\partial \mathbf{h}^{k+1}}\left\{\left[\theta_c\left(\mathbf{A}+\mathbf{B}+\mathbf{D}\right)^{k+\theta_c}+\frac{1}{\Delta t}\mathbf{C}\right]\mathbf{c}^{k+1}\right.
$$

$$
\left.+ \left[(1-\theta_c)\left(\mathbf{A}+\mathbf{B}+\mathbf{D}\right)^{k+\theta_c}-\frac{1}{\Delta t}\mathbf{C}\right]\mathbf{c}^k + q_c^{k+\theta_c}-b_c^{k+\theta_c}\right\}
$$

$$
= \theta_c\frac{\partial}{\partial \mathbf{h}^{k+1}}\left(\mathbf{A}+\mathbf{B}+\mathbf{D}\right)^{k+\theta_c}\mathbf{c}^{k+1}+(1-\theta_c)\frac{\partial}{\partial \mathbf{h}^{k+1}}\left(\mathbf{A}+\mathbf{B}+\mathbf{D}\right)^{k+\theta_c}\mathbf{c}^k
$$

$$
= \frac{\partial}{\partial \mathbf{h}^{k+1}}\left(\mathbf{A}+\mathbf{B}\right)^{k+\theta_c}\mathbf{c}^{k+\theta_c}
$$

The matrix $\mathbf{D}$ contains Cauchy boundary conditions and is thus independent of $\mathbf{h}$ or $\mathbf{c}$. The ij-th component of the contribution from matrix $\mathbf{A}$ is

$$
\left[\frac{\partial}{\partial \mathbf{h}^{k+1}}\left(\mathbf{A}^{k+\theta_c}\right)\mathbf{c}^{k+\theta_c}\right]_{ij} = \sum_s \frac{\partial}{\partial h_j^{k+1}}\left[A_{is}^{k+\theta_c}\right]c_s^{k+\theta_c}
$$

$$
= \sum_s \frac{\partial}{\partial h_j^{k+1}}\left\{\frac{1}{|V^e|}\left[\left(\alpha_L|v|^{k+\theta_c}+D_0 n\right)\varsigma_i'\varsigma_s'\right.\right.
$$

$$
+ \left(\alpha_T|v|^{k+\theta_c}+D_0 n\right)\zeta_i'\zeta_s'
$$

$$
\left.\left.+ \left(\alpha_T|v|^{k+\theta_c}+D_0 n\right)\xi_i'\xi_s'\right]\right\}c_s^{k+\theta_c}
$$

where ς', ζ', and ξ' are the basis functions calculated in a rotated frame of reference that has the x direction parallel to the velocity vector. This rotation transforms the dispersion tensor into a diagonal matrix. Differentiating velocities,

$$
\frac{\partial}{\partial h_j^{k+1}}|v^{k+\theta_c}| = \frac{\partial}{\partial h_j^{k+1}}\left(\sqrt{v_1^2+v_2^2+v_3^2}\right)^{k+\theta_c}
$$

$$= \left[\frac{1}{\sqrt{v_1^2 + v_2^2 + v_3^2}} \left(v_1 \frac{\partial v_1}{\partial h_j^{k+1}} + v_2 \frac{\partial v_2}{\partial h_j^{k+1}} + v_3 \frac{\partial v_3}{\partial h_j^{k+1}} \right) \right]^{k+\theta_c}$$

with

$$\frac{\partial}{\partial h_j^{k+1}} v_1^{k+\theta_c} = \frac{\partial}{\partial h_j^{k+1}} \left[-\frac{k_{11}}{V^e} \left(\varsigma_i h_i^{k+\theta_c} + \varsigma_l h_l^{k+\theta_c} + \varsigma_m h_m^{k+\theta_c} + \varsigma_n h_n^{k+\theta_c} \right) \right]$$

$$= -\frac{k_{11}}{V^e} \theta_c \varsigma_j$$

$$\frac{\partial}{\partial h_j^{k+1}} v_2^{k+\theta_c} = -\frac{k_{22}}{V^e} \theta_c \varsigma_j$$

$$\frac{\partial}{\partial h_j^{k+1}} v_3^{k+\theta_c} = -\frac{k_{33}}{V^e} \theta_c \xi_j$$

The contribution from matrix $\boldsymbol{A}$ is thus

$$\left(\frac{\partial}{\partial h^{k+1}} \left[\boldsymbol{A}^{k+\theta_c} \right] \boldsymbol{c}^{k+\theta_c} \right)_{ij} = -\frac{\theta_c}{\operatorname{sign}(V^e)(V^e)^2} \frac{1}{\sqrt{v_1^2 + v_2^2 + v_3^2}^{\,k+\theta_c}} \cdot$$

$$\left(v_1^{k+\theta_c} k_{11} \varsigma_j + v_2^{k+\theta_c} k_{22} \varsigma_j + v_3^{k+\theta_c} k_{33} \xi_j \right) \cdot$$

$$\sum_s \left(\alpha_L \varsigma_i' \varsigma_s' + \alpha_T \zeta_i' \zeta_s' + \alpha_T \xi_i' \xi_s' \right) c_s^{k+\theta_c}$$

The ij-th component of the contribution from matrix $\boldsymbol{B}$ is

$$\left(\frac{\partial}{\partial h^{k+1}} \left[\boldsymbol{B}^{k+\theta_c} \right] \boldsymbol{c}^{k+\theta_c} \right)_{ij} = \sum_s \frac{\partial}{\partial h_j^{k+1}} \left(B_{is}^{k+\theta_c} \right) c_s^{k+\theta_c}$$

where

$$B_{is}^{k+\theta_c} = \frac{\operatorname{sign}(V^e)}{4} \left(v_1^{k+\theta_c} \varsigma_s + v_2^{k+\theta_c} \zeta_s + v_3^{k+\theta_c} \xi_s \right)$$

so that

$$\frac{\partial}{\partial h_j^{k+1}} B_{is}^{k+\theta_c} = -\frac{\theta_c}{4|V^e|} \left(k_{11} \varsigma_j \varsigma_s + k_{22} \zeta_j \zeta_s + k_{33} \xi_j \xi_s \right)$$

and thus

$$\left(\frac{\partial}{\partial h^{k+1}} \left[\boldsymbol{B}^{k+\theta_c} \right] \boldsymbol{c}^{k+\theta_c} \right)_{ij} = -\frac{\theta_c}{4|V^e|} \sum_s \left(k_{11} \varsigma_j \varsigma_s + k_{22} \zeta_j \zeta_s + k_{33} \xi_j \xi_s \right) c_s^{k+\theta_c}$$

This component is independent of row index i.

Fourth term

$$\frac{\partial \boldsymbol{f}_2}{\partial \boldsymbol{c}^{k+1}} = \frac{\partial}{\partial \boldsymbol{c}^{k+1}} \left\{ \left[\theta_c \left(\boldsymbol{A} + \boldsymbol{B} + \boldsymbol{D} \right)^{k+\theta_c} + \frac{1}{\Delta t} \boldsymbol{C} \right] \boldsymbol{c}^{k+1} \right.$$

$$+ \left. \left[(1 - \theta_c) \left(\boldsymbol{A} + \boldsymbol{B} + \boldsymbol{D} \right)^{k+\theta_c} - \frac{1}{\Delta t} \boldsymbol{C} \right] \boldsymbol{c}^k + \boldsymbol{q}_c^{k+\theta_c} - \boldsymbol{b}_c^{k+\theta_c} \right\}$$

$$= \theta_c \left(\boldsymbol{A} + \boldsymbol{B} + \boldsymbol{D} \right)^{k+\theta_c} + \frac{1}{\Delta t} \boldsymbol{C} + \left[\frac{\partial}{\partial \boldsymbol{c}^{k+1}} \left(\boldsymbol{A} + \boldsymbol{B} \right)^{k+\theta_c} \right] \boldsymbol{c}^{k+\theta_c}$$

The ij-th component of the derivative term is

$$\left[\frac{\partial}{\partial c^{k+1}}\left(A+B\right)^{k+\theta_c}c^{k+\theta_c}\right]_{ij} = \sum_s \frac{\partial}{\partial c_j^{k+1}}\left(A_{is}+B_{is}\right)^{k+\theta_c}c_s^{k+\theta_c}$$

The contribution from matrix A is

$$\frac{\partial A_{is}^{k+\theta_c}}{\partial c_j^{k+1}} = \frac{\partial}{\partial c_j^{k+1}}\left\{\frac{1}{|V^e|}\left[\left(\alpha_L|v|^{k+\theta_c}+D_o n\right)\varsigma_i'\varsigma_s'\right.\right.$$
$$\left.\left.+\left(\alpha_T|v|^{k+\theta_c}+D_o n\right)\left(\zeta_i'\zeta_s'+\xi_i'\xi_s'\right)\right]\right\}$$

Differentiating velocities,

$$\frac{\partial}{\partial c_j^{k+1}}|v|^{k+\theta_c} = \frac{\partial}{\partial c_j^{k+1}}\left(\sqrt{v_1^2+v_2^2+v_3^2}\right)^{k+\theta_c}$$
$$= \left[\frac{1}{\sqrt{v_1^2+v_2^2+v_3^2}}\left(v_1\frac{\partial v_1}{\partial c_j^{k+1}}+v_2\frac{\partial v_2}{\partial c_j^{k+1}}+v_3\frac{\partial v_3}{\partial c_j^{k+1}}\right)\right]^{k+\theta_c}$$

and noting that $\partial v_1/\partial c = \partial v_2/\partial c = 0$, we get

$$\frac{\partial v_3^{k+\theta_c}}{\partial c_j^{k+1}} = -\frac{\partial}{\partial c_j^{k+1}}\left(k_{33}\epsilon\bar{c}\right) =$$
$$= -\frac{\partial}{\partial c_j^{k+1}}\frac{k_{33}\epsilon}{4}\left(c_i^{k+\theta_c}+c_l^{k+\theta_c}+c_m^{k+\theta_c}+c_n^{k+\theta_c}\right)$$
$$= -\frac{\theta_c k_{33}\epsilon}{4}$$

and thus

$$\sum_s \frac{\partial A_{is}^{k+\theta_c}}{\partial c_j^{k+1}}c_s^{k+\theta_c} = -\frac{\theta_c k_{33}\epsilon}{4|V^e|}\left(\frac{v_3}{\sqrt{v_1^2+v_2^2+v_3^2}}\right)^{k+\theta_c}\sum_s\left[\alpha_L\varsigma_i'\varsigma_s'+\alpha_T\left(\zeta_i'\zeta_s'+\xi_i'\xi_s'\right)\right]c_s^{k+\theta_c}$$

$$(18)$$

This term is independent of column index j. The contribution from matrix B is

$$\sum_s \frac{\partial B_{is}^{k+\theta_c}}{\partial c_j^{k+1}}c_s^{k+\theta_c} = \sum_s \frac{\partial}{\partial c_j^{k+1}}\left[\frac{\text{sign}(V^e)}{4}\left(v_1^{k+\theta_c}\varsigma_s+v_2^{k+\theta_c}\zeta_s+v_3^{k+\theta_c}\xi_s\right)\right]c_s^{k+\theta_c}$$
$$= \frac{\text{sign}(V_e)}{4}\frac{\partial v_3^{k+\theta_c}}{\partial c_j^{k+1}}\sum_s \xi_s c_s^{k+\theta_c}$$
$$= -\frac{\text{sign}(V_e)\theta_c k_{33}\epsilon}{16}\sum_s \xi_s c_s^{k+\theta_c}$$

$$(19)$$

This term is constant for each element.

REFERENCES

1. Henry, H. R., Salt intrusion into freshwater aquifers, *J. Geophys. Res.* 64(11), 1911–1919, 1959.

2. Henry, H. R., Effects of dispersion on salt encroachment in coastal aquifers. In: Cooper Jr., H. H., F. A. Kohout, H. R. Henry and R. E. Glover (eds.) *Seawater in Coastal Aquifers.* U. S. Geol. Surv. Water Supply paper, 1613-C, pp C70–C84, 1964.

3. Bear, J., *Hydraulics of Groundwater.* McGraw-Hill, New York, 1979.

4. Herbert, A. W., C. P. Jackson and D. A. Lever, Coupled groundwater flow and solute transport with fluid density strongly dependent upon concentration, *Water Resour. Res.* 24(10), 1781–1795, 1988.

5. Huyakorn, P. S., P. F. Andersen, J. W. Mercer and H. O. White, Saltwater intrusion in aquifers: Development and testing of a three-dimensional finite element model, *Water Resour. Res.* 23(2), 293–312, 1987.

6. Hassanizadeh, S. M. and T. Leijnse, On the modeling of brine transport in porous media, *Water Resour. Res* 24(3), 321–330, 1988.

7. Cooley, R. L., Some new procedures for numerical solution of variably saturated flow problems, *Water Resour. Res.* 19(5), 1271–1285, 1983.

8. Frind, E. O., Simulation of long-term transient density-dependent transport in groundwater, *Adv. Water Resources* 5, 73–88, 1982.

9. Gambolati, G., C. Paniconi and M. Putti, Numerical modeling of contaminant transport in groundwater. In: Petruzzelli, D. and F. G. Helfferich (eds.) *Migration and Fate of Pollutants in Soils and Subsoils.* Springer-Verlag, Berlin. Volume 32 of *NATO ASI Series G: Ecological Sciences*, pp 381–410, 1993.

10. Galeati, G. and G. Gambolati, On boundary conditions and point sources in the finite element integration of the transport equation, *Water Resour. Res.* 25(5), 847–856, 1989.

11. Aziz, K. and A. Settari, *Petroleum Reservoir Simulation.* Applied Science Publishers, London, 1979.

12. Gambolati, G., M. Putti and R. Rangogni, Saltwater contamination of a coastal Italian aquifer by a coupled finite element model of flow and transport, *Excerpta* 7 (1992-1993), 145–186, 1994.

13. Paniconi, C. and M. Putti, A comparison of Picard and Newton iteration in the numerical solution of multidimensional variably saturated flow problems, *Water Resour. Res.* 30(12), 3357–3374, 1994.

14. Fletcher, R., *Practical Methods of Optimization, Vol. 1: Unconstrained Optimization.* John Wiley & Sons, New York, NY, 1980.

MODIFIED EULERIAN LAGRANGIAN METHOD FOR FLOW AND TRANSPORT IN HETEROGENEOUS AQUIFERS

S. Sorek

Ben-Gurion University of the Negev, Beer Sheva, Israel

and

S. Lumelsky

Technion-Israel Institute of Technology, Haifa, Israel

Abstract

A Modified Eulerian Lagrangian (MEL) method is developed for the numerical solution of flow and transport in a heterogeneous aquifer. This is emerging from Lagrangian formulation of the governing equations written in terms of reduced velocities associated with the material derivatives. These velocities ,for forward and backward particle tracking techniques. depend on the heterogeneity of various solid matrix and fluid properties. Particles .in the transport problem. are shifted by a combination of fluid's velocity and the gradient of the hydrodynamic dispersion. In the case of the flow problem. we apply the MEL scheme with particle's velocity associated with the gradient of hydraulic conductivity.

Comparisons between the MEL. the Eulerian Lagrangian (EL) and the Eulerian Finite Elements (EFE) methods prove that the MEL scheme is superior in yielding almost no deviations from analytical solutions of 1-D flow and transport in a saturated aquifer.
Evolution of spatial accumulation of numerical mass balance errors. prove that such an assessment may be misleading.

The vertical unsaturated/saturated flow and transport problem is formulated by the MEL method as an example demonstrating the parabolic transformation of the governing equations.

1. Introduction

The solute transport equation is in general difficult to solve mainly because certain terms in this equation may conform it to become hyperbolic dominant. These are the cases when the advective flux or when the gradient of the dispersion factor becomes dominant. If the dispersive flux becomes dominant, the transport equation is of the parabolic type. Thus, the mathematical characteristics of the transport equation may vary between the hyperbolic and the parabolic type.

Because of this, much research was invested on numerical methods that aim at minimizing the problems of numerical dispersion and over/under shooting errors. Among such methods are the EL schemes that have been developed as special techniques for solving the advection dominated transport problems. All EL schemes are based on formal decomposition of the differential operator into "advection" along characteristic pathlines. In addition, Neuman [1, 2], Neuman and Sorek [3], Sorek and Braester [4] and Sorek [5, 6, 7] also decompose the dependent variable into advective and residual parts. The resulting advection problem is solved by methods applicable to the Lagrangian formulation. The "dispersive" problem may be solved by conventional finite elements or finite difference methods at a fixed grid.

An EL least-square collocation method was proposed by Bentley and Pinder [8]. The EL collocation method with the use of characteristic method for the advective problem is documented by Baptista [9] and by Allen and Khosravani [10]. The performance of the EL methods and sources of errors in such schemes is discussed by Bentley and Pinder [8].

Non of the above mentioned, incorporated the gradient of the dispersion factor as part of the velocity that drives the particle along its pathline.

Here in what follows, we will develop a Modified EL scheme (MEL) to solve for both 1-D flow problem and the migration of a solute in a saturated and heterogeneous porous domain. The numerical scheme is based on the decomposition of the differential operator and the dependent variables. Particle's velocity combines fluid's velocity and the gradient of the hydrodynamic dispersion, for the transport problem. In the case of the flow problem, this velocity is associated with the gradient of hydraulic conductivity.

In the framework of this paper the following objectives will be considered:

- comparison of the performance of the EL, MEL and EFE schemes for the transport problem with different spatial distributions of the dispersion coefficient,

- the use of the MEL scheme for the solution of certain flow problems.

- MEL formulation of the unsaturated/saturated flow and transport problem.

2. Mathematical Statement

Consider the general advection-dispersion problem, without source terms, given by

$$\frac{\partial u}{\partial t} = -\nabla \cdot [(1 - f)\mathbf{V}u - \mathbf{K}\nabla u]. \tag{1}$$

where $u(\mathbf{x}, t)$ denotes a dependent variable being a function of the spatial vector $\mathbf{x}$ and time t; $\mathbf{V}(\mathbf{x}, t)$ denotes the velocity vector; $\mathbf{K}(\mathbf{x})$ denotes a tensor associated with linear dependency between the dispersive flux and ∇u; f denotes a control coefficient describing:

the transport problem when $f = 0$

$$\frac{\partial u}{\partial t} = -\nabla \cdot [\mathbf{V} u - \mathbf{K} \nabla u].\tag{2}$$

the flow problem when $f = 1$

$$\frac{\partial u}{\partial t} = \nabla \cdot [\mathbf{K} \nabla u].\tag{3}$$

Note that in (2), u may be viewed as the concentration of a solute while $\mathbf{K}$ may describe the dispersion tensor. In (3), u may be viewed as the hydraulic head and $\mathbf{K}$ may describe the hydraulic conductivity tensor.

In order to investigate the different cases in which (1) may become hyperbolic dominant, let us rewrite it in a nondimensional 1-D form.
To do this, we define the following nondimensional $. (\,)_D$, variables,

$$x_D = x/\ell, \quad u_D = u/\dot{u}, \quad \dot{u} \equiv \frac{1}{\ell} \int_0^\ell u\,dx, \quad K_D = K/\dot{K}, \quad \dot{K} \equiv \frac{1}{\ell} \int_0^\ell K\,dx,$$

$$t_D = t/\dot{t}, \quad \dot{t} \equiv \frac{1}{\dot{K}}\ell^2, \quad V_D = V/\dot{V}, \quad \dot{V} \equiv \frac{\dot{K}}{\ell}.\tag{4}$$

where ℓ denotes a characteristic length.
In view of (4), the Eulerian 1-D form of (1) becomes,

$$\frac{1}{K_D}\frac{\partial u_D}{\partial t_D} = S_R \frac{\partial u_D}{\partial x_D} + \frac{\partial^2 u_D}{\partial x_D^2} - (1 - f)P_e \frac{\partial}{\partial x_D}(V_D u_D).\tag{5}$$

here P_e denotes the Peclet number given by

$$P_e \equiv \frac{\ell \dot{V}}{\dot{K}}.\tag{6}$$

and S_R a scalar ,proposed by sorek and Braester [4], which represents the ratio between the parabolic and hyperbolic terms of a flow equation. This is given by

$$S_R \equiv \frac{\ell \dfrac{dK}{dx}}{\dot{K}}.\tag{7}$$

From (6) and (7) we note that dK/dx is equivalent to $\dot{V}$.
The Lagrangian form of (5), reads

$$\frac{1}{K_D}\frac{Du_D}{Dt_D} = \frac{\partial^2 u_D}{\partial x_D^2} - (1 - f)P_e u_D \frac{\partial V_D}{\partial x_D}.\tag{8}$$

where the nondimensional hydrodynamic derivative $,\dfrac{D}{Dt_D},$ is given by

$$\frac{D}{Dt_D} \equiv \frac{\partial}{\partial t_D} + V_D^* \frac{\partial}{\partial x_D},\qquad(9)$$

and the nondimensional apparent velocity $,V_D^*,$ is given by

$$V_D^* \equiv K_D[(1 - f)P_e V_D - S_R].\qquad(10)$$

From (8) (or (5)) we note that the combination of both P_e and S_R numbers dominate the nature of the flow and transport problems.

The *transport* equation (f=0) becomes :

*   *hyperbolic* (advection) dominated when $(P_e, S_R) \Longrightarrow \infty$,
*   *parabolic* (dispersion) dominated when $(P_e, S_R) \Longrightarrow 0$.

For the *flow* equation (f=1), only the S_R number prevails, this becomes:

*   *hyperbolic* dominated when $S_R \Longrightarrow \infty$,
*   *parabolic* dominated when $S_R \Longrightarrow 0$.

Following the above nondimensional forms, we will aim at addressing also cases when the S_R number may dominate.

To do so, we note that we may further rewrite (1) in the form

$$\frac{Du}{Dt} = (1 - \varphi)\nabla \cdot [\mathbf{K}\nabla u] + \varphi \mathbf{K}\nabla^2 u - (1 - f + \lambda)u\nabla \cdot \mathbf{V} - \lambda \mathbf{V} \cdot \nabla u,\qquad(11)$$

where the hydrodynamic derivative is defined by

$$\frac{D}{Dt} \equiv \frac{\partial}{\partial t} + \mathbf{V}^* \cdot \nabla,\qquad(12)$$

and the apparent velocity $\mathbf{V}^*$ is defined by

$$\mathbf{V}^* \equiv (1 - f)\mathbf{V} - \varphi \nabla \mathbf{K},\qquad(13)$$

here φ and λ denote control coefficients.

Note that in (11), u is interpreted as being carried, say by a particle, along a pathline defined by

$$D\mathbf{x} = \mathbf{V}^* Dt.\qquad(14)$$

We may now construct various solution schemes all of which are described in table 1.

control coef.	EFE		EL		MEL	
	Flow	Transport	Flow	Transport	Flow	Transport
f	1	1	/	0	1	0
φ	0	0	/	0	1	1
λ	0	1	/	0	0	1

Table 1: Choice of control coefficients

Our aim in what follows is to investigate the performance of the MEL numerical scheme which is based on the modified hydrodynamic derivative, accounting also for $\nabla \mathbf{K}$.

2.1) Decomposition of the variable and the differential operator

Neuman and Sorek [3], proposed to split the dependent variable into an *advection part* $\bar{u}$ and a residual $\mathring{u}$ regarded as the *dispersion part*

$$u = \bar{u} + \mathring{u}. \tag{15}$$

Following Neuman and Sorek [3], we also decompose the differential operator in (11), to become formally a purely hyperbolic "*advection problem*" defined in terms of $\bar{u}$

$$\frac{D\bar{u}}{Dt} = 0, \tag{16}$$

and a predominantly parabolic "*dispersion problem*" defined in terms of $\mathring{u}$ and u,

$$\frac{D\mathring{u}}{Dt} = (1 - \varphi)\nabla\cdot[\mathbf{K}\nabla u] + \varphi\mathbf{K}\nabla^2 u - (1 - f + \lambda)u\nabla\cdot\mathbf{V} - \lambda\mathbf{V}\cdot\nabla u, \tag{17}$$

In what follows we will consider the case of a constant velocity and a 1-D domain. This, however, will not affect the development of the numerical scheme nor will it affect the implications resulting from the considered examples.

3.   Numerical Implementation

The numerical procedure is comprised of two main steps:
a. Solution of the "*advection problem*" by *particle tracking technique* along pathlines.
b. Solution of the residual "*dispersion problem*" using a fixed *Finite Element grid*.

a.   Partical tracking technique

Let a particle, p, located at point x_p at the $k-th$ time level (t^k), be associated with $\bar{u}_p^k \equiv u(x_p, t^k)$. At the end of the $(k+1)$ time level (t^{k+1}), each particle, p, reaches a new forward position, x_p^{k+1}, obtained from the solution of (14) by, say, the Runge-Kutta method.

This describes the *forward particles tracking* procedure, which projects $\bar{u}_p^{k+1}(=\bar{u}_p^k$, by virtue of (2.)) onto nodal points.
For nodes, that are not covered by these clouds, i.e., for any node (n) located at x_n such that

$$x_{p\ min}^{k+1} > x_n > x_{p\ max}^{k+1}. \tag{18}$$

we use *backward particles tracking* procedure to project $u \equiv {}^k u(= u({}^k x_p, t^k))$ from a backward location ${}^k x_p$ onto nodes obeying (18). The ${}^k x_p$ location is obtained by solving (14) with $x_p^{k+1} = x_n$ and ${}^k u$ is the interpolated value at the backward element.

b. Dispersion by Finite Elements

We approximate $u(x,t)$ and $K(x)$ by

$$u(x,t) \simeq \hat{u}(x,t) \equiv \sum_{j=1}^{N} u_j(t)\xi_j(x); \qquad K(x) \simeq \hat{K}(x) \equiv \sum_{l=1}^{N} K_l \xi_l(x). \tag{19}$$

where N denotes the total number of the grid nodes; K_l denotes the subscribed values of K at the nodal points and ξ_j denotes the shape functions satisfying

$$\xi_j(x_i) = \delta_{ij} = \begin{cases} 1, & \text{if } j = i \\ 0, & \text{if } j \neq i. \end{cases} \tag{20}$$

where δ_{ij} denotes the ij-th element of the Kroniker delta-function.

Substitution of (19) and (20) into (17), applying Galerkin's orthogonalization for the 1-D case spanning between $[x = 0, x = L]$, with $V = const$, reads (demonstrated for the case when $\lambda = 0$),

$$\int_{x=0}^{x=L} \left[\frac{D\hat{u}}{Dt} - (1-\varphi)\frac{\partial}{\partial x}\left(K\frac{\partial \hat{u}}{\partial x}\right) - \varphi\hat{K}\frac{\partial^2 \hat{u}}{\partial x^2} \right]\xi_i dx = 0, \qquad i = 1, 2, ..., N \tag{21}$$

where in view of (13) and (14), the 1-D hydrodynamic derivative is given by

$$\frac{D}{Dt} = \frac{\partial}{\partial t} + \left[(1-f)V - \varphi\frac{d\hat{K}}{dx} \right]\frac{\partial}{\partial x}. \tag{22}$$

The application of (22) over $\hat{u}$ for each time step $\Delta t(\equiv t^{k+1} - t^k)$, may be approximated by backward Finite Difference, to read

$$\frac{D\hat{u}}{Dt} \simeq \frac{\hat{u}^{k+1} - {}^k\hat{u}}{\Delta t}. \tag{23}$$

The global algebraic set after integration by parts of (21), in view of (23) and accounting for the forward particle projection, becomes

$$\sum_{j=1}^{N} u_j^{k+1} \int_{x=0}^{x=L} \left\{ \frac{1}{\Delta t}\xi_i\xi_j + \left[\sum_{l=1}^{N} K_l\xi_l \right]\frac{d\xi_i}{dx}\frac{d\xi_j}{dx} + \varphi\left[\sum_{l=1}^{N} K_l\frac{d\xi_l}{dx} \right]\xi_i\frac{d\xi_j}{dx} \right\}dx =$$

$$\frac{1}{\Delta t}\int_{x=0}^{x=L}\left(\sum_{j=1}^{N_1} u_j^k\xi(^k x_p) + \sum_{m=1+N_1}^{N} \bar{u}_m^{k+1}\delta_{im} \right)\xi_i dx + K\frac{\partial u}{\partial x}(\delta_{iN} - \delta_{i1}); \tag{24}$$

where N_1, denotes the nodes that are not covered by clouds of particles, and δ denotes the Kroniker delta-function.

In view of (22) and (24), for $\varphi = 1$, we note that the MEL scheme may be viewed as the EL scheme from which we subtract and add the product $\nabla K \cdot \nabla u$. This is implanted, respectively, in the velocity of the particles and in the orthogonalization procedure.

4. Examples

a. Numerical MEL simulation of saturated flow and transport

We set ourselves at solving (11) for constant velocity in a 1-D domain (under general units system) spanning between $x \in [0, 2.5]$ and during a time interval of $t \in [0, \tau]$. Spatial and time steps were chosen to be $\Delta x = 0.05$ and $\Delta t = 0.5$, respectively.

Adequate selection of the shape functions was found to be very important considering the discrete choice of K values (see, e.g., Fig.1) and the need to approximate their gradient to evaluate the apparent particle velocity as described in (13).

Use of linear polynomial functions essentially simplifies the matrix assembling procedure in a Galerkin FE scheme, but in our case it lead to numerical oscillations which was not experienced when quadratic polynomials were applied as the interpolating functions (Fig.2).

For the transport problem delineated in Figs.2 ($u \equiv c$ -solute concentration; $K \equiv D$ - hydrodynamic dispersion) we considered an initial concentration $c(x, 0) = 0$ and boundary conditions $c(0, t) = 1$ and $c(2.5, t) = 0$.

The numerical performance of the MEL scheme was examined in terms of its deviations from analytical solutions and compared to the numerical solutions of the EL and EFE schemes.

Analytical solution of (1) for the 1-D transport ($f = 0$) and flow ($f = 1$) problems were developed concerning $K(x)$ as a cubic polynomial function of space. The general analytical solution for (1) may be given by

$$u = e^{at}(1 - \frac{x}{2L} - \frac{x^2}{2L^2}). \tag{25}$$

where $a(= -0.075)$ denotes a constant time integration factor and $L(= 2.5)$ denotes the spatial span.
The analytical solution for the 1-D transport problem was developed using a dispersion coefficient distribution described by

$$D(x) = \frac{2ax^3 + (6V + 3aL)x^2 + (6VL - 12aL^2)x}{6(L + 2x)}, \tag{26}$$

where the constant fluid velocity was chosen to be $V = 0.5$.

Spatial distribution of $D(x)$ and the associated Pe Peclet number are described in Fig.3.

Comparisons between the solutions of MEL, EL and EFE schemes were performed and the relative deviation error

$$Relative \quad Error = \frac{u_{numerical} - u_{analytical}}{|u_{numerical} - u_{analytical}|_{max}}, \tag{27}$$

of these solutions from the analytical one is delineated in Fig.4.

In view of Figs. 3 and 4, we note that the MEL scheme results in less deviation in compare to the EFE and EL ones, throughout the entire range of the Pe number. This is valid not only at the range with high advection (at the vicinity of $x = 0$) but ,more significantly, also where dispersion dominates.

The maximum absolute deviation for the MEL scheme was found to be 0.0076, for the EL scheme 0.0289, and 0.0314 for the EFE.

Concerning the Courant number $Cr(\equiv \dfrac{V^*\Delta t}{\Delta x})$ based on particle's velocity V^*, we find that for the EL and EFE schemes this yields $Cr = 5$, while for MEL it decreases from $Cr = 72.25$ at $x = 0$ to $Cr = 3.77$ at $x = 2.5$.

To check the ability of the EFE, El and MEL methods to conserve mass ,we considered for this example the relative mass error for a unit cross section during each ,Δt, time step.

Hence, the accumulated ,M_B, mass crossing the x_0 $(= 0.)$ and x_L $(= 2.5)$ boundaries will be given by,

$$M_B = \int_{t^k}^{t^{k+1}} [(K\frac{\partial u}{\partial x} - Vu)_{x_0} - (K\frac{\partial u}{\partial x} - Vu)_{x_L}]dt \tag{28}$$

The change in mass ,ΔM, stored in the domain is obtained by,

$$\Delta M = \int_{x_0}^{x_L} [u^{k+1} - u^k]dx \tag{29}$$

The relative mass error now reads,

$$Relative \quad Mass \quad Error = \frac{M_B - \Delta M}{M_B} \tag{30}$$

The evolution of the accumulative relative mass error over the entire space for the EFE, EL and MEL methods, is described in Fig. 5.

In viewing both Figs. 4 and 5, it is evident that the criterion concerning mass balance error may be misleading. Yet, we note that the MEL scheme is producing a lower mass error than the EL scheme. This error is decreasing following the initial loading of the particles taking place at the beginning of the evolution.

In the case of the 1-D analytical solution for the flow problem ($f = 1$), we chose a conductivity coefficient $K(x)$ in the form

$$K(x) = \frac{2ax^3 + 3aLx^2 - 12aL^2x}{6(L + 2x)}. \tag{31}$$

The spatial distribution of K and S_R (the analogous to Peclet number as discussed in (7)) is described in Fig.6.

The relative error obtained by the MEL and EFE solutions is shown in Fig.7.
Maximum absolute deviation for the EFE scheme was 0.0021 and 0.0030 for MEL.
We note that in this case the performance of both schemes is similar.

Let us now reconsider the choice of the heterogeneous medium which in our case is
expressed by the random distribution of K in space (Fig.1).
We now apply also the EFE and EL schemes as numerical procedures to solve the 1-D
transport problem (11) using the same boundary and initial conditions, as in conjunction with Fig.2.
The comparison between the solutions obtained by the MEL, EL and EFE schemes
is depicted in Fig.8.
We note that the MEL scheme is superior in producing less numerical dispersion.
In comparing the relative mass balance error (Fig. 9), we note (as in Fig. 5) that the
MEL is reducing its mass error as time evolves, more than the EFE and EL schemes.

b. **MEL formulation for unsaturated/saturated flow and transport.**
Here we show an example of transforming a 1-D Eulerian representation of Richard's
flow and the advection - dispersion equations, into MEL forms.

The vertical unsaturated/saturated Eulerian balance equations are described by,
  Fluid's mass balance equation,

$$\sigma\frac{\partial \psi}{\partial t} = \frac{\partial}{\partial z}(\theta V) \, . \tag{32}$$

Where σ $(\equiv \dfrac{\partial \theta}{\partial \psi})$ denotes the soil water capacity. θ denotes the volumetric water
content and ψ denotes the pressure head .
  Fluid's linear (Darcy's) momentum balance equation,

$$\theta V = -K(\theta)(\frac{\partial \psi}{\partial z} - 1) \, . \tag{33}$$

Component's mass balance equation,

$$\frac{\partial(\theta c)}{\partial t} = \frac{\partial}{\partial z}[\theta(D\frac{\partial c}{\partial z} - Vc)] \, . \tag{34}$$

By sustituting (33) into (32), we obtain the flow equation in its Eulerian form.
This can now be transformed to read,
  the MEL flow equation,

$$\sigma\frac{D_{V_\psi}(\psi)}{Dt} = K\frac{\partial^2(\psi)}{\partial z^2} \, . \tag{35}$$

where the hydrodynamic derivative is defined by

$$\frac{D_{V_\psi}}{Dt} \equiv (\frac{\partial}{\partial t} + V_\psi \frac{\partial}{\partial z}) \ , \tag{36}$$

the velocity of a particle carrying pressure head along its path line, is given by,

$$V_\psi \equiv \eta(1 - \frac{\widetilde{\partial \psi}}{\partial z}) \ . \tag{37}$$

here $\eta \ (\equiv \frac{\partial K}{\partial \theta})$ is associated with the change of conductivity with respect to the water content. Note that V_ψ is also a function of the pressure head gradient . The numerical solution of (37), maybe resolved by iterations starting ,say, from the value of $\frac{\widetilde{\partial \psi}}{\partial z}$ at the previous time level.

In a similar way as for the flow equation, we now transform (34) to become,
the MEL transport equation ,

$$\frac{D_{V_c}(c)}{Dt} = D\frac{\partial^2 (c)}{\partial z^2} \ . \tag{38}$$

where the hydrodynamic derivative is defined by,

$$\frac{D_{V_c}}{Dt} \equiv (\frac{\partial}{\partial t} + V_c\frac{\partial}{\partial z}), \tag{39}$$

and the velocity of a particle carrying concentration is given by,

$$V_c \equiv V - \frac{\partial D}{\partial z} - D\frac{\sigma}{\theta}\frac{\widetilde{\partial \psi}}{\partial z} \ . \tag{40}$$

We note the reduction of fluid's velocity by spatial changes in the hydrodynamic dispersion and in the pressure head.

5.   Conclusion

The MEL scheme was developed specifically to address steep dispersion gradients in transport problems or steep conductivity gradients associated with flow problems. In such cases, the balance equation becomes hyperbolic dominated and simulations suffer from numerical dispersion problems.

The MEL scheme is based on a modified hydrodynamic derivative incorporating the dispersion (or conductivity) gradient as part of particle's velocity.

It was employed to solve 1-D flow and transport problems involving continuous or discontinuous distribution in space of the conductivity and dispersion coefficients, respectively.

Performance of the MEL scheme was compared against that of the EL and EFE schemes.

It was proven to be superior in producing much less deviation from analytical solution of the 1-D transport problem.

In the case of random spatial distribution of the dispersion coefficient, we note that both the MEL and EL schemes are practically free of numerical dispersion, in compare to the EFE scheme.

Yet, the MEL scheme produces a somewhat sharper front in better comply with the Peclet number.

The MEL and EFE schemes were compared to an analytical solution of a 1-D parabolic flow problem.

The MEL scheme yielded very small deviations, similar to the EFE scheme.

We conclude that the MEL scheme produced better results in terms of numerical efficiency and accuracy.

REFERENCES

[1] Neuman, S. P. A Eulerian-Lagrangian numerical scheme for the dispersion-convection equation using conjugate space-time grids, J. Comp. Phys., **41** (2), 270-294, 1981.

[2] Neuman, S. P. Adaptive Eulerian - Lagrangian finite element method for advection - dispersion, Int. J. Num. Math. Eng., **20**, 321-337, 1984.

[3] Neuman, S. P. and Sorek, S. Eulerian-Lagrangian methods for advection - dispersion, Proc. 4-th Int. Conf. F. E. W. R., Germany, 14.41-14.68, 1982.

[4] Sorek, S. and Braester, C. Eulerian-Lagrangian formulation of the equations for groundwater denitrification using bacterial activity, Adv. W. Resour., **11** (4), 162-169, 1988.

[5] Sorek, S. Eulerian-Lagrangian formulation for flow in soil, Adv. W. Resour., **8**, 118-120, 1985a.

[6] Sorek, S. Adaptive Eulerian-Lagrangian method for transport problems in soils, Scientific Basis for Water Water Resources Management, IASH Publication, 153, 393-403, 1985b.

[7] Sorek, S. Eulerian-Lagrangian method for solving transport in aquifers, Adv. W. Resour., **11** (2), 67-73, 1988.

[8] Bentley, L. R. and Pinder, G. F. Eulerian-Lagrangian solution of the vertically averaged groundwater transport equations, W. Resour. Res., **28** (11), 3011-3020, 1992.

[9] Baptista, A. E. Solution of advection dominated transport by Eulerian-Lagrangian method, using the backward method of characteristics, Ph.D. Dis., MIT, 1987.

[10] Allen, M. B. and Khosravani, A. Eulerian-Lagrangian method for finite-element collocation using the modified method of characteristic, Computational, Methods in Subsurface Hydrology Proc 8 Int. Conf. Comput. Method Water Resour, Springer-Verlag Publishers, Berlin, 375-379, 1990.

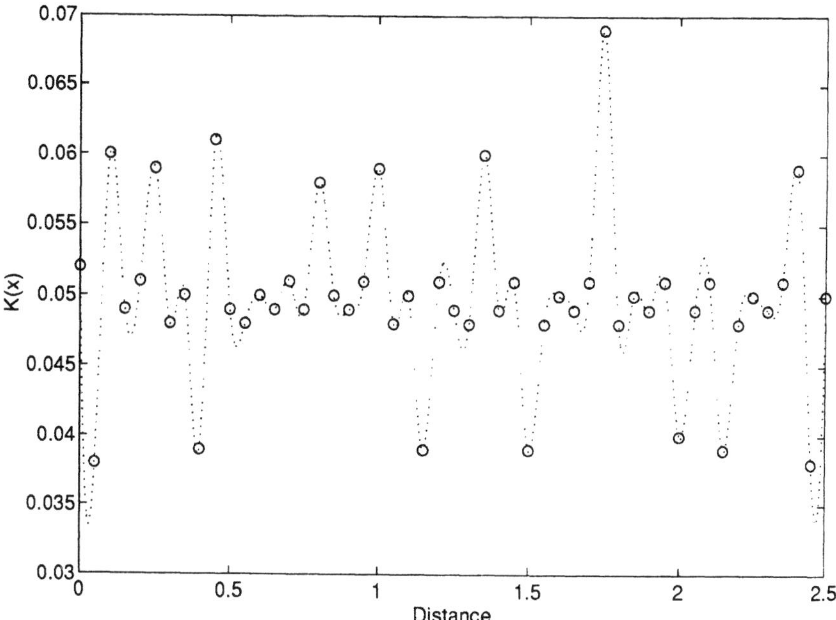

Figure 1: *Random spatial distribution of the K coefficient*

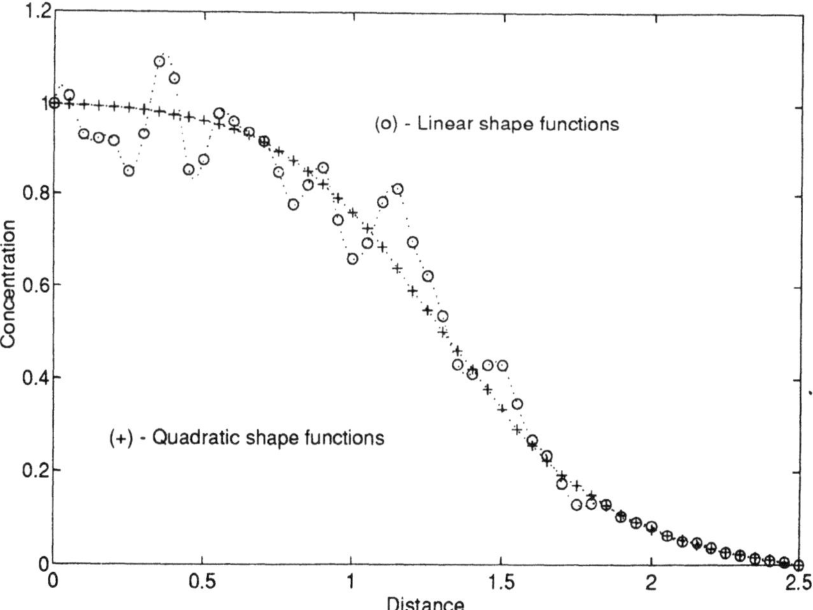

Figure 2: *Transport problem simulated by MEL for the above spatial distribution of K*

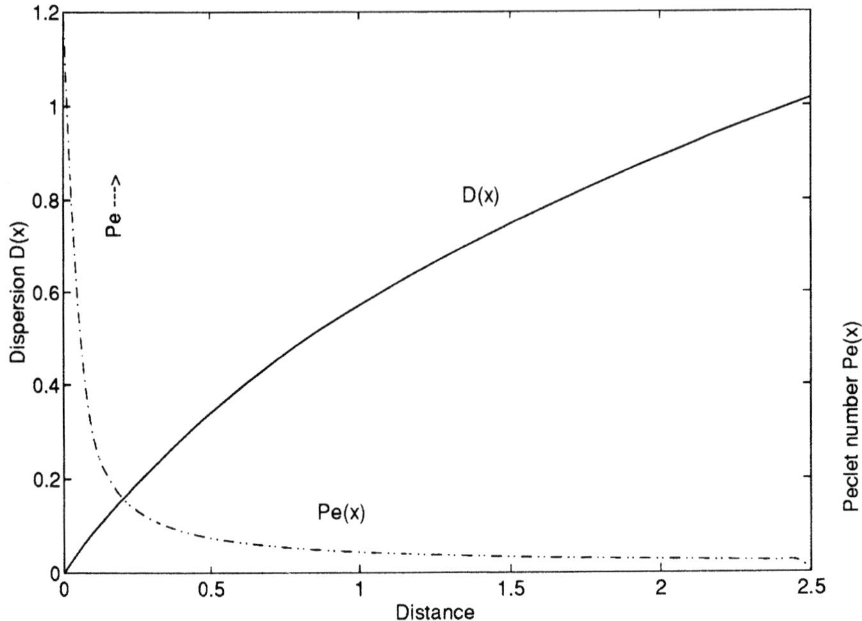

Figure 3: *Spatial distribution of the dispersion coefficient D and the associated Peclet P_e number*

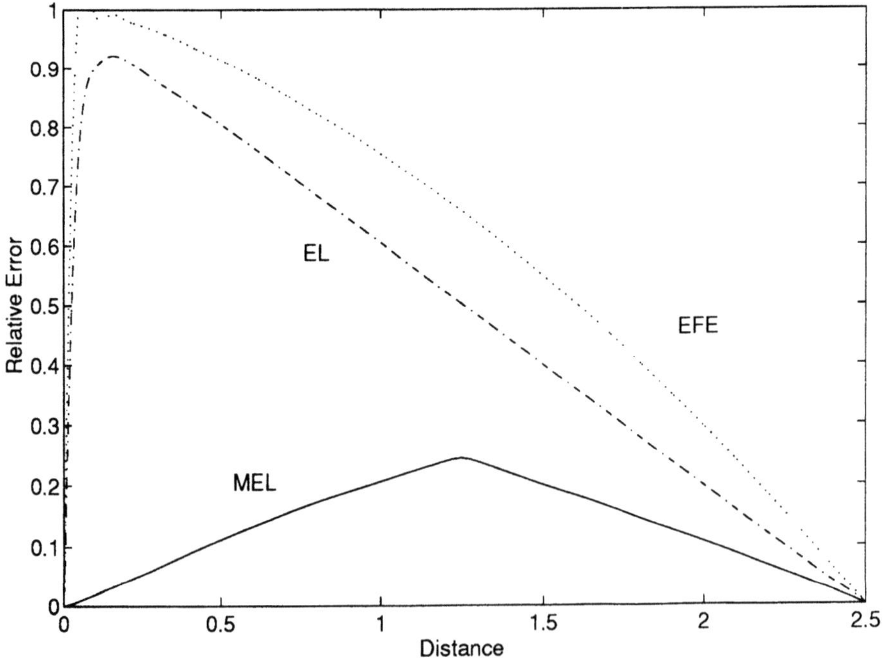

Figure 4: *Relative error distribution for the transport problem*

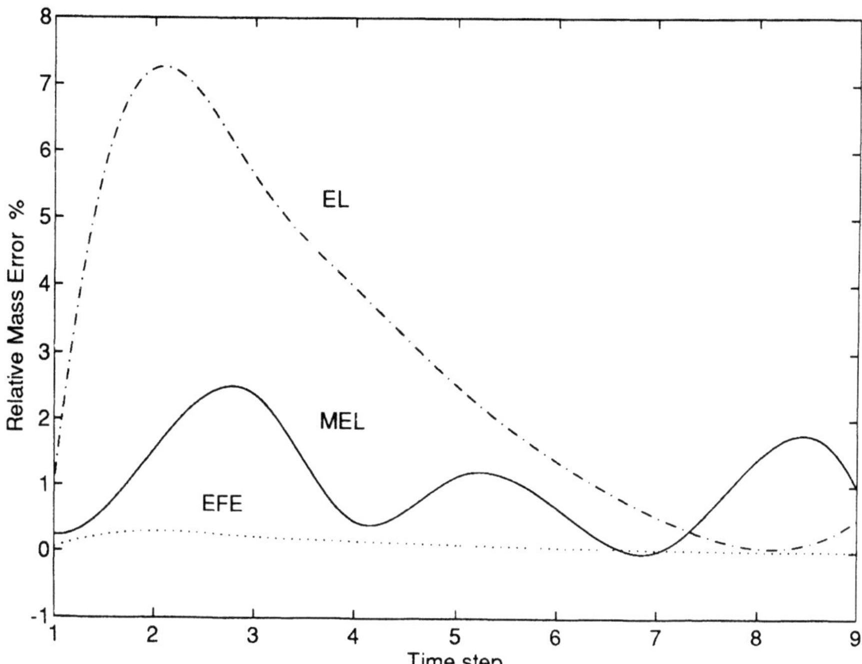

Figure 5: *Evolution of relative mass error for the transport problem*

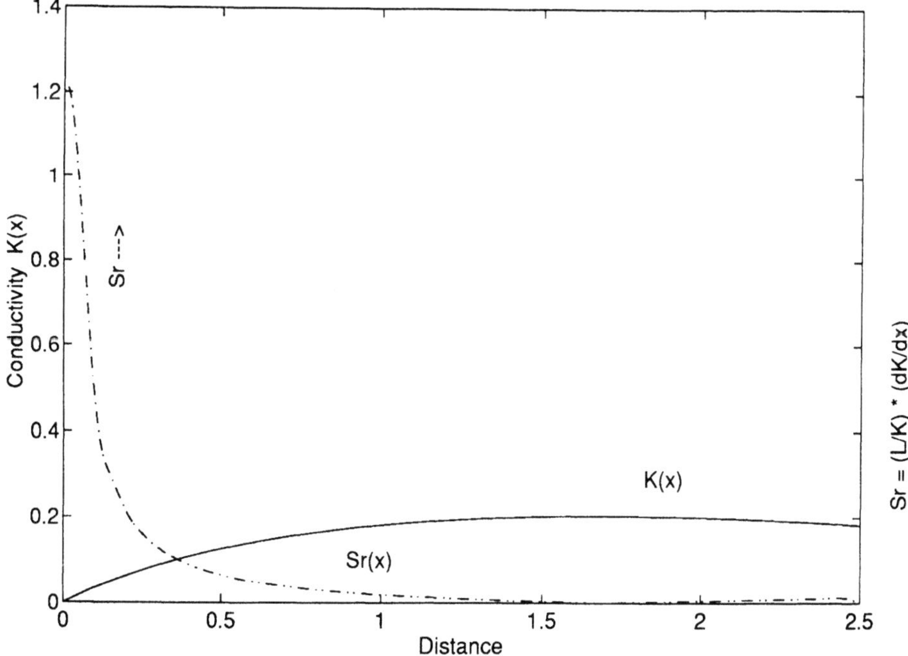

Figure 6: *Spatial distribution of the conductivity coefficient K and the associated S_R number*

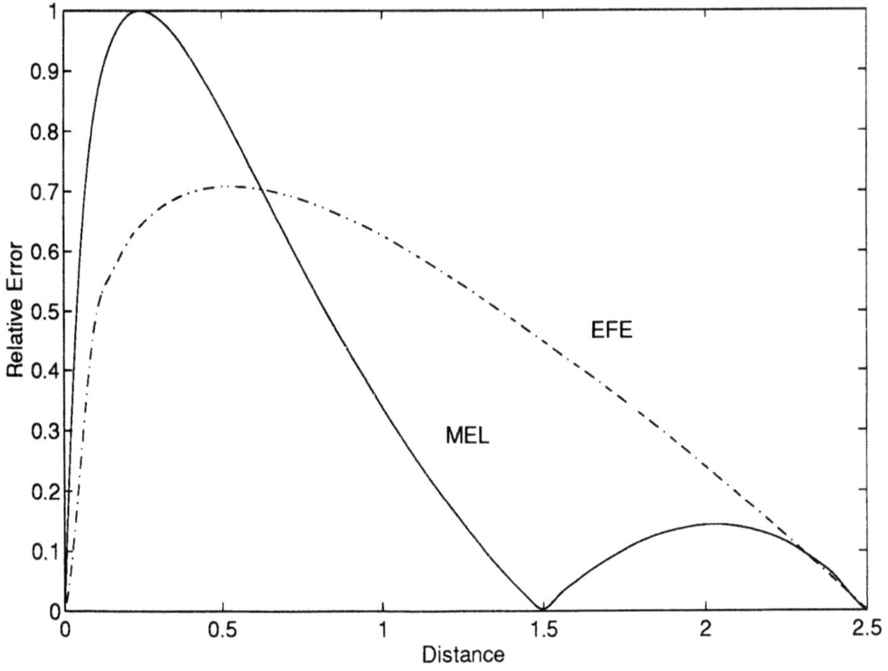

Figure 7: *Relative error distribution for the flow problem*

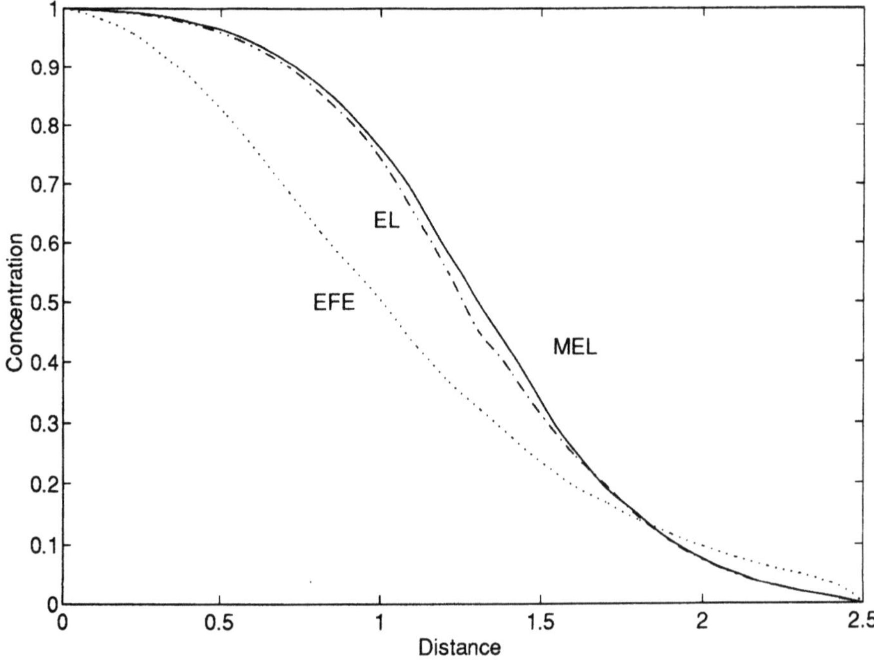

Figure 8: *Transport problem simulated by MEL, EL and EFE for the random spatial distribution of the dispersion coefficient*

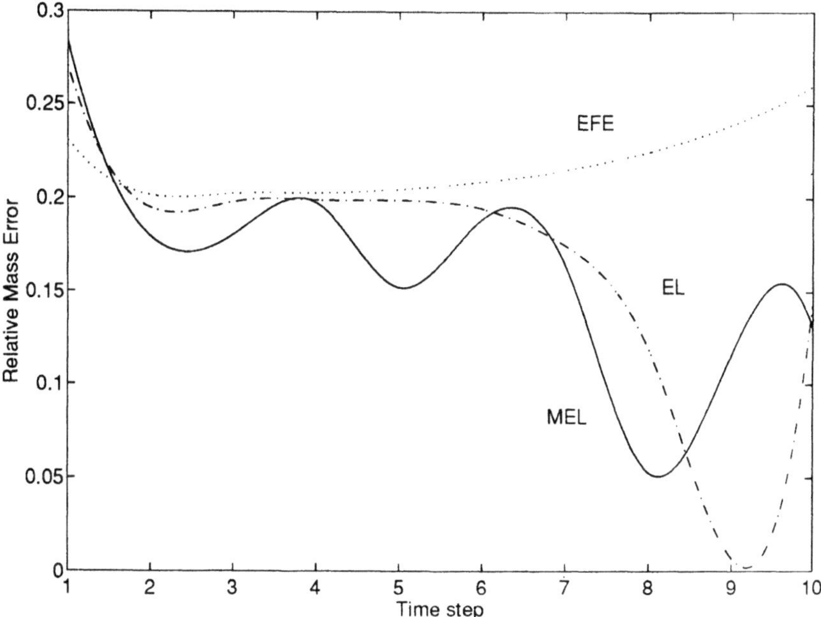

Figure 9: *Evolution of relative mass error when dispersion is randomly distributed*

RANDOM WALK MODELLING APPLICATION TO CONTAMINANT TRANSPORT IN THE UNSATURATED ZONE

I.F. Kontur

Technical University of Budapest, Budapest, Hungary

ABSTRACT

Examples of the application of a random walk model for the description of the transport of contaminants in the unsaturated upper soil cover zone are presented. In sections 2.2, 2.3 and 2.4 the water household of the cover layer, the transport processes of conservative- and non-conservative contaminating substances are described, respectively.
Application of the particle approach to the modelling of pollutant transport allows the use of simple and favourable computer simulation technique. The probabilistic formulation enhances the application of the "armoury" of the probability theory.

1. INTRODUCTION

The paper describes the application of the random walk techniques for the modelling of transport, migration, of pollutants in the three-phase zone, utilizing the concept of particle movement. Different levels of modelling are presented in the light of selecting time- and spatial scale. No preference will be given to specific models but rather the options of applying the random walk models, on the basis of the "particle" concept, will be given.

1.1 SELECTION OF THE SPACETIME

In describing the transport process to which a particle moving in the three-phase zone is subjected macro- and micro scales should be distinguished. In the case of the macro scale the three-phase soil matrix (having no relation with the term matrix as used in mathematics) is considered a unit or it will be split to parts of the size where soil particles and the pore space between them can not be separated. The micro scale means that one is to investigate processes taking place in between the soil particles and thus the space between the particles shall be cut into parts. Consequently the macro scale means a soil block of a couple of centimetres size in soils, while in fissured rock it may be as large as a couple of metres. I will not discuss here the transition between micro and macro scale, although this may be an extremely interesting question.

Selection of the time scale is of importance in the case of discrete modelling techniques only. Neither in-time nor in-space discretization should disrupt the continuity of events. Only the structure and form of the transition probability matrix changes with time scale of discretization.

Discretizing in space means the simplification of the space in the form of a set of discrete points of the space. In terms of the stochastic concept these points represent states. The transition between these states is described by the matrix of state transition probability. In the case of discretely modelling in time the parameters are discrete ones while in the case of modelling in continuous time we consider stochastic processes with continuous parameters, since time is also a parameter in this approach. The "state" is then the space, modelled as a set of discrete points, which means that we consider a stochastic process of discrete state, characterized by the transition matrix of the state. Stochastic processes of the continuous state will not be discussed here.

1.2 THE CONCEPT OF WATER/POLLUTANT PARTICLES

In subsurface waters the movement of both water and pollutant is of interest. The propagation, transport, of a polluting substance can not be separated from the motion of water. The random walk particle modelling concept refers to both water and substance particles in a similar way. However, the transport processes of conservative substances, moving along with the water but not reacting or decaying, should be distinguished from the non-conservative ones which are subject to dissolution-absorption, decay, settling, chemical reactions etc. In the former case the very same laws of motion -and transition probability matrix- refer to the water and to the pollutant particle. This approach will not hold in the case of non-conservative substances and the walking pathways of the pollutant particle should be analyzed separately.

Considering a particle, either water or pollutant, it is of a given mass, a kvantum, which can not be further subdivided. In the case of chemical reactions one might consider this a special restriction, but it is not, since the mass as an extensive quantity can be subdivided, split, in an unlimited way. Finally we obtain probabilities and can calculate with them: what is the probability of a particle being in one or another state, that is place, or what is the probability that it will decay, settle or be transformed in any other way. Thus the particles (water and pollutant) are considered kvantums that can not be further subdivided.

In selecting the proportions of discrete mass and discrete space (which means, in the case of water, the selection of the discrete unit of water and the discrete unit of the space in which the water moves) one should take care that the unit of water be not larger than a small fraction of the spatial unit. That is ones should be able to state that a water/pollutant particle is found in one or other space unit with a specified probability. This is especially important in the case of non-conservative polluting substances.

1.3 CONSERVATION OF MASS AND CONTINUITY

The mass of the above described particles must not disappear and must not be generated in course of the movement of the particle. This means that the particle must be found somewhere if it was within the system in a previous point of time. Consequently all states must be specified, including the state where the particle enters the system (by dissolution), where it leaves the system (absorption, settling) and where it is transformed, also specifying the states of transformation (from what into what). The full specification of states, that is places, means that if in one point of the time the particle was in one state/spatial-unit (with the probability of 1, that is with full certainty) then in the next point of time one must be able to find it elsewhere. The probability of finding a particle in a state is varying, but the sum of these probabilities must equal to one, the full certainty. In the matrix of transition probabilities it means that the sum of a raw must be one (if one writes the probabilities in the form of a line vector). These matrices are termed stochastic matrices. Consequently the stochastic character of a transition probability matrix means that the law of the conservation of mass holds.

Eventually the system may include sources and sinks. They should be considered the states of entrance and exit and the transition probability matrix should include such states.

In the case when random walk modelling also involves chemical transformation processes then the method to be followed is that the probability of the transformation of the given substance within a time step of the calculation should be determined and the respective value of the transition probability matrix should be multiplied with this. This method involves the utilization of the assumption that transformation and transport processes are independent ones. This means that in-time and in-space discretization should be made in such a way as to secure the conditions at which the above assumption of independency holds with a specified accuracy.

2. THE RANDOM WALK MODEL

2.1 THE GENERAL APPROACH

After discretizing the space is represented by a set of points. The probability of transition of the particles (water particles) from j to k during time delta t be p(j,k), where j= 1....n and k=1....n, provided the number of points that describes the space is n. Among these points there are internal points (n1) and boundary points (n2), thus n1+n2=n. The matrix of transition probability (being a square matrix of n*n dimension) can be subdivided into four blocks: a square matrix of n1*n1 size, representing transition from internal point to internal point; two rectangle matrices, n1*n2 and n2*n1, representing transitions from internal point to boundary point and from boundary point to internal point; and a square matrix n2*n2 which will not be dealt with here, since they represent the interconnection of the boundary points.

Considering a given space of volume V and mean residence (retention) time $\bar{t}$, the probability of a particle leaving the space within time step Δt is

$$q = 1 - e^{-\Delta t/\bar{t}}$$

when it was inside the space at time zero. Thus the probability that the particles is still inside the space is $1-q=e^{-\Delta t/\bar{t}}$. This equation, however, can also be obtained from the differential equation of a single linear reservoir, in a deterministic way.

In the case of a system which is continuous in time (and discrete in space) the matrix of transition probability can be obtained as the exponent of the so called infinitesimal matrix (Karlin and Taylor [1])

$$P(t) = e^{A\,t}$$

where

P(t) is the matrix of transition probability considering time step t and A is the infinitesimal matrix, having time^{-1} dimension. Thus element $a(j,k)$ is just the reciprocal of the mean time of transition from j to k. The sum of these elements is in the main diagonal with negative sign. Consequently the sum of the rows of the infinitesimal matrix is zero and thus

$$P(t) = I + \frac{A*t}{1!} + \frac{(A*t)^2}{2!} + \frac{(A*t)^3}{3!}$$

where I is the identity matrix. Since any power term of a matrix of zero sum of row is also of the zero sum of row type, then the sum of the rows of matrix P(t) is one.(Nevertheless the elements of matrix P are not necessarily numbers between zero and one, that is they are probabilities; hypothesis: If matrix A is negative definite then P(t) is a matrix of transition probabilities). If from the above equation of P(t) one considers 1, 2, 3, etc terms, then the result becomes ever more accurate. Considering the first term only we obtain an approximation of the probability of transition, which is nothing else but the ratio of time t to the mean residence (retention) time.

The elements of the infinitesimal matrix form the coefficient matrix of the storage differential equation system of the individual parts of the space. The effects of spatial and temporal inhomogeneity, variance and non-linearity, will not be discussed here.

2.2 WATER BUDGET MODEL OF THE SOIL-MOISTURE ZONE

The mathematical model of the water household can be formulated as

$$\frac{\partial S}{\partial t} + K(S)\frac{\partial S}{\partial z} - D(S)\frac{\partial^2 S}{\partial z^2} + f = 0 \qquad (1)$$

where

s[-] -	is the saturation of the soil, varying along the depth z
f $[T^{-1}]$-	is a source term,
K(S) $[L\ T^{-1}]$ -	is the seepage coefficient,
D(S) $[L^2\ T^{-1})$ -	is the diffusion coefficient,
z [L] -	is the vertical coordinate
t [T] -	is the time.

The dependence of seepage coefficient K(S) and diffusion coefficient D(S) on the level of saturation are described by the formulae by Irmay [2] and Gardner and Mayhugh [3], respectively:

$$K(S) = K_0(z)\left(\frac{S(z) - S_0(z)}{1 - S_0(z)}\right)^3$$

$$D(S) = D_0(z)\ e^{B(z)\ [S(z) - S_0(z)]}$$

where $K_0(z)$, $S_0(z)$, $D_0(z)$ and B(z) are soil physical parameters which vary with the depth but are constants in time.

Two different ways of using the particle approach will be presented below:

The first is the classical random walk model which dccrics the rearrangement of continuum packages, concentrated in the nodes of a fixed grid, in discrete time steps. here the transition can be made from one node to another one and this shown an analogy with the Brownian motion [5]. The computation procedure is given by the Markovian-chain model of transition probabilities, where the state-space is discrete in terms of both time and space. A simple version of this model can be constructed in the following way:

$$N_i^{j+1} = N_i^j * r + N_{i-1}^j * p + N_{i+1}^j * q - Nf_i\ , \qquad (2)$$

where

i-	is the moving subscript of the spatial coordinates of the fixed nodes (i=0...n, z: =Δz.i);
j-	is the moving subscript of the time coordinates of the nodes (j=0...m, t: =Δt.j)
N -	is the number of particles found in the given node (i,j);
Nf -	is the number of source particles;
r -	is the probability that the particle stays where it is;
p -	is the probability that the particle moves forward;
q -	is the probability that the particle moves backward;
Δt [T] -	is the time step (can be constant or varying);
Δz [L] -	is the spatial step (can be constant or varying);
K [L T^{-1}] -	is the seepage coefficient, inhomogeneous, non-linear
D [L^2 T^{-1}] -	is the diffusion coefficient, inhomogeneous, non-linear

It should be noted that the dependence of seepage- and diffusion coefficients on the level of saturation is non-linear and thus the Markovian process to be modelled is inhomogeneous, which means that transition probabilities r, p, and q vary both in time and space (subscripts i,j will be omitted below in order to maintain lucidity). The same refers to Courant number Cr and Peclet number Pe. None of the particles can walk over more than one grid step during a single time step. This condition can be met only if quantities r, p and q are probabilities, that is their value vary between zero and 1.0 over the entire space and time range. This requirement is met when the following conditions are complied with, for the Courant and Péclet numbers:

$$0 \leq Cr \leq 1, \text{ and } 0 \leq Pe \leq 1$$

These conditions limit the size of time- and space steps. It could be demonstrated that the solution of differential equation 1. can be obtained with the method of finite differences only when the above conditions hold.

The _other method_ of calculation, which was used in the case of this study, is the spatially continuous and in-time discrete model, the so called random flight model. The algorithm of the solution is provided by the following formula (Kinzelbach [4])

$$z_k(t+\Delta t): =z_k(t)+K1\Delta t+\zeta(2D\Delta t)^{1/2} \tag{3}$$

where

ζ -	is a probability variable of normal distribution, with zero expectable value and 1.0 standard deviation;
z_k -	is the spatial coordinate of the kth particle;
K1 -	is the virtual seepage coefficient (K1= K+∂D/∂z)

Boundary condition are handled in the following way. At the spatial boundaries of the space of calculation time-varying boundary conditions are considered as source/sink terms,allowing the entrance or exit of particles. Three different boundary conditions can be considered: fully permeable; fully impermeable and semi-permeable. At the upper (aerial) boundary of the three-phase zone a fully impermeable (reflecting) boundary can be considered in the case of no sources or sinks (no infiltration and no evaporation). At the other, lower, boundary of the space considered, that is at the groundwater table, a semi-permeable boundary is considered, since water can enter the space from the groundwater but only to the extent of saturation.

For the computation one should first determine the unit element and the number of particles (n_s) that would cause saturation in that element. The requirement of accuracy will determine this selection. It is important to note that, owing to the non-linearity of the problem, seepage- and diffusion coefficients depend on the number of particles considered in a spatial unit element, that is on the distribution of the particles that will characterize the saturation. There is a random error, the so called statistical noise, in a distribution when it is characterized by a finite number of particles. With the increasing number of the time steps of calculation the error of the distribution will be further cumulated owing to the effects of coefficients that depend on it.

The accuracy can be significantly improved if instead of a fixed grid of spatial elements one applies a flexible system of cells, tailored to the actual distribution of the particles, and this system is re-generated in every time step (Gáspár and Szél [6]). Nevertheless the example to be shown in this study, did not involve the application of such "flexible grids".

2.3 TRANSPORT OF CONSERVATIVE   SUBSTANCES

In groundwater the transport of a dissolved conservative substance can be described by the following differential equation

$$\frac{\partial C}{\partial t} + Kc\frac{\partial C}{\partial z} - Dc\frac{\partial^2 C}{\partial z^2} + F = 0 \tag{4}$$

where

    C -   is the concentration of the contaminant at saturation;
    Kc -   is the actual speed of motion of the conveying medium;
    Dc -   is the dispersion coefficient, depending also on the concentration of the contaminant;
    F -   the source term of the contaminant

Comparing equations (1) and (4) it is seen that the structures of the mathematical model of water- and pollutant transport are similar to each other. Consequently algorithm 3. can be applied for this case as well.

Below an example of the infiltration process, carrying conservative contaminants, will be presented. For the sake of simplicity let us assume that the liquid and the contaminant move together, which means that the velocity and diffusion coefficients can be calculated identically [Kc=K(S), Dc=D(S)].

2.4 TRANSPORT OF NON-CONSERVATIVE SUBSTANCES

The transport of non-conservative substances, dissolved in the groundwater, can be described by the following differential equation:

$$\frac{\partial C}{\partial t} + Kc\frac{\partial C}{\partial z} - Dc\frac{\partial^2 C}{\partial z^2} + F + G(C) = 0 \qquad (5)$$

Where

G(C) -   is the general indication of internal reactions and/or internal sources and sinks;

Let us now assume, for the sake of simplicity, that the changes of the concentration, as described by Eq. 5, do not depend on reactions with other constituents of the liquid, that is the reaction kinetic term is a function of the concentration of the substance in concern only.

The parameters of computation were the same as before and first order reaction kinetics (exponential decay) was assumed as the internal reaction term in the form of

G(C) $=\lambda$ C

where

λ - is the decay rate constant [$\lambda = 1.0$ day^{-1} was assumed]

Results of the calculation are shown in Figure 1 and 2. It is to be noted that two different methods can be followed in calculating the reaction processes. In the first case the number of particles are changed in such a way as to account for the rate of decay, as specified by the rate constant. In the second case, as applied in this study, the number of particles is left unchanged and their mass is varied to the necessary extent.

It should be also noted that when the non-conservative reaction process is described to the mathematical details of particle-collision kinetics then the case can be handled as a fully Lagrangian particle transport process (Kontur, Szél and Józsa [7]).

Finally in Figures 1,2 examples are shown for the application of the models described in sections 2.3 and 2.4 . The figures show a representation of the particles, the degree of saturation and the distribution of the pollutant concentration, with the indication of the parameters assumed.

Acknowledgment
The author wishes to express his gratitude towards Dr. Sándor Szél, who has prepared the computer models and carried out the computations.

REFERENCES

[1] <u>Karlin, S.-H.M. Taylor:</u> A First Course in Stochastic Processes, Academic Press, New York 1975.

[2] <u>Irmay, S.:</u> On the hydraulic conductivity of unsaturated soils, Trans Am. Geophys. U.N. 35. (1954), 463-468.

[3] <u>Gardner, W.R. and Mayhugh, M.S.:</u> Solutions and tests on the diffusion equation for the movement of water in soil, Proc.Soil.Soc. Am.22 (1958), 197-201.

[4] <u>Kinzelbach, W.:</u> Groundwater Modelling an Introduction with Sample Programs in Basic. Elsevier. Amsterdam (1985)

[5] <u>Kontur, I.:</u> Random walk model of water movement in nusatured zones, Proc.of the Int. Symposium RIZA München Vol1 (1984) 365-373.

[6] <u>Gáspár, Cs. and Szél, S.:</u> Application of unstractured grids in Monte-Carlo simulations, Proc. of the XXIV IAHR Congress, Madrid (1991)

[7] <u>Kontur, I., Szél S. and Józsa, J.:</u> Mixing of reacting substances by simple random walk. Proc. of the XXIV IAHR Congress, Madrid (1991)

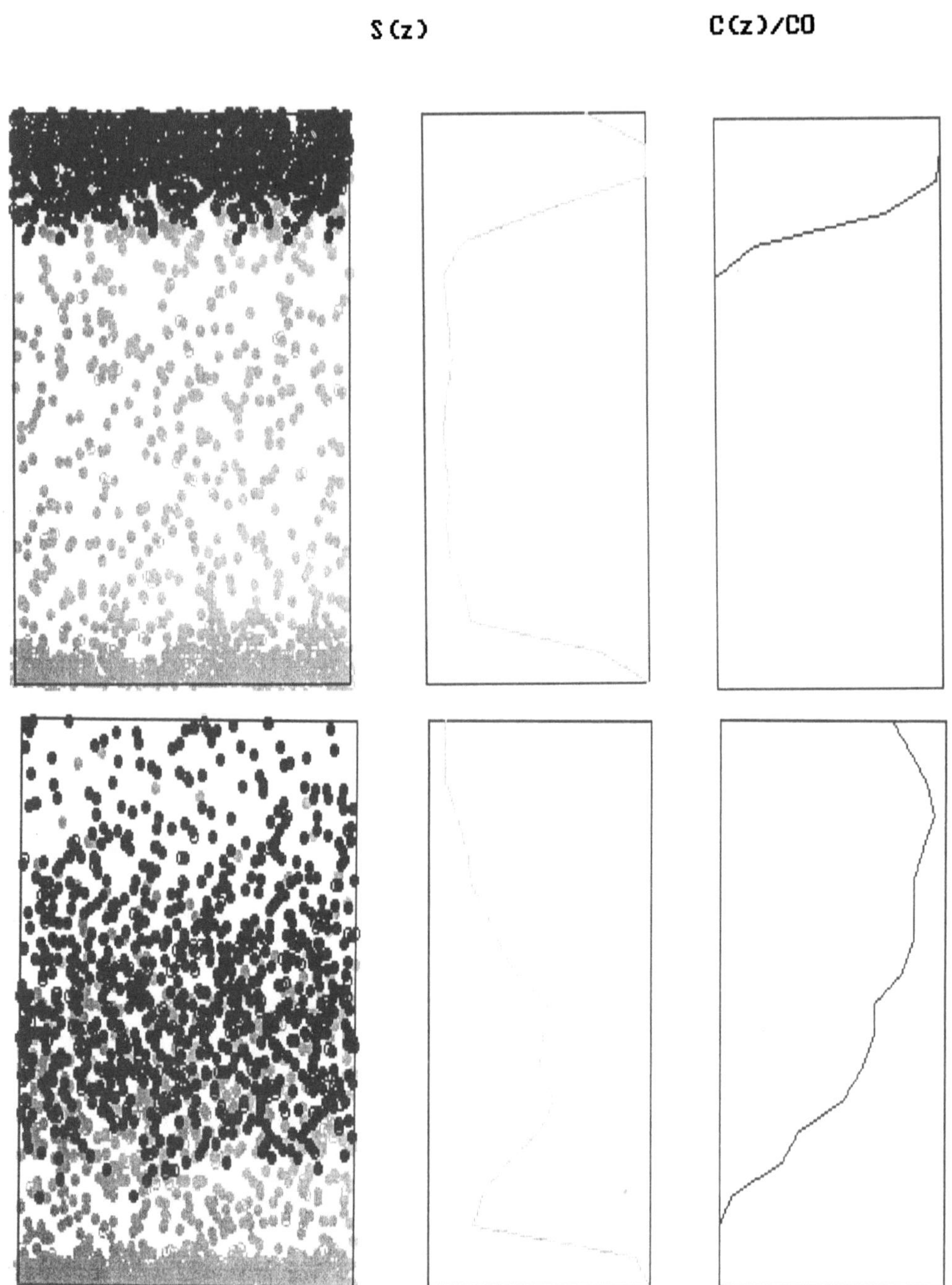

Figure 1.   Transport of dissolved pollutant infiltrated to the unsaturated zone. Transport of conservative subtances. Time=0.12 and 0.56 day.

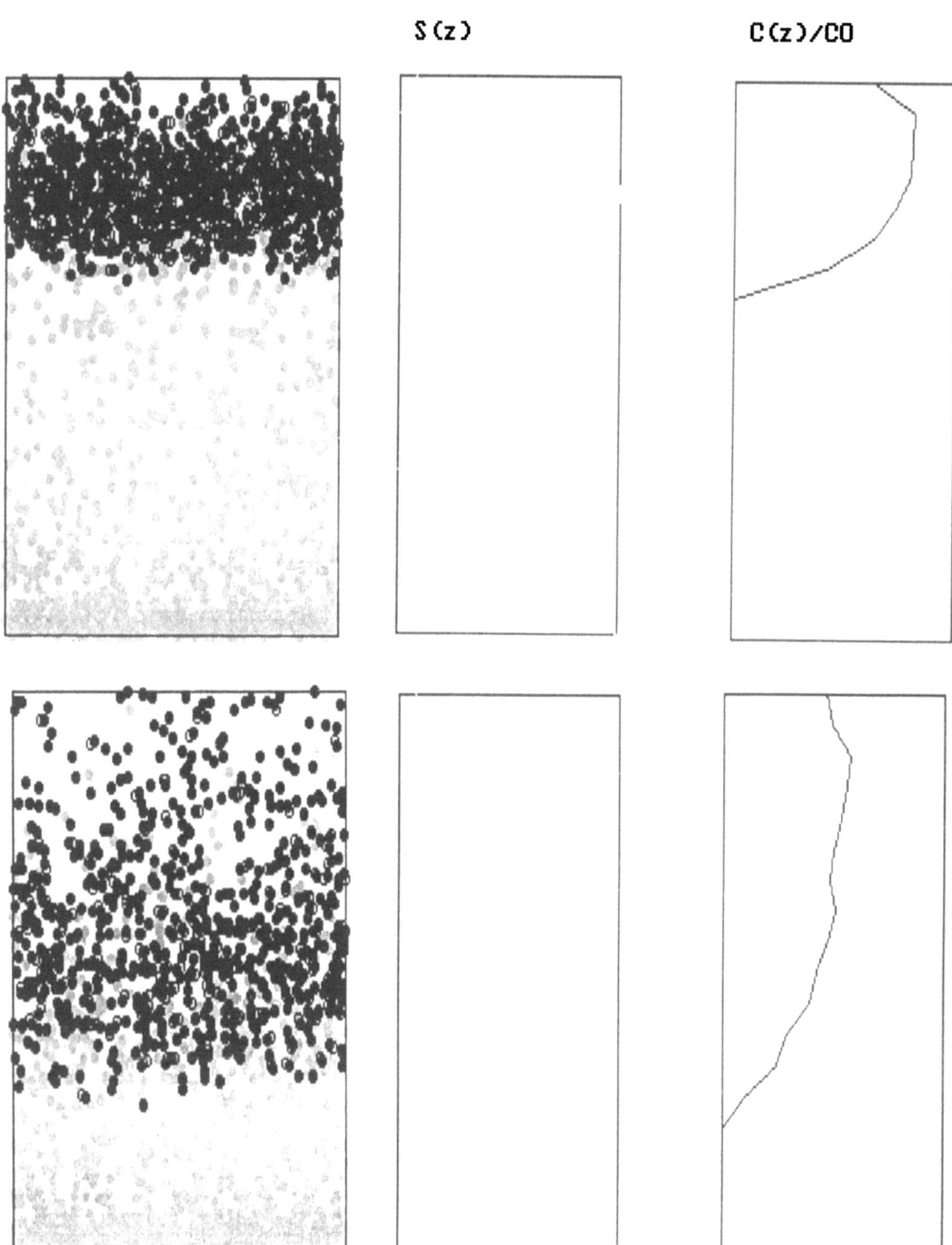

Figure 2. Transport of dissollved pollutant infiltrated to the unsaturated zone. Transport of non-conservative subtances. Decay coefficient = 1.0 (1/day). Time = o.19 and o.49 day.

SOLVING GROUNDWATER MANAGEMENT PROBLEMS USING A NEW METHODOLOGY

G.P. Karatzas and G.F. Pinder
University of Vermont, Burlington, VT, USA

ABSTRACT

Recently the problem of groundwater management has been approached by several optimization techniques including the classical linear/nonlinear programming methods, simulated annealing, neural networks, genetic algorithms and the outer approximation method. The 'outer approximation' method is a global optimization technique for the minimization of a concave function over a compact set of constraints. The concept of this method, as well as applications of the method to groundwater management problems, was first presented by the authors for problems with a convex set of constraints (Karatzas and Pinder, [1993]), and in a later work for a non-convex set of constraints (Karatzas and Pinder, [1994]).

Herein, a brief description of the theoretical concept of the method is introduced, followed by applications of the method to groundwater management problems using a 2-D and a 3-D numerical simulator. First, the methodology is applied to a hypothetical contaminated aquifer problem using a 2-D numerical simulator. A remediation scheme using two pumping wells is proposed for an easy representation of the concept of the outer approximation method in a 2-D space. Next, a hypothetical aquifer system is considered, which is represented by the 3-D numerical simulator PTC (Princeton Transport Code); four scenarios are examined, locating the potential pumping wells either on the first, second, or third layer and then by distributing them among the three layers. For each of the above scenarios the total remediation cost is computed, and conclusions drawn from the comparison.

1   INTRODUCTION

The outer approximation method combined with a 2-D numerical simulator was first presented by the authors to solve groundwater management problems formulated as minimization problems involving a continuous concave function over a convex set of constraints. The objective function included treatment and installation costs. The optimization theory of the proposed methodology was based on the work presented by Thieu et al., [1983].

Since convexity does not occur in all groundwater quality management problems the aforementioned research was extended to consider a non-convex feasible region. This concept was presented by the authors in 1994. The theory was based on the work presented by Hillestad and Jacobsen [1980] and Horst and Tuy [1990].

The method takes advantage of the basic property of a concave function f,that is that the minimum of the function over a compact set of constraints D is always attained in at least one extreme point of the set. The feasible region D (constraint set) is approximated by a simpler set D_1 (relaxed set) containing D, and the objective function is minimized over the relaxed set (defined by the rectangular area in fig. 1a). If the solution to this simpler problem is in D (i.e. all the original constraints are satisfied), then this is a global optimum. Otherwise at the computed minimum a suitable cutting hyperplane is introduced, such that a portion of the relaxed set D_1 is cut off and a new relaxed set D_2 is defined (containing D) (fig. 1b). The process is successively repeated until a solution in D is obtained (fig. 1c).

In geometrical terms, any relaxed set D_k in n-dimensional space will be defined as a *polytope* and since $D_k \supset D$, i.e. D_k contains D, D_k is termed an *enclosing polytope*. A hyperplane is a 'flat' geometric shape in n-dimensional space analogous to the line in two-dimensional space and to the plane in three-dimensional space.

The optimization process starts by determining the set of vertices by which the original enclosed polytope is defined; then specifies the vertex that minimizes the objective function (if there are more than one, any of them can be selected); if the selected vertex satisfies all the constraints then this is an optimal solution, otherwise the most violated constraint (the most positive value) at this vertex is determined. Since this vertex is located in the infeasible region, it can be eliminated of the present vertex set by introducing a cutting hyperplane between the vertex and the feasible region. The role of the cutting hyperplane, at any step, is to eliminate some part of the infeasible region, and to determine a new enclosed polytope that is a better approximation of the feasible region than in the previous step. The new set of vertices is determined and the process is repeated as described above. For linearly constrained problems, the cutting hyperplane is defined as one of the original constraints. For problems with nonlinear constraints, the cutting hyperplane is the "linearized" form of one of the original nonlinear constraints. The appropriate procedure for the determination of the cutting hyperplane is dependent on the behavior of the most violated constraint, whether convex or non-convex (Karatzas and Pinder[1993 and 1994]).

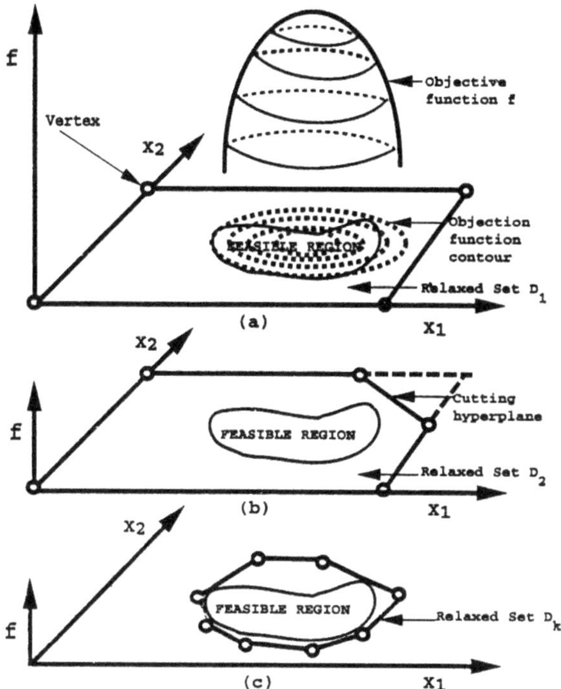

Figure 1: The theoretical concept of the outer approximation method

In the present work, applications of the outer approximation method to groundwater management problems will be demonstrated by the following examples: First, a hypothetical groundwater management problem of a contaminated aquifer is presented, using two pumping wells (decision variables), four observation points (constraints) and a 2-D numerical simulator. The purpose of this example is to demonstrate the performance of the outer approximation method in a 2-D space. Second, a hypothetical 3-D problem is presented, with several potential wells pumping from different layers of the aquifer and with constraints distributed among the layers. Finally, the last section contains the conclusions and the discussion.

2 PROBLEM:1. AN APPLICATION OF A GROUNDWATER QUALITY MANAGEMENT PROBLEM USING THE OUTER APPROXIMATION METHOD AND A 2-D NUMERICAL SIMULATOR

A hypothetical homogeneous, isotropic, unconfined aquifer is considered and represented by a 2-D finite element coupled model for areal flow and transport of a non-decaying, adsorbing-desorbing contaminant in a groundwater aquifer system. The model first was presented by Bredehoeft and Pinder [1973] and modified later by Ahlfeld [1986]. It consists of the following equations:

1. Flow equation

$$\nabla \cdot (\mathbf{K} b \nabla h) + q_l - q \delta(x_k, y_k) = 0 \tag{1}$$

2. Darcy's law

$$\mathbf{V} = -\frac{\mathbf{K}}{\theta} \nabla h \tag{2}$$

3. Mass transport equation

$$\nabla \cdot (b \theta \mathbf{D} \nabla c) - \theta b \mathbf{V} \nabla c + q_l(c - c_l) - q(c - c_k) \delta(x_k, y_k) = \frac{\partial c b R}{\partial t} \tag{3}$$

where

h = hydraulic head, $h(x, y)$ (L);
b = saturated aquifer thickness (L);
$\mathbf{K}$ = hydraulic conductivity tensor (L/T);
q_l = fluid flux into the aquifer from top or bottom boundaries (L/T);
q = fluid sinks(+), sources(-) (L/T);
$\delta(x_k, y_k)$ = dirac delta function evaluated at point (x_k, y_k).
c = solute concentration, $c(x, y, t)$ (M/L^3);
θ = effective porosity (dimensionless);
$\mathbf{D}$ = hydrodynamic dispersion tensor (L^2/T);
$\mathbf{V}$= average pore velocity vector, $\mathbf{V}(\mathbf{x}, \mathbf{y}, \mathbf{t})$ (L/T);
c_l = solute concentration in leakage fluid (M/L^3);
c_k = solute concentration in source/sink fluid (M/L^3);
R = retardation coefficient defined as $\theta + \rho_s k_D$ (dimensionless);
ρ_s = bulk density of the soil (M/L^3);
k_D = solute partition coefficient (L^3/M);

The dimensions of the aquifer are 870 m by 870 m with physical and numerical parameters presented in Table 1 and initial and boundary conditions as shown in fig. 2. It is assumed that two contaminant sources, located at nodes 425 and 485 (indicated by the black circles), contaminated the aquifer for 15 years prior to remediation.

The representative contours show contamination levels after normalizing the concentration, i.e, the normalized concentration at the sources was considered equal to 1. It is assumed that the contaminant sources are eliminated when the remediation action starts.

In fig. 3 a simple remediation scheme is presented, with two pumping (extraction) wells, indicated by w1 and w2, and four arbitrary observation points, indicated by o1,o2,o3,o4. At the end of a five year remediation period, points 1, 2, and 4 are required to achieve a normalized concentration less than or equal to 0.05 and point 3 must achieve a concentration less than or equal to 0.1.

The behavior of the constraints for the above remediation scheme is shown in fig. 4. Each individual curve represents one concentration constraint with all the points

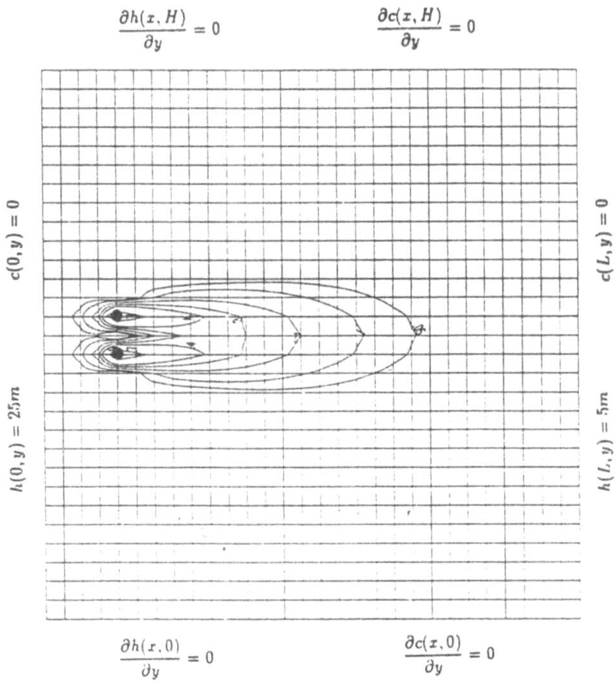

Figure 2: A Plan View of the Example Problem with the Initial and Boundary Conditions

Number of nodes	900
Number of elements	841
Hydraulic conductivity, $\mathbf{K}$	0.125 m/hour
Longitudinal dispersivity, α_L	18 m
Transverse dispersivity, α_T	1.8 m
Diffusion coefficient, d	0.00001
Porosity, θ	0.2
Time step, Δt	4 months (2928 hours)
Number of time steps	15

Table 1: Physical and numerical parameters of the hypothetical aquifer.

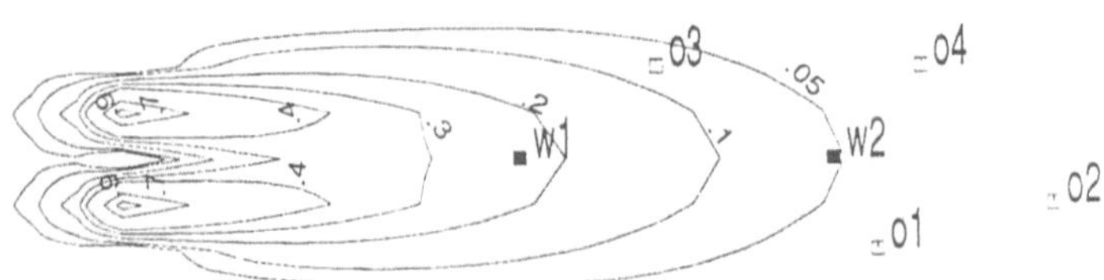

Figure 3: A remediation scheme of two wells and four constraints for the example problem.

along the curve satisfying the constraint with equality, i.e $c_j = c_j^*$ and $j = 1, 2, 3, 4$. The response of each individual constraint to the pumping rates is examined as follows: a constant pumping rate is assigned for well 1. By running the simulator the pumping rate of well 2 is determined such that the concentration at the observation point is equal to the constraint value. The feasible region is defined by constraints ($C1$), ($C2$), $C(3)$ and the upper bounds of the pumping rates at the two wells.

Figure 5, shows the above described behavior of the constraints in a 3-D space. The pumping rates, $\mathbf{q_1}$ and $\mathbf{q_2}$, are represented by the x and y axes and the evaluation of the constraints on the z axis. The feasible region, projected on the x-y plane passing through the 0 of the z axis, is defined as the area of the plane where all the constraints, in the form $c_j(\mathbf{q}) - c_j^*$, are evaluated to be less than or equal to zero and the pumping rates of the wells are less than the upper bounds of the pumping rates (i.e the area defined by the intersection of the constraint hyperplanes and the horizontal plane). The shape of the feasible region for this particular remediation scheme is a closed convex set (figs. 4 and 5).

The problem is formulated as:

$$min \quad \sum_{i=1}^{n} \alpha_i q_i + \alpha_i^0 (1 - e^{-bq_i}), \quad i \in I \tag{4}$$

$$s.t. \quad c_j(\mathbf{q}) \leq c_j^*, \quad j \in J \tag{5}$$

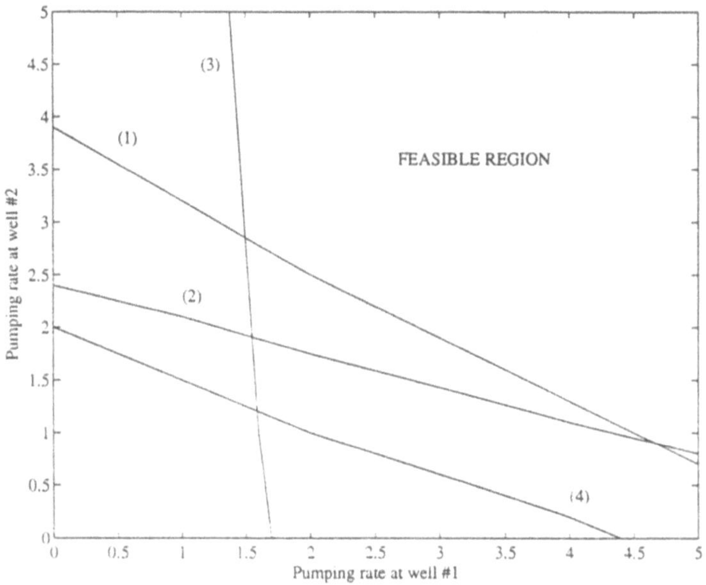

Figure 4: Representation of the feasible region and the constraints in a 2-D space.

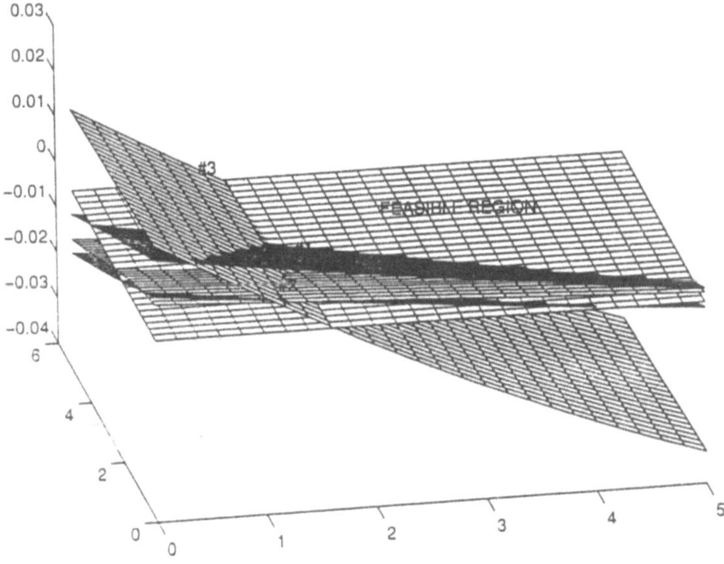

Figure 5: Representation of the constraint hyperplanes and the feasible region in a 3-D space.

$$0 \leq q_i \leq q_i^*, \qquad i \in I \tag{6}$$

where:

α_i is the unit pumping cost,

α_{0_i} is the fixed cost of installing a well at node i,

q_i is the pumping rate at node i,

$c_j(\mathbf{q})$ is the concentration at node j at the end of the remediation period,

c_j^* is the upper bound of c_j,

q_i^* is the fixed upper bound of q_i,

b is a scaling factor(=1000 for this problem);

I is the set of points with the number of elements
    equal to the number of parameters (pumping wells),

J is the set of points with the number of elements
    equal to the number of constraints (observation nodes).

Because the concentration of contaminant in a groundwater system is a nonlinear function of the pumping rates, constraints posed in (5) are nonlinear functions of $\mathbf{q}$. In the above optimization formulation the implicit representation of the nonlinear convective-dispersive simulation model is used (Ahlfeld [1986]).

The problem was solved first using the outer approximation Mmethod where the optimal solution was obtained by introducing four cutting hyperplanes (each represented by a line for a 2-D problem). A graphical representation of the delineation of the segmants of the infeasible region by the cutting hyperplanes for each algorithmic step is given in figs. 6 through 10. Figure 6 is the initial step of the optimization technique, where the infeasible region is indicated by the darker color; the rest of the figures show the part of the infeasible region eliminated at each optimization step by introducing a new cutting plane, and in figure 10 the optimal solution is indicated by the black dot.

This example, is a typical groundwater remediation problem with a nonlinear convex set of constraints. Only two wells were proposed, to provide an easy representation of the feasible domain and the 'cutting hyperplane' concept in two dimensional space. A comparison of the results, obtained by using the outer approximation method and the graphical representation of the exact feasible region (figs. 4 and 6), verified the validity of the computed optimal solution.

3   PROBLEM:2, A GROUNDWATER QUALITY MANAGEMENT PROBLEM USING THE OUTER APPROXIMATION METHOD AND A 3-D NUMERICAL SIMULATOR

The previous 2-D example was extended to 3-D. A three-layer hypothetical homogeneous, isotropic, unconfined aquifer was considered with dimensions: Length $= 870m$, Width $= 870m$ and Depth $= 30m$. The 3-D numerical simulator PTC (Princeton Transport Code) Babu et al.,[1993]) was employed for the representation of the aquifer. PTC

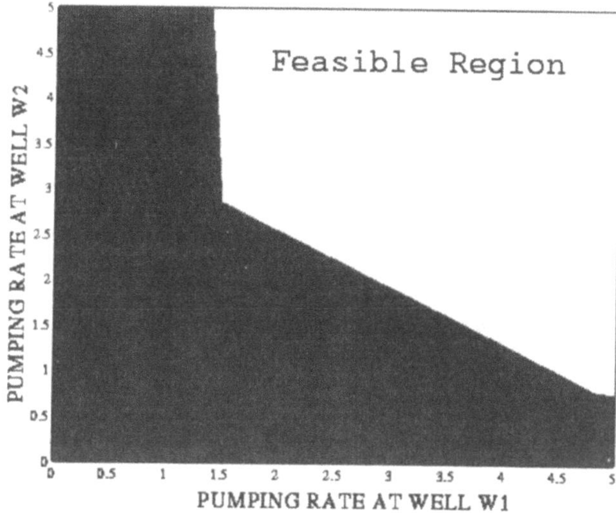

Figure 6: Initial optimization step for the example problem

is a hybrid finite element/finite difference groundwater flow and contaminant transport simulator. The physical and numerical parameters used in the simulations are presented in Table 2.

As in the previous example, it was assumed that two contaminant sources, located at nodes 425 and 485 in the second layer, contaminated the aquifer for 15 years prior to remediation. The boundary conditions for the flow and transport are that there is no water flux or dispersive flux across the boundaries and no applied fluxes occur anywhere within the domain. A plan view of the aquifer with the initial and boundary conditions is presented in fig. 2.

There were twenty three observation points (constraints), in each of the three layers, whereat it was requested that the normalized concentration becomes less or equal to .04 by the end of a remediation period of five years, and three points where the concentration had to remain less than or equal to .1 (points close to the original contaminant source). All these points are indicated by a black triangle in figures 11 through 22. Also, it was assumed that the sources had been removed when the remediation started. Also, constraints (upper bounds) were imposed on the pumping rates.

Four different remediation scenarios were examined: The first three scenarios were: (1) Ten potential wells pumping from the first layer (bottom layer) (fig. 11), (2) Ten potential wells pumping from the second layer (middle layer) (fig. 12), (3) Ten potential wells pumping from the third layer (top layer) (fig. 13). In all the above scenarios the potential well locations were considered at the same nodal locations. Figures 11 through 13 show the optimal solution at the end of the proposed remediation period

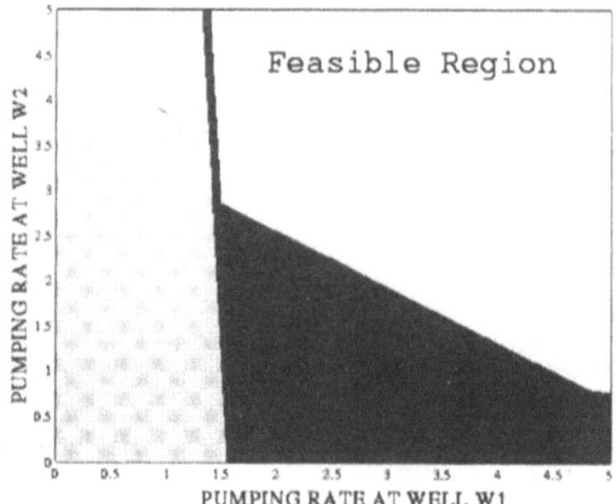

Figure 7: Optimization step:1; Introduction of the first cutting hyperplane

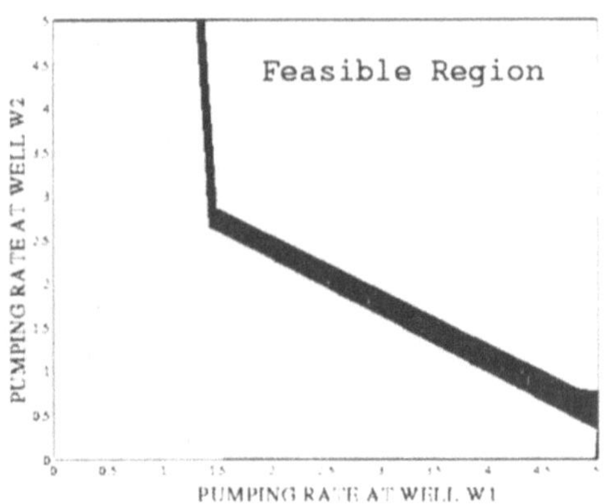

Figure 8: Optimization step:2; Introduction of the second cutting hyperplane

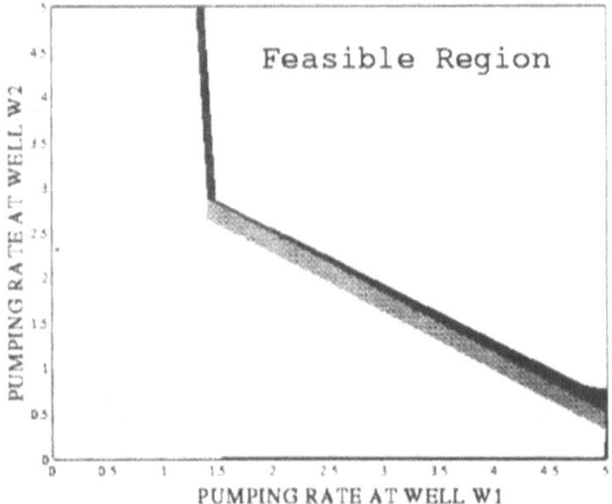

Figure 9: Optimization step:3; Introduction of the third cutting hyperplane

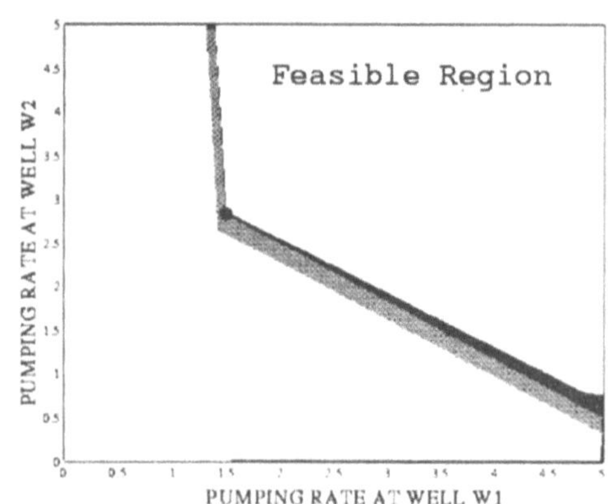

Figure 10: Optimization step:4; Introduction of the last(forth) cutting hyperplane - Optimal solution

Number of layers	3
Number of nodes	900
Number of elements	841
Hydraulic conductivity, $\mathbf{K}_x$	0.125 m/hour
Hydraulic conductivity, $\mathbf{K}_y$	0.125 m/hour
Hydraulic conductivity, $\mathbf{K}_z$	0.0125 m/hour
Longitudinal dispersivity, α_L	18 m
Transverse dispersivity, α_T	1.8 m
Vertical dispersivity, α_V	0.18 m
Diffusion coefficient, d	0.00001
Porosity, θ	0.2
Time step, Δt	4 months (2928 hours)
Number of time steps	15

Table 2: Physical and Numerical Parameters of the Hypothetical Aquifer

for each scenario, as well as the total cost related to the optimal remediation scheme. Well locations are indicated by the black squares and activated wells are shown by the black bars which are propotional to the pumping rates. (4) The fourth scenario was developed by studying the results from the previous three scenarios. Five wells were selected as potential pumping wells, and these were the most often activated pumping wells in the previous three scenarios. These five well locations (nodes) were proposed as potential wells on each layer; therefore the total number of the proposed wells was fifteen, distributed in the three layers. Figures 14 through 22 show all the optimization steps for the fourth scenario; figure 14 represents the initial optimization step and figure 22 indicates the optimal solution as well as the total cost related to this remediation scenario. The term 'optimization step' is defined as a certain scenario of pumping rates, applied at the proposed well locations, selected by the optimization algorithm and for which the constraints are evaluated. If all the contraints are satisfied, the optimization algorithm ends, otherwise it continues for the next optimization step. For all the scenarios the problem was formulated as in the previous example. Active wells are indicated by a black bar with its size to be proportional to the pumping rate. The upper bound of the pumping rate at any well location was $20m^3/hour$.

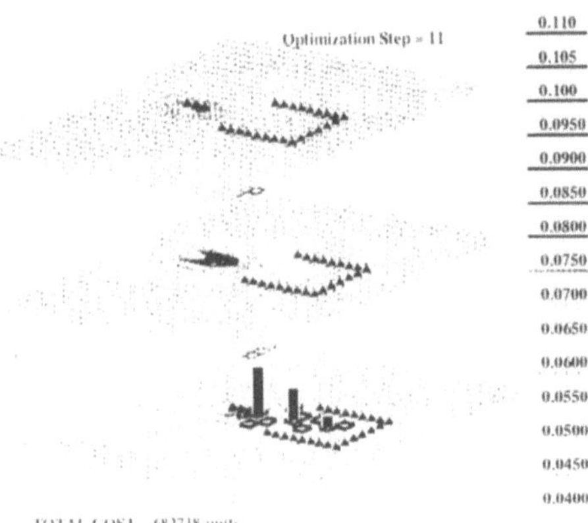

Figure 11: Scenario 1: Ten potential wells pumping from the first layer - Optimal Solution

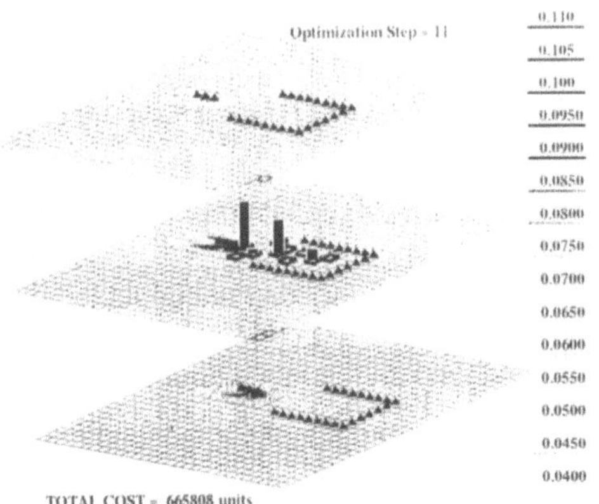

Figure 12: Scenario 2: Ten potential wells pumping from the second layer - Optimal Solution

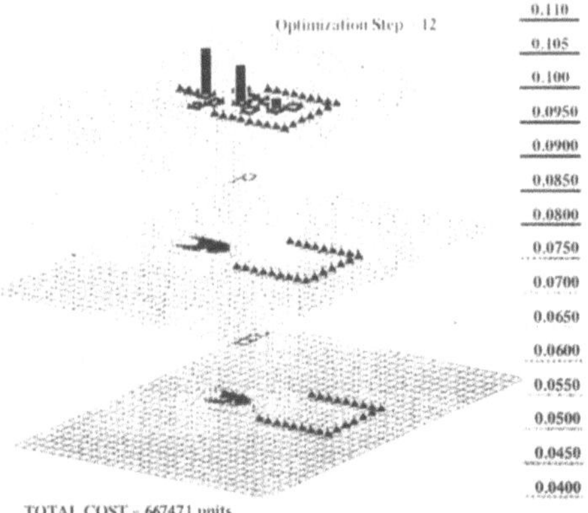

Figure 13: Scenario 3: Ten potential wells pumping from the third layer - Optimal Solution

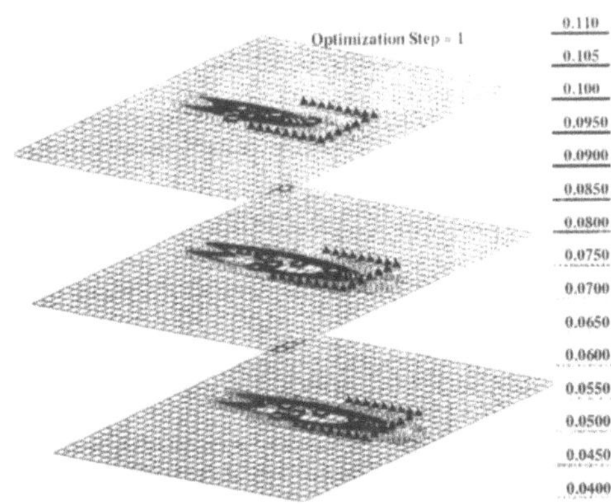

Figure 14: Scenario 4: Fifteen wells distributed in the three layers; Optimizastion step = 1

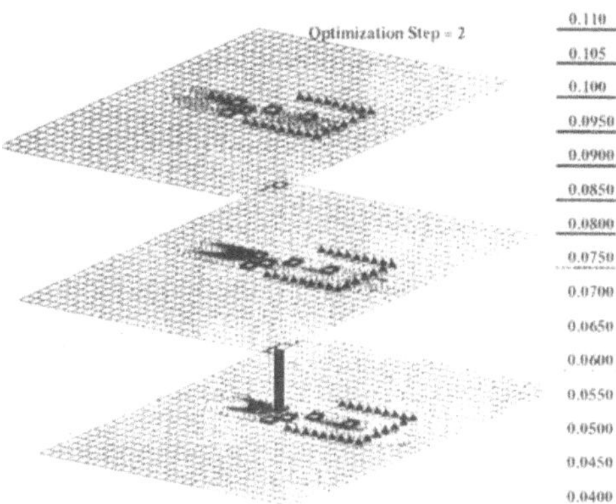

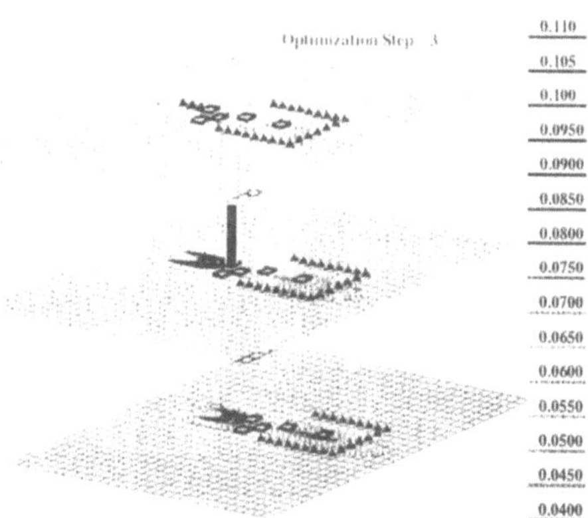

Figure 15: Scenario 4: Fifteen wells distributed in the three layers; Optimizastion step = 2

Figure 16: Scenario 4: Fifteen wells distributed in the three layers; Optimizastion step = 3

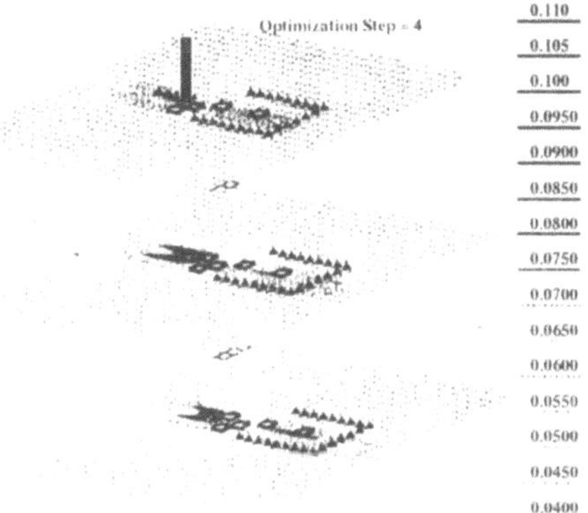

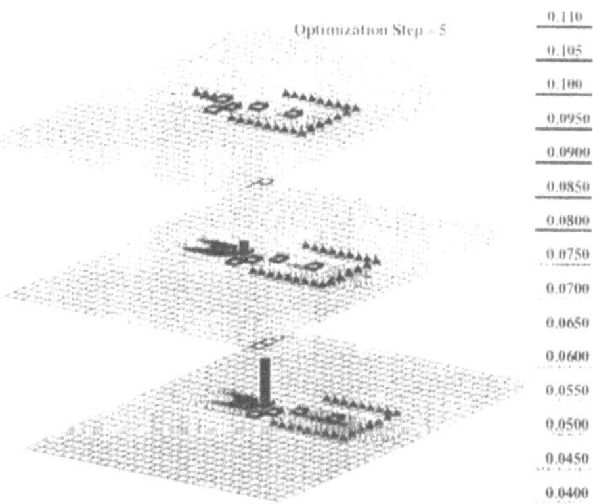

Figure 17: Scenario 4: Fifteen wells distributed in the three layers; Optimizastion step = 4

Figure 18: Scenario 4: Fifteen wells distributed in the three layers; Optimizastion step = 5

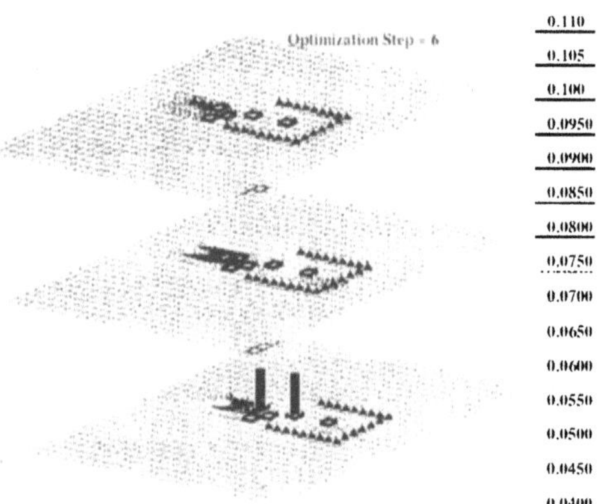

Figure 19: Scenario 4: Fifteen wells distributed in the three layers; Optimizastion step = 6

Figure 20: Scenario 4: Fifteen wells distributed in the three layers; Optimizastion step = 7

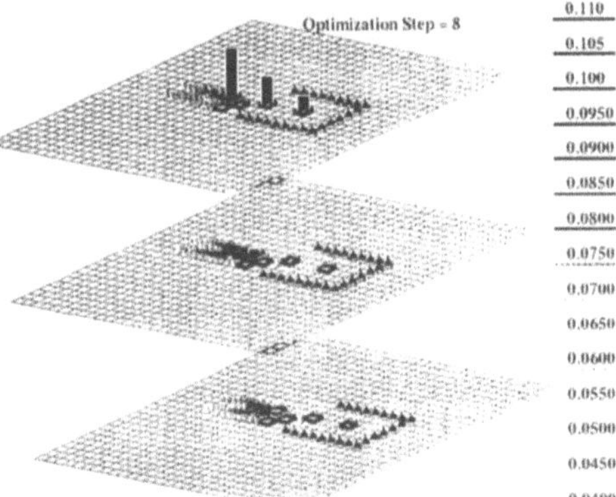

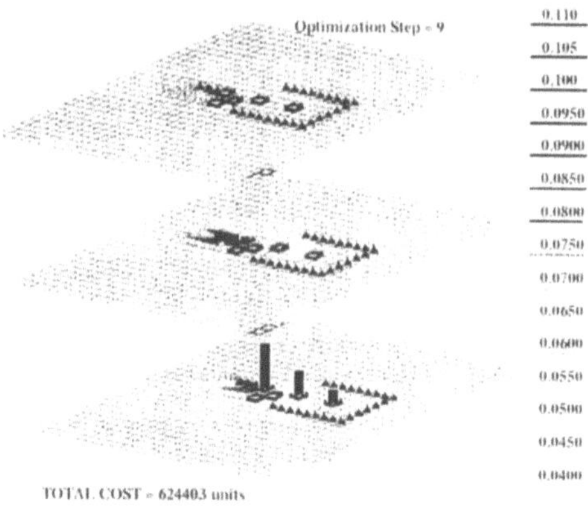

Figure 21: Scenario 4: Fifteen wells distributed in the three layers; Optimizastion step = 8

Figure 22: Scenario 4: Fifteen wells distributed in the three layers; Optimizastion step = 9 - Optimal Solution

CONCLUSIONS AND DISCUSSION

The concept of, and two applications of the outer approximation method for problems with convex or non-convex feasible domains herein is presented. Since the concept of the method is based on combinatorial geometry, the computational effort increases as the number of decision variables increases.

Conclusions regarding the use of the 3-D numerical simulator can be obtained by studying figures 11,12,13 and 22. For the first three scenarios the computation of the total costs involves selectio not only of different well locations, but also different pumping rates. A significant difference is observed for the fourth case scenario where pumping was allowed from all the layers. For this study, it was assumed that the well installation cost was the same for all the layers. For cases where the installation cost is different for each layer, the difference in total cost will depend more uppon which layer is being pumped. The importance of using a 3-D rather than 2-D simulator is in agreement with previous studies (Tiedeman and Gorelick [1993]), wherein was emphasized the fact that this is a better representation of the physical system (aquifer).

REFERENCES

1. Karatzas, G.P. and G.F Pinder (1993), 'Groundwater Management Using Numerical Simulation and the Outer Approximation Method for Global Optimization', *Water Resources Research*, vol.29, no.10,3371-3378.

2. Karatzas, G.P. and G.F Pinder (1994), 'Groundwater Management Using a 3-D Numerical Simulator and a Cutting Plane Optimization Technique', Proceedings of the X Conference on Computational Methods in Water Resources, Heidleberg, Germany, Kluwer Academic Publishers, vol.2, 841-848.

3. Thieu, T.V., B.T. Tam, and V.T. Ban (1983), 'An Outer Approximation Method for Globally Minimizing a Concave Function over a Compact Convex Set', *Acta Mathematica Vietnamica*, 8, 21-40.

4. Hillestad, R.J. and S.E. Jacobsen (1980), 'Reverse Convex Programming', *Applied Mathematics and Optimization*, 6, 63-78.

5. Horst, R. and H. Tuy (1990), *Global Optimization*, Springer-Verlag, Berlin. Heidelberg.

6. Bredehoeft, J.D. and G.F. Pinder (1973), 'Mass Transport in Flowing Groundwater', *Water Resources Research*, vol.9, no.1, February

7. Ahlfeld, D.P. (1986), 'Designing Contaminated Groundwater Remediation Systems Using Numerical Simulation and Nonlinear Optimization', Ph.D dissertation, Princeton University.

8. Babu, D.K., G.F. Pinder, A. Niemi, D.P. Ahlfeld, S.A. Stothoff (1993), 'Princton Transport Code, Report 84-WR-3 (revised)', University of Vermont, Burlington, VT, USA.

9. Tiedeman, C. and S. M. Gorelick (1993), 'Analysis of of Uncertaintly in Optimal Groundwater Contaminant Capture Design', *Water Resources Research*, vol.29, no.7, 2139-2153.

THE GROUNDWATER AND
THE GROUNDWATER QUALITY MANAGEMENT PROBLEM:
RELIABILITY AND SOLUTION TECHNIQUES

T. Tucciarelli

University of Reggio Calabria, Reggio Calabria, Italy

ABSTRACT

The groundwater and the groundwater quality management models are defined as an engineering decision problem. The solution of this problem is not trivial and can be reached with the use of deterministic or stochastic techniques. The determistic approaches (Primal Method) compute only local minima, or the global minimum (Outer Approximation Methods) only according to strict hypothesis. The stochastic methods (Simulated Annealing, Genetic Algorithms or Neural Networks) find the global minimum, but only in statistical sense. The most recent approaches to the problem treat the physical parameters (transmissivity) as a random field. The uncertainty of the transmissivity makes impossible to guarantee the feasibility of the optimal strategy. A reliability level is defined as the ratio between the number of realizations that mantain feasible the optimal solution out of the total number of realizations. Some of the algorithms proposed to reach a fixed reliability are discussed. The idea of linking the goal of the management problem with the number and location of new field measurements is introduced.

1. AN ENGINEERING DECISION PROBLEM

The simulation models give us a mathematical tool to evaluate the state variables (e.g. concentrations, velocities and piezometric heads) as functions of both the natural parameters (e.g. conductivity, storativity, retardation coefficients) and the artificial parameters (e.g. extraction and injection rates, well locations). These models can be simply used to predict the effect of different choices of the artificial parameters on the state variables, but can also be part of an engineering decision problem, casted in optimization form.

In this engineering decision problem the simulation model is the main tool to relate three quantities:

1) The objective function(s) (OF)
2) The constraint functions (CF)
3) The decision variables (DV)

1) The OF is a measure of the performance of our decision. This performance can be represented by a single number, usually a cost, or by a multiobjective function [1] if the final decision has to be made according to some parameters not known 'a priori'. In the first case the most common OF are:

a) the cost of the extraction, the injection, the treatment and the disposal of the groundwater. The injection and the treatment costs [2] are proportional to the pumping rates q; the extraction cost depends on the energy required, and is therefore a function of the hydraulic head at the extraction wells. Because the drawdown immediately around the well is mainly a function of the pumping rate of the well, it is common practice to compute all these costs with a polynomial function of q, where the polynomial has a degree varying from one to two.

b) the installation cost of the extraction/injection wells [3]. This is an integer (0-1) variable. There is no compromise between the decision of installing or not installing a well. In long-period (15-20 years) restoration projects this cost is usually negligible. When this cost is not neglected, it implies a much larger computational effort for the solution of the problem.

2) The CF are a measure of the changes that affect the enviroment due to our decision (for example an elevation of

the piezometric head or a reduction of the concentration).
These functions are constrained by some values given by the
manager according to sanitary, sociological and political
issues.

These constraints can also be replaced by a recourse
cost [4], added to the OF. The recourse cost is the cost
predicted at the end of the restoration period to reduce the
CF inside the range allowed. This approach can be usefull
when the degree of violation of the constraints at different
control points is thought to be important and the uncertainty
of the parameter changes the problem from deterministic to
stochastic.

In many groundwater quality management problems the
constraints on the concentrations are replaced with much
easier constraints on the piezometric gradients [5]. For
example, if the migration of a plume has to be contained in a
given area, the gradients around this area can be constrained
in the inward direction.

3) The DV are the artificial parameters of our simulation
model, usually the pumping rates in different time periods
(continuous variables) and the well locations (integer
variables), but they can also be the location of slurry
walls, drains or impermeable caps.

An example of groundwater quality management problem is the
following one:

$$\text{Minimize } f(\mathbf{q}) \tag{1}$$

$$\text{s.t. } c_i(\mathbf{q}) \le c^{max}_{i} \quad i=1,k \tag{2}$$

$$0 \le q_j \le q^{max}_{j} \quad j=1,n \tag{3}$$

where the CF in (2) are the concentrations at the end of the
restoration period at selected observation points and the
pumping rates in (3) can be either extraction or injection
rates according to the location of the well.

2. SOLUTION TECHNIQUES

The solution of problem (1)-(3) is not a trivial task,
specially if the installation cost is part of the objective
function. The main problem in solving (1)-(3) is that the

feasible domain, the set of decision variables that satisfy
all the constraints, is not convex. See in Figg. 1a and 1b an
example of convex and not convex feasible domain.

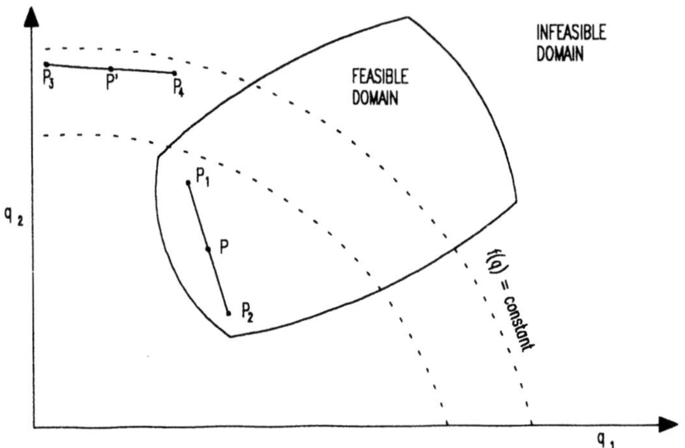

Fig. 1a: Example of convex objective function and convex
feasible domain.

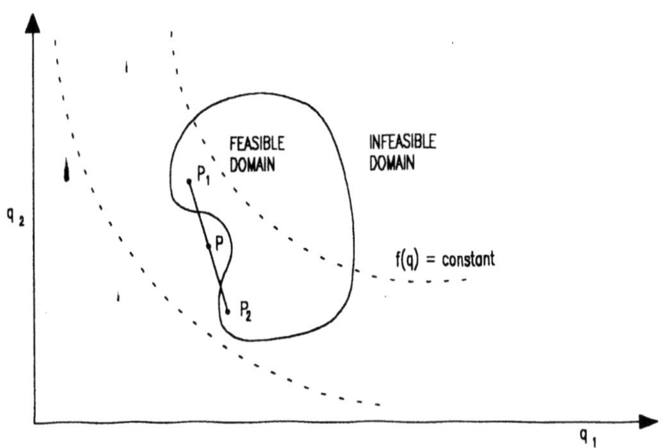

Fig. 1b: Example of concave objective function and not convex
feasible domain.

In Fig. 1a any point P between the feasible points P_1 and P_2 is also feasible; in Fig. 1b the point P is not feasible. In the same way, the OF is said to be convex if any point P' between two points with the same OF has a smaller OF and concave if any point P' between two points with the same OF has a larger OF. The OF in Fig. 1a is convex, and the OF in Fig. 1b is concave. It can be shown that the convexity of the feasible region and the OF implies the uniqueness of the local minimum and, therefore, that the local minimum is also the global minimum.

2.1 Deterministic Methods

All the deterministic methods compute only a local minimum, that depends on the location of the initial point. The most common strategy used to deal with this problem is to start the minimization process from different initial points and to select the best local minimum. For the location of each minimum universal codes for nonlinear constrained problems are generally used. These codes have shown poor convergence properties when applied to groundwater quality problems (problems with few tens of decision variables are likely to miss convergence or convergence is very slow). This is due to the flat shape of the boundary between the feasible and the infeasible domain (a problem similar to the one encountered in the solution of the inverse problem,[6]).

A new minimization algorithm has recently been tested [7], that allows for these problems the use of hundreds of decision variables. The methodology can be classified as a 'primal' method, because the solution is found by moving from a feasible point P_i to an other feasible point P_{i+1} (Fig. 2). At each point P_i the hyperplane tangent to the boundary between the feasible and the infeasible domain is computed. The direction search from point P_i to point P_{i+1} is a linear combination of two directions, $\mathbf{d}^1$ and $\mathbf{d}^2$. Direction $\mathbf{d}^1$ moves from point P_i to the point P_i^*, solution of subproblem i.

Subproblem i has the same OF of the main problem, but linear constraints, given by most of the hyperplanes previously computed. In the case of Fig.2, if i=1, the subproblem number 2 is

$$\text{Minimize } f(\mathbf{q}) = \mathbf{cost\ q} \qquad (4)$$

$$\text{s.t. } \nabla c_r(P_j)(\mathbf{q} - \mathbf{q}(P_j)) \leq 0 \quad j=1,2 \qquad (5)$$

$$0 \leq q_j \leq q_j^{max} \quad j=1,2 \qquad (6)$$

where r is the index of the constraint that is satisfied as an equality in point P_j. Direction $\mathbf{d}^2$ is aimed to move from P_i into the feasible domain. A simple way to choose $\mathbf{d}^2$ is to use the direction normal to the hyperplane with index i. It can be proved that a maximum number of N+K hyperplanes have to be saved in the subproblem constraints, where N and K are respectively the number of DV and the number of constraints.

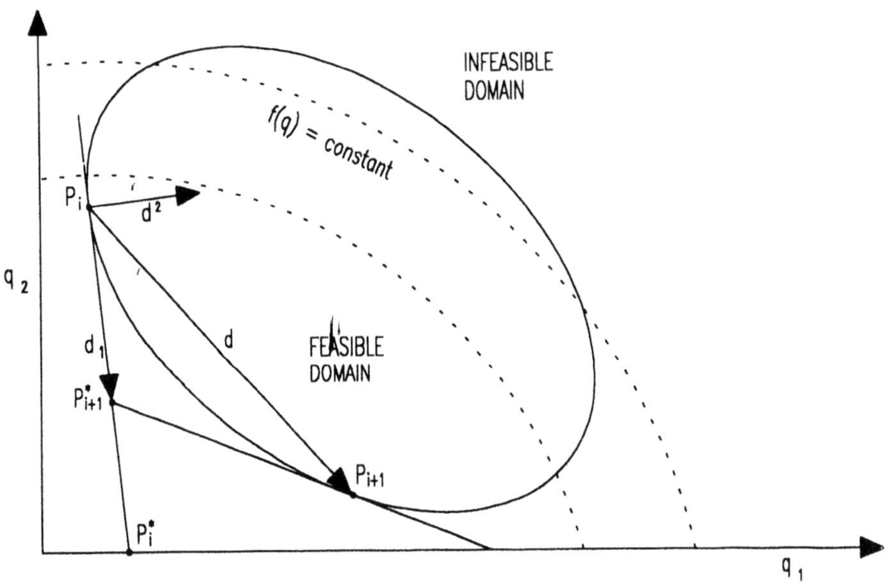

Fig.2: Location of point P_{i+1} from point P_i in the primal method.

The problem (1)-(3) can also be solved in the framework of the optimal control theory [8], viewing the pumping rates at time t as the control vector at the same time, and the simulation model as the transition equation of the state variables from time t to time $t + \Delta t$.
To deal with the nonuniqueness of the local minimum, alternative methods have recently been proposed. A first group of methods is dedicated to problems with a special form of the objective function. Among them there are the linear

mixed integer programs, for the solution of linear problems where the OF is increased with the installation cost [9] and the outer approximation methods (OAM), for problems with concave OF [10]. The OF is said to be concave if it is not convex in all the feasible points.

The OAM (Fig.3) solve a sequence of subproblems with linear constraints. It can be shown that the minimum of a problem with linear constraints and a concave OF lies in one of the vertices V_i, intersections of the hyperplanes that represent the linear constraints. At each subproblem a new hyperplane (called cutting plane) with equation

$$\nabla c_r(V_p)\,(q-q(V_p))\ +\ c_r(V_p)\ -\ c_r^{max}\ =\ 0 \tag{7}$$

is computed and added to the ones computed in the previous subproblems. Index p and r are the ones that maximize the difference

$$g(V_p)\ =\ c_r(V_p)\ -\ c_r^{max} \tag{8}$$

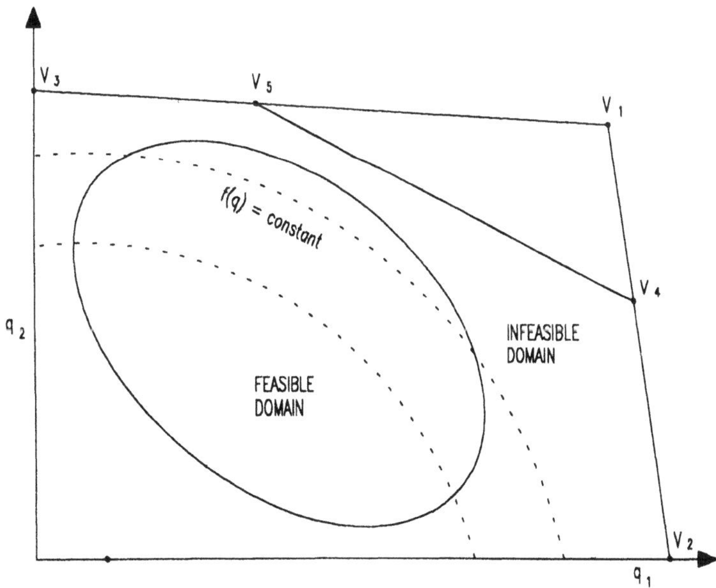

Fig.3: Example of cutting planes in the OAM.

It can be shown that all the points on the hyperplane defined by equation (7) are infeasible and form an approximating polytope around the feasible region. The new vertices and the corresponding OF are computed. Most of the new vertices have to be saved to compute the vertices of the new hyperlanes, but some of them can be discarded (for example the vertex V_1 in the example of Fig.3). It can be shown that the vertex with the minimum OF converges to the global minimum. The OAM are very usefull when the remediation period is very short, the installation cost is important and the energy cost is negligible. In fact, it is easy to see that a problem where the OF is the sum of a linear cost plus the installation cost is equivalent to a problem with a continuous, concave objective function (see Fig. 4).

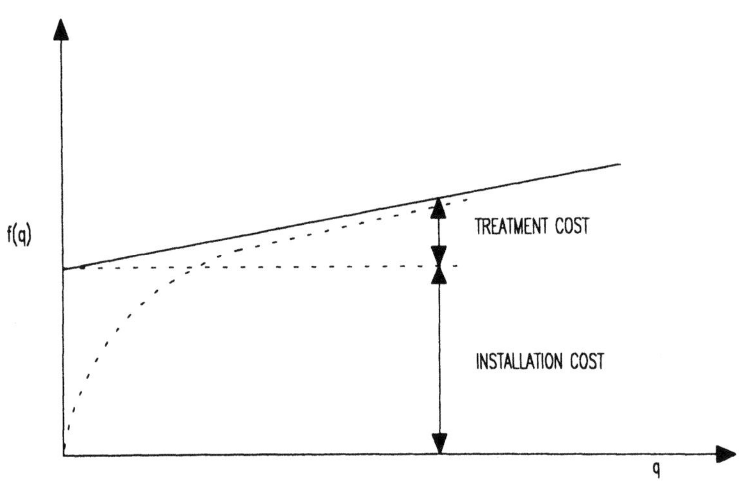

Fig.4: Approximation of the remediation cost by means of a concave objective function.

The inconvenience of the OAM and of all the deterministic methods for the solution of non-convex problems is that the numerical effort and the memory requirement increase dramatically along with the number of DV. In the OAM methods the number of vertices to store (and to handle) in order to guarantee convergence becomes overwhelming after few tens of DV. To avoid this inconvenience stochastic methods have to be used.

2.2 Stochastic Methods

The stochastic methods give up the idea of finding a vector of DV that can be proved to be the global minimum. Assuming a much more 'humble' aptitude, they apply procedures, based on the generation of random numbers, that converge exactly to the global minimum only in statistical sense. The final result is, therefore, only an approximation of the global minimum, in the same way as the statistics computed from a sample of realizations are an approximation of the unknown population parameters. All these methodologies have strong analogies with natural processes; the most important are the genetic algorithms [11], the simulated annealing [12] and the neural networks [13].

The genetic algorithms maximize the OF starting from a set (called population) of decision variable vectors. Like a living pupulation, this set of vectors changes from one generation to the next one. The individuals (DV vectors) of each generation survive according to a likelihood that is proportional to their OF. If they survive, they randomly mate and they reproduce other individuals that have components chosen from both the parents. Moreover, in some generations there are mutations that randomly change the components of the individuals. It can be shown that the winning population is finally composed by vectors that maximize the objective function. In the example of Table I a set of DV is written in binary form. Assume that the OF is equal to the sum of the unit digits and that the survived individuals mate by changing the first three and the last two digits. Observe that the new generation will have the second and the sixth individuals with a larger OF.

Table I

Old Individuals	Survival Likelihood	Survived Individuals	New Individuals
1) 01011	3/14	01011	01001 (1-2)
2) 00100	1/14	10101	10111 (1-2)
3) 10101	3/14	10101	10110 (3-5)
4) 10001	2/14	10001	10001 (3-5)
5) 10010	2/14	10010	10000 (4-6)
6) 11100	3/14	11100	11101 (4-6)

The simulated annealing starts from a single vector, where each component is thought as the position of a fluid particle in the space (Fig. 5). To each vector, called configuration, it corresponds an objective function, called internal energy. According to the temperature of the fluid, the configuration of the particles will change in time. For high temperatures, the particles will easily move from an old configuration to a new one, regardless the energy of the new configuration. When the temperature is lowered, the particles will be more likely to move to new configurations with a lower energy.

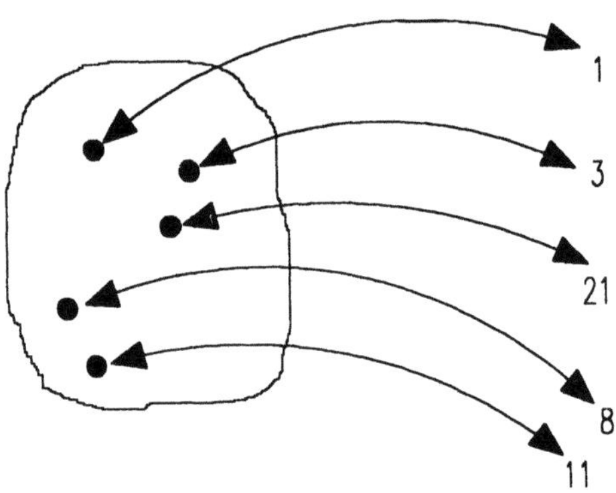

Fig.5: Analogy between the location of particles and the DV values.

In the algoritm the temperature is a parameter that, starting from a high value, is lowered to zero. For each temperature, the configuration of the particles is randomly changed. If the new objective function is lower than the old one, the new configuration is always accepted. Otherwise, the new configuration is accepted with a likelihood that is proportional to

$$p = \exp\left(-\frac{\Delta f}{T}\right) \qquad (9)$$

where Δf is the increase of the OF and T is the temperature. After a large enough number of trials (say 100n, if n is the vector dimension), the temperature is dropped. If we assume that the configuration $F_{old} = (1,3,21,8,11)$ of Fig. 5 is the old one and the OF is the sum of the vector components, we can see in Table II the probability p of three possible configurations corresponding to two different temperatures

Table II

	F1 = 2,6,5,24,14 $\Delta f = 7$	F2 = 1,2,15,10,11 $\Delta f = -5$	F3 = 2,6,18,13,7 $\Delta f = 2$
T = 10	0.49	1	0.81
T = 5	0.24	1	0.67

It is evident that the probability of a new configuration to be accepted when its OF is greater than the OF of the old one goes to zero along with the temperature. This allows the running configuration to stall into a local minimum. See in Fig. 6 the thermodynamic analogy between the simulated annealing process and the lava solidification.

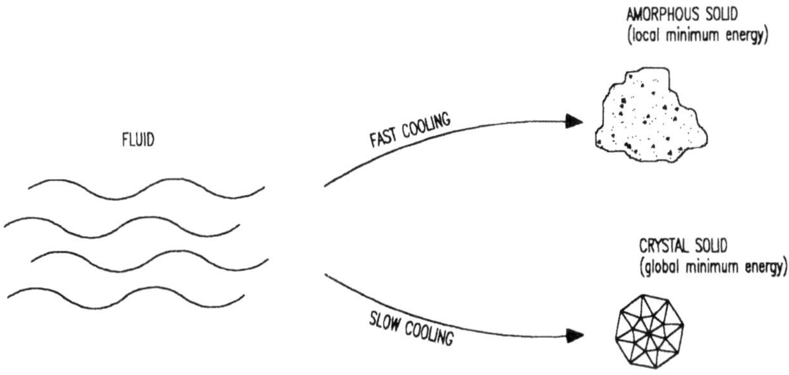

Fig.6: Analogy between the lava solidification and the simulated annealing algorithm.

The neural networks are used when the solution of several management problems, each one for a different set of natural parameters, is required. It is a learning method that allows the user to find more approximated, but easier representations of the relationship between the natural parameters and the solution of a management problem like (1)-(3). This relationship is based on the use of a somehow arbitrary number of intermediate units. To each natural parameters vector (its elements are called input unit) it corresponds an activation level of the intermediate units. If a_i is the value of each input unit, the activation level of the intermediate unit j is defined as

$$b_j = \frac{1}{1 + \exp(-I_j)} \tag{10}$$

where

$$I_j = \Sigma_i \; w_{ij} \; a_i \tag{11}$$

,and the minimum value of the objective function is computed as

$$f = \Sigma_i \; w_i b_i \tag{12}$$

The coefficients w_{ij} and w_i are calibrated with a training process that uses the solutions of a limited sample of management problems.

3. RELIABILITY OF THE MANAGEMENT MODELS

The reliability of the management models depends on the reliability of the simulation model and the reliability of the constraints, as well as the costs and the boundary conditions forecasted in the future [14]. Much emphasis has been given in the last years on the uncertainty of the natural parameters, especially of the transmissivities.

Moreover, the most recent advances in groundwater quality modeling suggest that the natural parameters cannot be assumed piece-wise constant and must be treated as a random variable, conditioned to assume measured values in a restricted number of locations. This ' random' perspective implies that the constraints (2) cannot always be guaranteed, at least for some parameter realizations.

Wagner and Gorelick [15] assume piece-wise constant parameters, but use a chance-constrained technique to take into account the uncertainty of the parameters. Later on the same authors [15] suggest to solve the problem (1)-(3) assuming spatially variable parameters. They define a reliability level R_1 of an optimal pumping strategy as

$$R_1 = \frac{N_f}{N_t} \tag{13}$$

where N_f is the number of transmissivity realizations that mantain feasible the optimal pumping strategy and N_t is the total number of realizations.

Two different strategies are proposed. In the first strategy, termed the multiple realization management model, they solve the management problem by changing equations (2) with a whole set of constraints

$$c_i(\mathbf{q}, \mathbf{T}_1) \leq c^{max}_i \quad i=1,k$$

$$c_i(\mathbf{q}, \mathbf{T}_2) \leq c^{max}_i \quad i=1,k$$

$$\ldots\ldots$$

$$c_i(\mathbf{q}, \mathbf{T}_s) \leq c^{max}_i \quad i=1,k \tag{14}$$

where $\mathbf{T}_j$ is the transmissivity vector corresponding to realization j and s is the number of realizations. This implies that the optimal pumping strategy is feasible for all the transmissivity realizations included in the management model. The choice of the number s remains somehow empirical. With 'a posteriori' Monte Carlo experiments they find a reliability level of more than 90% by using 30 realizations only.

In the second strategy, termed the Monte Carlo Management Model, several problems (1)-(3) are solved, each one with a different transmissivity realization. A different optimal pumping strategy corresponds to each transmissivity realization and a p.d.f of the pumping rates correspondes to a p.d.f. of the random field parameters. If only one reclamation well is present in the model and the feasibility of the computed state variables increases along with the pumping rate, there is also a monotonic relationship between the pumping rate and the reliability level. For example, the pumping rate associated with the 80th percentile of the optimal pumping rate p.d.f. corresponds to a 80% reliability level.

Wagner et al. [4] use the recoursive cost approach to include the degree of violation of the constraints into the final decision. They assume that the only constraints are set on the head gradients, in order to avoid the contaminant to enter the cleanup area. This would change the constraints (14) in the linear form

$$R_i(\mathbf{T}_1)\mathbf{q} \geq g_i \quad i=1,k$$

$$R_i(\mathbf{T}_2)\mathbf{q} \geq g_i \quad i=1,k$$

$$\ldots\ldots$$

$$R_i(T_s)q \geq g_i \quad i=1,k \tag{15}$$

where R_i is the ith row of a matrix with k rows and n columns. The problem is solved by changing constraints (15) with a recoursive cost, that is a monotonically increasing function of the differences

$$S_{ij} = g_i - R_i(T_j)q \tag{16}$$

The objective function becomes the sum of the recoursive costs associated to each transmissivity realization, multiplied by the probability of each realization and by a penalty coefficient, plus the pumping cost. The penalty coefficient has to be given according to economic value given to the constraints violation. It can be shown that the objective function is still convex, unless the benefit of the pumped water is subtracted to the objective function.

Morgan et al. [17] use a chance-constrained technique to solve a non linear mixed-integer problem where a given number of sets in (15) is allowed to be violated. This allows to assign a fixed reliability to the solution of the management problem. The same semplification of Wagner et al. [4] is used, and constraints are given on the head gradients only. The inequalities (15) are changed in the form

$$R_i(T_1)q + My_1 \geq g_{i1} \quad i=1,k$$

$$R_i(T_2)q + My_2 \geq g_{i2} \quad i=1,k$$

$$\ldots\ldots$$

$$R_i(T_s)q + My_s \geq g_{is} \quad i=1,k$$

$$\sum_{j=1}^{s} y_j \leq V \tag{17}$$

where M is a very large positive number, V is the maximum number of realizations allowed to be infeasible ($V = R_1 s$), y_j is equal to zero when all the constraints (15) for realization j are satisfied for all i and equal to one otherwise. The OF (1) and the constraints (17) form a mixed-integer problem with a not convex feasible domain. The authors suggest a special technique, called the Mixed Integer Chance Constrained Programming, that is a branch and bound technique specially suited for this problem.

4. THE MANAGEMENT MODEL AS A TOOL FOR THE MEASUREMENTS LOCATION

The groundwater quality management problem, when the natural parameters are assumed to be random quantities, can also be used to locate new measurements for a better calibration of the simulation model [18],[19].
The basic idea is to relate the number and the location of new field measurements with the goal of the management problem. Most of the pumping test design techniques are in fact based on the precision required 'a priori' to the state variables of the simulation model [20].
Tucciarelli and Pinder [19] assume that the transmissivities are piece-wise constant, and change the deterministic problem (1)-(3) in the chance-constrained form

$$\text{minimize } \textbf{cost } \textbf{q} + f(x^O) \tag{18}$$

$$\text{s.t. } c_i + F^{-1}(\pi) \sqrt{\text{diag}_i(\textbf{r}C_T\textbf{r}^t)} \le c_i^{max} \qquad i=1,k \tag{19}$$

where $f(x^O)$ is the cost of the measurement at location x^O, $F^{-1}(\pi)$ is the inverse of the standard normal cumulative distribution corresponding to reliability π, C_T is the covariance matrix of the log-transmissivities, $\textbf{r}$ is the sensitivity matrix, defined as

$$r_{ij} = T_j \frac{dc_i}{dT_j} \tag{20}$$

It can be shown that, according to the hypothesis of normal distribution of the unknown log-transmissivities and concentrations around the expected values, the square root on the left-hand side of (19) is the first order approximation of the concentration standard deviation at the control point i. Therefore, the optimal solution of (18)-(20) will satisfy the deterministic constraints (2) with reliability π. Assuming that the new measurement affects the second order moments of the transmissivities, without any meaningfull change of the simulation model state variables, it is possible to predict the change of the log-transmissivity covariance matrix C_T before performing the measurement. This allows to compare the optimal OF of several problems (18)-(20), each one run with a different C_T matrix.

CONCLUSIONS

The definition of the groundwater quality management problem and some of the most important techniques for its solution have been discussed. The most popular methodologies used to deal with the stochastic behaviour of the transmissivities and guarantee a given reliability to the optimal pumping strategy have been presented.

A very common criticism to the management models is that the uncertainty of the results can be reduced only with expensive field measurements. The idea to decide about new measurements according to the goal and the constraints of the management model has been introduced.

REFERENCES

1. Male, J.W. and F.A. Mueller: Model for prescribing ground-water use permits, J. Water Resour. Plann. Manag. Div. Am. Soc. Civ. Eng., 118(5), 543-561, 1992.

2. Lehr, J.H. and D.H. Nielsen: Aquifer restoration and groundwater rehabilitation, Groundwater, 20(6), 650-656, 1982.

3. Galeati, G. and G. Gambolati: Optimal dewatering schemes in the foundation design of an electronuclear plant, Water Resour. Res., 24(4), 541-552, 1988.

4. Wagner, J.M. U. Shamir and H.R. Nemati: Groundwater quality management under uncertainty: stochastic programming approaches and the value of information, Water Resour. Res., 28(5), 1233-1246, 1992.

5. Tiedeman, C., and S.M. Gorelick, Analysis of uncertainty in optimal groundwater contaminant capture design, Water Resour. Res., 29(7), 2139-2153, 1993.

6. Carrera, J. and S.P. Neuman: Estimation of aquifer parameters under transient and steady state conditions:

1. Maximum likelihood method incorporating prior information, Water Resour. Res., 22(2), 199-210, 1986.

7. Karatzas, G.P. ,T. Tucciarelli and G.F. Pinder: Groundwater quality management using numerical simulation and a primal optimization technique, Proceedings of Computational Methods in Water Resources X, Kluwer Academic Publisher, July 1994.

8. Culver, T.B. and C.A. Shoemaker: Dynamic optimal control for groundwater remediation with flexible management periods, Water Resour. Res., 28(3), 629-641, 1992.

9. Aguado E. and I. Remson: Groundwater management with fixed charges, Water Resour. Plann. Manag. Div. Am. Soc. Civ. Eng., 106(2), 375-382, 1980.

10. Karatzas, G.P. and G.F. Pinder: Groundwater management using numerical simulation and the outer approximation method for global optimization, Water Resour. Res., 29(10), 3371-3378, 1993.

11. Goldberg, D.E.: Genetic algorithms, Addison-Wesley, 1989.

12. Dougherty, D.E. and R.A. Marryott: Optimal groundwater management:1. Simulated annealing, Water Resour. Res., 27(10), 2493-2509, 1991.

13. Ranjithan S., J.W. Eheart and J.H. Garrett: Neural network-based screening for groundwater reclamation under uncertainty, Water Resour. Res., 29(3), 563-574, 1993.

14. McLaughlin, D. and E.F. Wood: A distributed parameter approach for evaluating the accuracy of groundwater model predictions. 1. Theory, Water Resour. Res., 24(7), 1048-1060, 1988.

15. Wagner, B.J. and S.M. Gorelick: Optimal groundwater quality management under parameter uncertainty, Water Resour. Res., 23(7), 1162-1174, 1987.

16. Wagner, B.J. and S.M. Gorelick, Reliable aquifer remediation in the presence of spatially variable hydraulic conductivity: from data to design, Water Resour. Res., 25(10), 2211-2225, 1989.

17. Morgan, D.R., J.W. Eheart and A.J. Valocchi: Aquifer remediation design under uncertainty using a new chance

constrained programming technique, Water Resour. Res., 29(3), 551-561, 1993.

18. Maddock, T.: Management model as a tool for studying the worth of data, Water Resour. Res., 9(2), 270-280, 1973.

19. Tucciarelli, T. and G. Pinder: Optimal data acquisition strategy for the development of a transport model for groundwater remediation, Water Resour. Res., 27(4), 577-588, 1991.

20. McCarthy, J.M. and W. W-G. Yeh: Optimal pumping design for parameter estimation and prediction in groundwater hydrology, Water Resour. Res., 26(4), 779-791, 1990.

SOME CONSIDERATIONS ABOUT UNIQUENESS IN THE IDENTIFICATION OF DISTRIBUTED TRANSMISSIVITIES OF A CONFINED AQUIFER

M. Giudici

University of Milan, Milan, Italy

and

G. Morossi

Consorzio Milano Ricerche, Milan, Italy

and

G. Parravicini and G. Ponzini

University of Milan, Milan, Italy

ABSTRACT

The identification of the transmissivity of a confined aquifer can be achieved by the solution of a generally ill-posed inverse problem when measurements of piezometric head and source term are available. Herewith some classical results are considered; the most promising approach consists of the simultaneous utilisation of several sets of data, namely piezometric heads and source terms relative to different steady hydraulic conditions of the aquifer. The main advantage of this approach is that the required data are the easiest to measure in hydrogeological field applications.

Two discrete formulations of the inverse problem, within the framework of finite difference schemes, are reviewed; the first one considers the internode transmissivities as the unknowns of the inverse problem, whereas the second considers the node transmissivities as unknowns. The relationship between internode and node transmissivities together with advantages and drawbacks of these two formulations of the inverse problem are discussed.

1. INTRODUCTION AND DEFINITIONS

The equation that governs the two-dimensional hydraulic flow through a heterogeneous isotropic confined aquifer in steady conditions under appropriate approximations (see, e. g., [1]) reads:

$$\operatorname{div}(T\,\operatorname{grad}H) = Q \tag{1.1}$$

where T is the hydraulic transmissivity of the aquifer, H is the piezometric head and Q is the source term. In the particular case of a one-dimensional domain the equation is reduced to:

$$D_x(TD_xH) = Q \tag{1.2}$$

where D_x denotes $\dfrac{d}{dx}$.

Equations (1.1) and (1.2) are called balance equations because they are obtained by using Darcy's law and by imposing the mass conservation principle.

Once T, Q and appropriate boundary conditions on H are given, the determination of the function H that satisfies the balance equation is called the forward problem which is the central point of a management model.

Note that appropriate boundary conditions on H (Dirichlet) are often available in real applications and that it is generally possible to obtain an evaluation for the source term Q. On the contrary the estimate of the phenomenological coefficient T is not easily obtained in field applications. There are several methods devoted to the determination of T, among which we recall slug tests, pumping tests and laboratory tests on samples. In the last twenty years researchers have also tried to determine the transmissivity by the solution of an inverse problem. For a thorough review on these subjects we refer the reader to [2] and [3].

The inverse problem, also called the identification problem, consists of determining T given H and Q such that H, T and Q satisfy the balance equation.

It is evident from equation (1.2) that the solution of the inverse problem in one-dimensional domains is given by the formula:

$$T(x) = \frac{C + \int Q(x)dx}{D_x H}. \tag{1.3}$$

where C is an integration constant to be determined in ways to be discussed later.

The computed $T(x)$ are inserted into management models which simulate scenarios for any Q and for any boundary conditions. Therefore to obtain meaningful forecasts with management models it is necessary that the inverse problem is well posed, that is to say that T must be determined uniquely and in a stable way. This is not the case since in general the inverse problem is an ill posed problem for several reasons:

1) with noisy data the existence of a solution is not ensured;

2) for the uniqueness of the solution additional information is generally required;

3) the transmissivity depends on the derivatives of H and not on H itself and this leads to instability.

In this paper we focus on the uniqueness of the solution for both the continuous case and the discrete case; we always assume that a solution to the inverse problem exists and we disregard the issue of instability.

We elaborate the continuous one-dimensional case for which we can compute the analytical solution of the problem. This fact allows for an easy examination of the numerical models which have to be used in real applications.

In section 2 uniqueness for the continuous case is discussed. In section 3 some finite difference schemes are introduced.

We stress that results useful in real applications are mainly numerical; thus some considerations on the applicability are discussed in the sections devoted to the discrete cases (sections 4 and 5).

2. UNIQUENESS FOR THE CONTINUOUS INVERSE PROBLEM

The problem of the uniqueness of the solution appears immediately from formula (1.3), in which case things are easily settled. In fact we have the ordinary differential equation of first order (1.2) whose solution $T(x)$ is determined unique once an initial value of transmissivity is assigned. The solution to equation (1.2) reads

$$T(x) = \frac{T(x_1)D_x H(x_1) + \int_{x_1}^{x} Q(x)dx}{D_x H}, \tag{2.1}$$

where $T(x_1)D_x H(x_1)$ is the flow density at the point x_1 belonging to the one-dimensional domain where the problem is defined. If $T(x_1)$ is known then the solution is found unique.

For two-dimensional domains the problem of uniqueness has been studied by many authors; it has been shown in [4], [5], [6] and [7] that the solution of the inverse problem in a two-dimensional domain is unique when the transmissivity is assigned on the inflow or outflow boundary surface of the domain or even at one point for each flow line. However in practice it is very difficult to obtain these Cauchy data, so it is desirable to use different information easier to achieve. The rest of this section deals with these issues.

2.1 Stationary transmissivity

In a one-dimensional domain if there exists a point x_1 where $D_x T(x_1) = 0$ and $Q(x_1) \neq 0$ then it is possible to compute $T(x_1)$ from equation (1.2) according to the formula

$$T(x_1) = \frac{Q(x_1)}{D_{xx} H(x_1)}. \tag{2.2}$$

Therefore the initial value for the Cauchy problem is determined and the solution is found unique all over the one-dimensional domain.

In a two-dimensional domain the knowledge of a point $\mathbf{x}_1$ where $\mathrm{grad}\, T(\mathbf{x}_1) = 0$ still allows to compute $T(\mathbf{x}_1)$ by means of equation (1.1) as:

$$T(\mathbf{x}_1) = \frac{Q(\mathbf{x}_1)}{D_{xx} H(\mathbf{x}_1) + D_{yy} H(\mathbf{x}_1)}, \tag{2.3}$$

provided that $Q(\mathbf{x}_1) \neq 0$.

The transmissivity is hence found unique along the flow line of H passing through the point $\mathbf{x}_1$; however this is clearly not enough to compute a unique solution over the whole domain.

2.2 Zero gradient point

If in a one-dimensional domain there exists a point x_1 where $D_x H(x_1) = 0$ the knowledge of T in x_1 is irrelevant and the solution is given by formula (2.1) as follows:

$$T(x) = \frac{\int_{x_1}^{x} Q(x)dx}{D_x H}.$$ \hfill (2.4)

In two-dimensional domains the knowledge of the existence and of the location of an isolated point where the gradient of the potential vanishes is relevant from a mathematical point of view. In fact it is possible to compute the transmissivity along all the flow lines starting from or converging to such point (see [5] and [8]). In real applications, however, such an information is seldom obtained, because it is very difficult to distinguish experimentally between very small gradients and vanishing gradients. This fact leads to instability in the transmissivity identification. For these reasons we do not deal with this technique in the discrete cases and in the applications.

2.3 The approach of multiple independent sets of data

We have outlined that the knowledge of T and of H necessary for the solution of the inverse problem, as treated so far, are difficult to obtain in field applications. We now deal with a way of solving the inverse problem using a kind of information easier to collect in field applications.

Suppose that two sets of steady-state data relative to the same one-dimensional aquifer - i. e., two piezometric heads $H^{(1)}, H^{(2)}$ and two source terms $Q^{(1)}, Q^{(2)}$ - are known. Then we can write the following system:

$$\begin{cases} D_x(T D_x H^{(1)}) = Q^{(1)} \\ D_x(T D_x H^{(2)}) = Q^{(2)} \,. \end{cases}$$ \hfill (2.5)

System (2.5) can be rewritten as a system of linear algebraic equations for the two unknowns $T(x)$ and $D_x T(x)$. In matrix form the following equation holds:

$$\begin{pmatrix} D_x H^{(1)}(x) & D_{xx} H^{(1)}(x) \\ D_x H^{(2)}(x) & D_{xx} H^{(2)}(x) \end{pmatrix} \cdot \begin{pmatrix} D_x T(x) \\ T(x) \end{pmatrix} = \begin{pmatrix} Q^{(1)}(x) \\ Q^{(2)}(x) \end{pmatrix}.$$ \hfill (2.6)

At a point x_0 where the following condition holds:

$$\det \begin{pmatrix} D_x H^{(1)}(x_0) & D_{xx} H^{(1)}(x_0) \\ D_x H^{(2)}(x_0) & D_{xx} H^{(2)}(x_0) \end{pmatrix} \neq 0$$ \hfill (2.7)

the solution of the system (2.6) is given by the formula:

$$\begin{pmatrix} D_x T(x_0) \\ T(x_0) \end{pmatrix} = \begin{pmatrix} D_x H^{(1)}(x_0) & D_{xx} H^{(1)}(x_0) \\ D_x H^{(2)}(x_0) & D_{xx} H^{(2)}(x_0) \end{pmatrix}^{-1} \begin{pmatrix} Q^{(1)}(x_0) \\ Q^{(2)}(x_0) \end{pmatrix}.$$ \hfill (2.8)

Once $T(x_0)$ has been computed, the initial datum for a Cauchy problem is determined and it is possible to compute the solution at every point of the domain. If condition (2.7) holds at every point of the domain then the system (2.6) is directly solvable and it is not necessary to integrate the balance equation (1.2) with a Cauchy datum.

In two-dimensional problems two sets of data are not enough to determine the three unknowns $T(\mathbf{x}_0)$, $D_x T(\mathbf{x}_0)$ and $D_y T(\mathbf{x}_0)$ at any point $\mathbf{x}_0$ of the domain; three sets of data are required which satisfy the following independence condition:

$$\det\begin{pmatrix} D_x H^{(1)} & D_y H^{(1)} & \Delta H^{(1)} \\ D_x H^{(2)} & D_y H^{(2)} & \Delta H^{(2)} \\ D_x H^{(3)} & D_y H^{(3)} & \Delta H^{(3)} \end{pmatrix}(\mathbf{x}_0) \neq 0. \tag{2.9}$$

Condition (2.9) renders the system

$$\begin{pmatrix} D_x H^{(1)} & D_y H^{(1)} & \Delta H^{(1)} \\ D_x H^{(2)} & D_y H^{(2)} & \Delta H^{(2)} \\ D_x H^{(3)} & D_y H^{(3)} & \Delta H^{(3)} \end{pmatrix}(\mathbf{x}_0) \cdot \begin{pmatrix} D_x T \\ D_y T \\ T \end{pmatrix}(\mathbf{x}_0) = \begin{pmatrix} Q^{(1)} \\ Q^{(2)} \\ Q^{(3)} \end{pmatrix}(\mathbf{x}_0). \tag{2.10}$$

uniquely solvable at the point $\mathbf{x}_0$ (see also [9]).

As in the one-dimensional case, we can uniquely determine the transmissivity in all the region of the domain where condition (2.9) holds by directly inverting system (2.10). In real applications it is difficult to verify whether condition (2.9) holds in the whole domain. In addition it is desirable to obtain information from the least possible experimental data. A new approach that takes into account the above arguments has been presented in [10] and [11] where the issue of determining a unique solution of the two-dimensional inverse problem from two sets of data and the value of transmissivity at a single point is analysed from both the mathematical and the numerical point of view.

The main result presented in those papers relies upon the fact that the system

$$\begin{cases} \operatorname{div}(T \operatorname{grad} H^{(1)}) = Q^{(1)} \\ \operatorname{div}(T \operatorname{grad} H^{(2)}) = Q^{(2)} \end{cases} \tag{2.11}$$

can be written in the normal form

$$\operatorname{grad} T(\mathbf{x}) = -T(\mathbf{x})\mathbf{a}(\mathbf{x}) + \mathbf{b}(\mathbf{x}) \tag{2.12}$$

where $\mathbf{a}(\mathbf{x})$ and $\mathbf{b}(\mathbf{x})$ are vector functions depending on the known functions $H^{(1)}(\mathbf{x})$, $H^{(2)}(\mathbf{x})$, $Q^{(1)}(\mathbf{x})$ and $Q^{(2)}(\mathbf{x})$. If the following independence condition holds

$$\left[(D_x H^{(1)})(D_y H^{(2)}) - (D_x H^{(2)})(D_y H^{(1)}) \right](\mathbf{x}) \neq 0 \tag{2.13}$$

at every point $\mathbf{x}$ of the domain and if the true value of the transmissivity at one point only of the domain is known then there is a unique solution of the inverse problem. For a rigorous proof of this theorem and for a thorough discussion of the applicability and other subjects we refer the reader to [10]; there it is shown that condition (2.13) is easy to check even in real cases when measurements of piezometric head are available at a limited number of points. Numerical case studies based on this approach have been developed and discussed in [11]; both noiseless and noisy data have been considered.

3. FINITE DIFFERENCES CONSERVATIVE SCHEMES

Let us consider a one-dimensional domain discretized with a regular lattice of N nodes (see figure 1); it is possible to write a discrete balance equation for each internal node starting with the integral form of equation (1.2) (see, e. g., [12]).

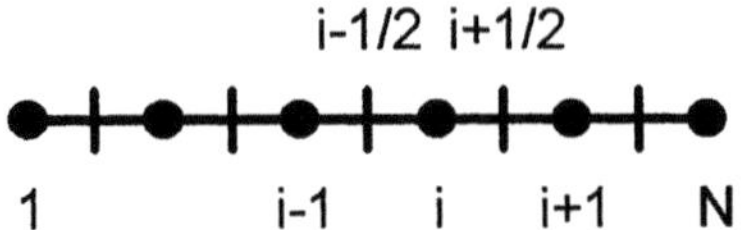

Figure 1: Discretized one-dimensional domain:
the bullets denote the nodes.

The discretized equation reads:

$$t_{i-1,i}\frac{H_{i-1}-H_i}{\Delta x}+t_{i,i+1}\frac{H_{i+1}-H_i}{\Delta x}=Q_i, \quad i=2,...,N-1 \tag{3.1}$$

where:

Δx is the distance between two adjacent nodes of the lattice;

$H_j, j=1,...,N$ are the piezometric heads at the N nodes;

$$Q_j = \int_{x_{j-1/2}}^{x_{j+1/2}} Q(x)dx, j=2,...,N-1 \text{ are the discrete source terms;}$$

$t_{j,j+1}, j=1,...,N-1$ are called internode transmissivities and are referred to the intervals $[x_j, x_{j+1}]$; under assumptions that are natural in the finite difference schemes, it is shown (see, e. g., [12]) that the internode transmissivities are related to the transmissivity function $T(x)$ by the relation

$$\frac{1}{t_{i,i+1}} = \frac{1}{\Delta x}\int_{x_i}^{x_{i+1}}\frac{1}{T(x)}dx. \tag{3.2}$$

Note that the internode transmissivities link linearly the discrete hydraulic gradients $(H_{i+1}-H_i)/\Delta x$ with the discrete flow densities $q_{i,i+1}$ between the nodes x_i and x_{i+1}; the relation

$$q_{i,i+1} = -t_{i,i+1}\frac{H_{i+1}-H_i}{\Delta x} \tag{3.3}$$

is the discrete counterpart of Darcy's law.

In a two-dimensional discrete domain (figure 2), fixed the indices i and j, the nodes (i,j), $(i,j+1)$, $(i,j-1)$, $(i-1,j)$ and $(i+1,j)$ are denoted with the capital letters C, N, S, W and E respectively, for the sake of shortness.

From the integral equation corresponding to (1.1) one obtains the following two-dimensional discrete balance equation for the node C:

$$\Delta x\left(t_{N,C}\frac{H_N-H_C}{\Delta y}+t_{S,C}\frac{H_S-H_C}{\Delta y}\right)+$$
$$\Delta y\left(t_{W,C}\frac{H_W-H_C}{\Delta x}+t_{E,C}\frac{H_E-H_C}{\Delta x}\right)=Q_C \tag{3.4}$$

where Q_C is the integral of the source term $Q(\mathbf{x})$ over the block centred in the lattice point C, and is hence given by the formula

$$Q_C = \int_{-\Delta x/2}^{+\Delta x/2} \int_{-\Delta y/2}^{+\Delta y/2} Q(x_C + \xi, y_C + \eta)\, d\xi\, d\eta \ . \tag{3.5}$$

Note that equation (3.4) describes not only the case of heterogeneous isotropic aquifer but also the case of heterogeneous anisotropic aquifer when the reference axes are oriented along the principal directions of anisotropy all over the domain.

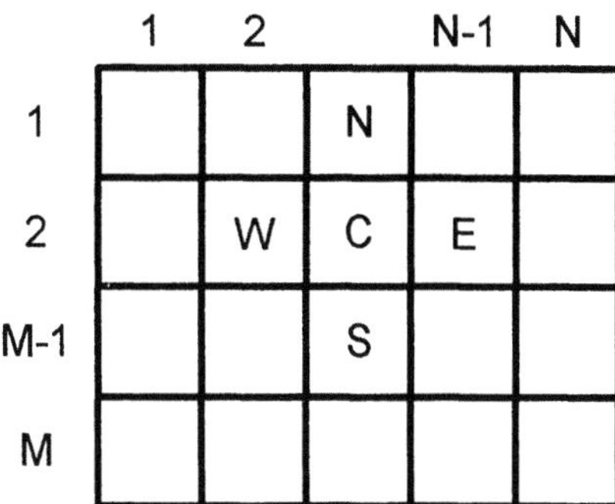

Figure 2: Two-dimensional discrete domain; the lattice nodes
are located at the centre of each square.

4. THE DISCRETE FORWARD AND INVERSE PROBLEMS

Equations (3.1) and (3.4) can be cast in the synthetic form:
$$\mathbf{AH} = \mathbf{Q} + \mathbf{B} \tag{4.1}$$
where $\mathbf{A}$ is the coefficient matrix whose elements linearly depend on the internode transmissivities, $\mathbf{H}$ is the vector of the piezometric heads, $\mathbf{Q}$ is the vector of the source terms and $\mathbf{B}$ is a vector linked to the boundary conditions.

As an example for the one-dimensional case, if $N=4$ and the value of the piezometric heads is known and fixed at the two boundary nodes, we can write the following system:

$$\begin{aligned}
t_{1,2}\,\frac{H_1 - H_2}{\Delta x} + t_{2,3}\,\frac{H_3 - H_2}{\Delta x} &= Q_2 \\[2mm]
t_{2,3}\,\frac{H_2 - H_3}{\Delta x} + t_{3,4}\,\frac{H_4 - H_3}{\Delta x} &= Q_3
\end{aligned} \tag{4.2}$$

or else:

$$\begin{pmatrix} -(t_{1,2}+t_{2,3}) & t_{2,3} \\ t_{2,3} & -(t_{2,3}+t_{3,4}) \end{pmatrix} \begin{pmatrix} H_2 \\ H_3 \end{pmatrix} = \Delta x \begin{pmatrix} Q_2 \\ Q_3 \end{pmatrix} - \begin{pmatrix} H_1\, t_{1,2} \\ H_4\, t_{3,4} \end{pmatrix} . \tag{4.3}$$

By comparing equation (4.3) with equation (4.1) it is apparent how $\mathbf{A}$, $\mathbf{Q}$ and $\mathbf{B}$ are related to the internode transmissivities, the source terms and the boundary conditions.

Given the internode transmissivities (and hence given $\mathbf{A}$), the discrete source terms $\mathbf{Q}$ and the boundary conditions (and hence given $\mathbf{B}$) the discrete forward problem consists of finding $\mathbf{H}$ that satisfies system (4.1).

Discrete sources terms, **Q**, and Dirichlet boundary conditions on **H** are often available in real case studies. Therefore the main obstacle in the determination of **H** in a unique way is the knowledge of the internode transmissivities. They can be determined by solving a discrete inverse problem in analogy with the continuous case.

Once the piezometric heads **H** and the source terms **Q** have been given, we define the discrete inverse problem of type 1 as the determination of the internode transmissivities that satisfies system (4.1).

As the continuous inverse problem, the discrete inverse problem of type 1 is generally an ill posed problem, due to the fact that the uniqueness of the solution is seldom granted; the difficulty in assigning boundary conditions on the transmissivity in real case studies causes the linear algebraic system (4.1) to be underdetermined. We will examine this issue and some techniques to overcome the underdetermination in the next section.

5. UNDERDETERMINATION OF THE DISCRETE INVERSE PROBLEM

As the discrete inverse problem consists of determining the unknown internode transmissivities by the solution of the linear algebraic system (4.1), it is necessary to count the number of equations and of unknowns.

The number of equations N_e is equal to the number of nodes for which it is possible to write the discrete mass balance; in the case of Dirichlet boundary conditions on the whole boundary N_e is equal to the number of internal nodes. The number of unknowns N_u is given by the number of internode transmissivities appearing in system (4.1).

For a one-dimensional discrete domain with N nodes (see figure 1) N_e and N_u are given by the relations:
$$N_e = N - 2, \; N_u = N - 1. \tag{5.1}$$
For a two-dimensional discrete domain with $N \times M$ nodes (see figure 2) N_e and N_u are given by the relations:
$$\begin{aligned} N_e &= (N-2)(M-2), \\ N_u &= 2NM - 3(N+M) + 4. \end{aligned} \tag{5.2}$$
While in the one-dimensional discrete case the number of unknowns exceeds the number of equations always by one, in the two-dimensional case the discrete inverse problem is generally highly underdetermined. It is therefore necessary either to reduce the number of unknowns or to increase the number of equations in order to obtain a unique solution.

5.1 Reducing the number of unknowns

Different approaches have been analysed in the current literature to reduce the number of unknowns. We shall briefly examine the most common ones.

The approach of stationary transmissivity presented in subsection 2.1 for the continuous case has been transferred to the discrete case with the approach called zonation (see, e. g., [2] and [13]).

Zonation is based on the introduction of additional information on the spatial distribution of the transmissivity. The aquifer is divided into a number of subdomains or zones, each of them characterised by a constant value of transmissivity. Thus the number of unknowns becomes equal to the number of zones. Such information about transmissivity can be derived either from the geological surveys of the aquifer or from remote geophysical prospecting, vertical electrical soundings or other techniques. In real case studies, however, both geological and geophysical methods provide only rough indications about transmissivity and this reduces the applicability of zonation.

Another widely used technique consists of considering the node transmissivities instead of the internode transmissivities as the unknowns of the inverse problem. In the classical finite difference approach the internode transmissivity $t_{1,2}$ is either the arithmetic mean or the harmonic mean of the transmissivities T_1 and T_2 of two adjacent nodes, that is to say it is given by the relation

$$t_{1,2} = \frac{T_1 + T_2}{2} \quad \text{(arithmetic mean)} \tag{5.3}$$

or else by the relation

$$t_{1,2} = \frac{2T_1 T_2}{T_1 + T_2} \quad \text{(harmonic mean).} \tag{5.4}$$

Equations (5.3) and (5.4) are not the only possible choices; in general we can express the relationship between node transmissivities and internode transmissivities as $t_{1,2} = f(T_1, T_2)$ provided that the function f is continuous, positive and satisfies the conditions: $f(T_1, T_2) = f(T_2, T_1)$, $f(T, T) = T$.

The discrete balance equation (3.4) becomes:

$$\Delta x \left(f(T_N, T_C) \frac{H_N - H_C}{\Delta y} + f(T_s, T_C) \frac{H_s - H_C}{\Delta y} \right) +$$
$$\Delta y \left(f(T_W, T_C) \frac{H_W - H_C}{\Delta x} + f(T_E, T_C) \frac{H_E - H_C}{\Delta x} \right) = Q_C \ . \tag{5.5}$$

The same equation should still be written for each internal node of the lattice thus giving rise to a system of equations similar to system (4.1).

It is now possible to provide a different definition of the inverse problem that we call inverse problem of type 2. This consists of determining the <u>node</u> transmissivities that satisfy equation (5.5) when the piezometric heads **H**, the discrete source terms **Q** and the function f have been assigned.

We stress that in real hydrogeological applications it is an almost impossible task to know which function f best approximates the real internode transmissivity especially when different flow conditions have to be modelled. The choice of that function is then arbitrary and its introduction causes a further approximation for the simulation of the flow.

It is right and proper to add that only when the function f is equal to the arithmetic mean (5.3) the inverse problem of type 2 is linear.

The number of unknowns for the inverse problem of type 2 is simply given by the number of node transmissivities appearing in the system of balance equations (5.5).

For a one-dimensional discrete domain with N nodes (see figure 1) N_e and N_u appearing in the system (5.5) of balance equations are given by the relations:

$$N_e = N - 2, \quad N_u = N. \tag{5.6}$$

In the case of a two-dimensional heterogeneous isotropic domain with $N{\times}M$ nodes (see figure 2), N_e and N_u are given by the relations:

$$N_e = (N-2)(M-2),$$
$$N_u = NM - 4 \, . \tag{5.7}$$

For a two-dimensional discrete domain with $N{\times}M$ nodes (see figure 2), in the case of heterogeneous anisotropic aquifer with the reference axes oriented along the principal directions of anisotropy, if the principal directions of anisotropy have a constant orientation, N_e and N_u are given by the relations:

$$N_e = (N-2)(M-2),$$
$$N_u = 2(N-2)(M-2) + 2(N+M-4) \, . \tag{5.8}$$

In fact for this case to each internal node of the discrete lattice there are associated the transmissivities referred to the two directions x and y. For the boundary nodes only the transmissivity referred to the direction pointing toward the internal part of the lattice is requested.

It is important to note that the definition of the inverse problem of type 2 increases the number of unknowns with respect to that of type 1 for one-dimensional domains and for two-dimensional domains in the case of heterogeneous anisotropic aquifer.

Only in the case of heterogeneous isotropic aquifer the number of unknowns in the inverse problem of type 2 is reduced with respect to that of type 1. Nevertheless, also in this case, the inverse problem of type 2 is still highly underdetermined and additional information is therefore necessary.

5.2 Increasing the number of equations

For the purpose of increasing the number of equations it is always necessary to obtain additional information about the real system. Of course it is desirable to use that information which is as simple as possible to collect.

If the flow through the boundary is known, additional discrete balance equations can be considered. However this information is seldom obtained in field applications, but for the case of an aquifer whose transmissivity at the border is so low that the flow through the boundary can be approximated as zero. This is the case, for instance, if we survey outcrops of thick impervious geological formations. However these geological situations are seldom met in real case studies and moreover they are not enough to ensure the uniqueness of the solution to the inverse problem.

Our approach aims to increase the number of balance equations through the use of several independent sets of data. This idea was originally proposed in [14] for the steady state case and in [9], [13], [15] and [16] for both steady and transient conditions.

Let us remark that in hydrogeological practice the piezometric heads, the sinks due to pumping wells, the levels of lakes and rivers - which make up the boundary condition for the aquifer - and the ratios of the leakage rates are quantities that are measurable in a relatively easy way.

Next it is necessary to understand how several data sets may be obtained. Considering the system of equations (4.1) it is apparent that, given $\mathbf{A}$, the piezometric heads $\mathbf{H}$ may be modified either by changing $\mathbf{Q}$, due to a change in the flow rates from the wells or to a change in the leakage, or by changing $\mathbf{B}$, due to a change in the boundary conditions.

It is therefore possible to observe different states of the aquifer induced artificially by modifying the flow rates from wells or naturally by a change in the boundary conditions or in the leakage. We must stress that the influence of a well does not exceed few decametres and that the leakage is not easily measurable; however different steady state conditions of the aquifer may happen in long drought or rain spells and in flood or drought regime of the lakes and rivers that make up the boundary conditions of the aquifer.

Somebody could rise the question whether it is reasonable to observe steady state conditions in real aquifers. With respect to this objection we may refer the reader to [1] and quote [17] where the authors reported their experience and concluded that "the pumping rate from an aquifer is momentarily stabilized or is increasing slowly, and a steady rate may be approximately observed. Or if shallow aquifers are considered, the yearly fluctuation brings the water state to a low level at which steady state is approximately observed". Moreover steady state conditions may also be met in aquifers where the sources are due to an intensive irrigation schedule which may happen in rice cultivation.

Let us now suppose that two sets of data have been measured. The number of balance equations is doubled, while the number of unknowns is unchanged.

In a one-dimensional domain the problem is overdetermined as the number of equations and of unknowns is now given by the relations $N_e = 2(N-2)$, $N_u = N-1$.

In a two-dimensional domain the number of unknowns $N_u = 2NM - 3(N+M) + 4$ still exceeds the number of equations $N_e = 2(N-2)(M-2)$ and more information is necessary. Yet in reference [11] a technique to determine a unique solution to the discrete inverse problem starting from two sets of data and the value of the transmissivity given at one point only of the domain is presented.

CONCLUSIONS

Some techniques leading to the uniqueness of the solution of the inverse problem have been reviewed. Among these we sketched how and why the several sets of data approach leads to the uniqueness of the solution of the inverse problem with the minimum information on transmissivity.

Actually the direct measurements of transmissivity or information on its spatial distribution are difficult and expensive to obtain with the up to date techniques. For this reason we think that the several independent sets of data approach, based on measurements of piezometric heads and of source terms, is a solid step for the solution of real world aquifer management problems.

REFERENCES

1. Bear, J.: Dynamics of fluids in porous media, American Elsevier, New York 1972.
2. Yeh, W.-G.W.: Review of parameter identification procedures in groundwater hydrology: the inverse problem, Water Resour. Res., 22(1986), 95-108.
3. Carrera, J.: State of art of the inverse problem applied to the flow and solute transport equations, in Groundwater flow and quality modeling (Eds. E. Custodio et al.), Reidel, Dordrecht 1988, 549-583.

4. Chavent, G.: Analyse fonctionnelle et identification de coefficients répartis dans les équations aux dérivées partielles, Thése d'Etat, Facultè des Sciences de Paris, 1971.

5. Richter, G.R.: An inverse problem for the steady state diffusion equation, SIAM J. Math. Anal., 41(1981), 210-221.

6. Richter, G.R.: Numerical identification of spatially varying diffusion coefficient, Mathematics of Computation, 36(1981), 375-386.

7. Emsellem, Y. and G. de Marsily: An automatic solution for the inverse problem, Water Resour. Res., 7(1971), 1264-1283.

8. Chicone, C. and J. Gerlach: A note on the identifiability of distributed parameters in elliptic equations, SIAM J. Math. Anal., 18(1987), 1378-1384.

9. Sagar, B., Yakowitz, S. and L. Duckstein: A direct method for the identification of the parameters of dynamic nonhomogeneous aquifers, Water Resour. Res., 11(1975), 563-570.

10. Parravicini, G., Giudici, M., Morossi, G. and G. Ponzini: Minimal assignment of phenomenological coefficients and uniqueness for an inverse problem, Preprint IFUM 480/FT, Dipartimento di Fisica, Università di Milano, Milan, Italy, (1994).

11. Giudici, M., Morossi, G., Parravicini, G. and G. Ponzini: A new method for the identification of distributed transmissivities, Preprint IFUM 481/FT, Dipartimento di Fisica, Università di Milano, Milan, Italy, (1994).

12. Samarskij, A. and V. Andreev: Méthodes aux différences pour equation elliptiques, Mir, Moskow 1978.

13. Carrera, J. and S.P. Neuman: Estimation of aquifer parameters under transient and steady-state conditions: 3. application to synthetic and field data, Water Resour. Res., 22(1986), 228-242.

14. Scarascia, S. and G. Ponzini: An approximate solution for the inverse problem in hydraulics, L'Energia Elettrica, 49(1972), 518-531.

15. Carrera, J. and S.P. Neuman: Estimation of aquifer parameters under transient and steady-state conditions: 1, maximum likelihood method incorporating prior information, Water Resour. Res., 22(1986), 199-210.

16. Carrera, J. and S.P. Neuman: Estimation of aquifer parameters under transient and steady-state conditions: 2. uniqueness, stability and solution algorithms, Water Resour. Res., 22(1986), 211-227.

17. Emsellem, Y. and G. de Marsily: Reply to "Comments on 'An authomatic solution for the inverse problem' by Y. Emsellem and G. de Marsily" by D. Kleinecke, Water Resour. Res., 8(1972), 1130-1131.

VERIFICATION OF ACTIVE AND PASSIVE GROUND-WATER CONTAMINATION REMEDIATION EFFORTS

M.J. Barcelona

University of Michigan, Ann Arbor, MI, USA

ABSTRACT

The verification of ground-water contamination remediation efforts requires thorough documentation of subsurface conditions before, during and after cleanup efforts have ceased. The documentation include include proof of: reduction of risk to human or environmental health, achievement of regulatory cleanup concentration goals in soil, gas or liquid media, or verification of continued approach to background environmental quality conditions. Meeting any one or all of these requirements calls for a comprehensive approach to the design and operation of remediation efforts with an emphasis on the monitoring of environmental conditions. These tasks are most challenging for in-situ remediation efforts which employ active (i.e., pumping or vacuum application) rather than passive (i.e., natural water and vapor gradient) conditions.

1. INTRODUCTION

The practice of site characterization for potential organic contaminants has evolved slowly in the past decade. Early guidelines [1,2,3], for minimal ground-water contamination detection monitoring (i.e., monitoring wells upgradient and downgradient) have been applied to many sites of potential concern from detection through remedial action selection phases.

This minimal approach has often been applied regardless of the physicochemical characteristics of contaminant mixtures or the complexity of the hydrogeologic setting. For soluble inorganic constituents, this approach may be adequate for detection. Assessment efforts require substantially more comprehensive approaches. For organic contaminant detection and assessment (i.e., determination of the nature and extent of contamination) efforts, wells alone have been found to be inadequate monitoring tools. This paper focuses on the monitoring needs for remediation principally by in-situ biological methods for volatile organic compounds present in hydrocarbon fuels and organic solvents. Recognition of the value of subsurface soil vapor surveys for volatile organic components of fuel and solvent mixtures have generated a flurry of modified, monitoring well-based site characterization approaches [4]. However, these approaches to site characterization and monitoring network design suffer from the failure to identify the total mass of contaminant in the subsurface for three main reasons.

First, although volatile organic compounds (VOC's) are mobile in ground-water and frequently early indicators of plume movement [5], their detection in vapor or well samples and apparent aqueous concentration distribution does not identify the total mass distribution of organic contaminant [6]. Secondly, efforts to correlate observed soil vapor or ground water VOC concentrations with those in subsurface solid cores have often been unsuccessful. This is because current bulk jar collection/refrigeration at 4°C guidelines for solid core samples for VOC analyses lead to gross negative errors [7]. Thirdly, "snapshots" (i.e, one-time surveys) of background and disturbed ground-water chemistry conditions have been interpreted as "constant" ignoring temporal variability in subsurface geochemistry.

The result of the slow improvement in site-characterization and monitoring practices has often been the very low probability detection of the source of mobile organic contaminants. This outcome may be followed by the misapplication of risk-assessment and remediation models.

Nonetheless, there exist good reasons for a more optimistic view for the future reliability of site characterization and monitoring efforts.

2.   ACTIVE AND PASSIVE REMEDIATION APPROACHES

Active in-situ remediation efforts generally involve the control of subsurface ground-water or vapor flow. Also they rely on the application of a suite of physical, chemical, and microbiological processes to destroy or transform contaminants to less harmful or less mobile chemical constituents. The most effective active remediation schemes sustain hydraulic (ground-water) or pneumatic (vapor) control within the zone of treatment. This facilitates contaminant removal or transformation but calls for careful design of an active remediation-based monitoring system. Such levels of control have most often been achieved in "closed-loop" treatment designs where extracted fluids are returned to the subsurface treatment zone. In these instances it is necessary to monitor both the process stream and in-situ environmental conditions for concentrations of parent compounds and transformation products. Linked to net flow and the volume of the treatment zone, these monitoring data provide the basis for estimates of the net removal/transformation of the original contaminants. The reliability of these data is critical to the verification of cleanup performances.

Passive remediation efforts rely on intrinsic biological and/or chemical processes to mediate the destruction or transformation of contaminants. Though they may take more time to achieve acceptable levels of contaminant removal than active methods, the existing monitoring design from detective or assessment phases of the project may need only slight modification as to sampling location, frequency and selection of monitoring parameters. This approach may significantly reduce the cost of remediation.

The shortcomings of previous contaminant detection and assessment monitoring efforts have been recognized. New guidelines and recommendations on network design and operations will lead to more comprehensive, cost-effective site characterization [7,9] in general. Also, excellent reviews of characterization and long term monitoring needs and approaches in support of in-situ remediation efforts should guide us in this regard [10,11]. Site characterization efforts provide a basis for long term monitoring design and actually continue throughout the life of a remediation project.

Active and passive in-situ bioremediation approaches have been applied frequently to subsurface cleanups of organic contaminants (e.g. fuels, solvents, pesticides, etc.). The monitoring measures for verification of bioremediation performances have been identified as: the documented decline of contaminant concentrations, identification of favorable conditions (e.g. substrate, nutrient, pH, electron-acceptors) for microbial activity, demonstration of an active microbial population capable of transforming the major contaminants, and the identification of intermediate break-down or end-products in the subsurface. The supply of suitable electron acceptors (e.g. O_2, NO_3^-, FeIII, $SO_4^=$, etc.) may be the crucial element in successful in-situ remediation efforts.

While these measures are necessary, they are not sufficient to establish the remedial effectiveness or performance of in-situ methods. The minimal measures noted above must be integrated into a mass-balance for contaminants and transformation products. A number of inorganic and organic indicators of subsurface transformation can be used to permit the approach to mass balance for specific organic contaminants. Table 1 shows various example monitoring indicators appropriate for solvent and fuel contamination situations where transformations occur under known limits of oxidation-reduction conditions.

General Conditions	Contaminant Mixture	Inorganic	Organic General	Organic Specific
aerobic (oxic)	Gasoline (Benzene, Toluene, Xylene, Alkylbenzenes	CO_2 O_2 NO_3^- NO_2^- FeII	low- molecular wt. organic acids	aromatic acids
anaerobic (anoxic)	Tetrachloro- ethylene Trichloro- ethylene	CO_2 NH_3 FeII	low- molecular wt. organic acids CH_4-methane	Trichloro- ethylene, Trichloro- ethane Dichloro- ethylenes Vinyl chloride C_2H_4-ethylene C_2H_6-ethane

TABLE 1 General Monitoring Indicators for Organic Contaminant Mixtures

It should be noted that subsurface redox conditions are not in chemical equilibrium and that transitional environments exist where intermediate transformation product stability may be significant [8].

The use of these indicators along with monitoring the concentration of the original compounds provides a more comprehensive approach to verifying remediation performance. There are relatively few examples of the mass balance approach, since there may be multiple pathways for field microbial transformation and the reaction products may be unknown. The pathways for microbial transformation are being delimited by a combination of field and laboratory experiments. Also the suite of reaction products are being determined by advanced analytical methods which will support the mass balance approach.

3.   ADVANCED SITE CHARACTERIZATION AND MONITORING

How do we proceed to estimate the potential for subsurface intrinsic bioremediation success and track its performance into the future? Clearly, we should seek to design technically-defensible characterization and monitoring networks which will provide reasonable estimates of the in-place contaminant distribution over time. A dynamic, ongoing site characterization effort therefore includes objectives to:

1)   identify the spatial distribution of contaminants, particularly their relative fractionation in subsurface solids, water, and vapor, along potential exposure pathways recognizing that the mass of contaminants frequently resides in the solids;
2)   determine the corresponding spatial distribution of total organic matter since overall microbial activity and disruptions in subsurface geochemical conditions (and bioremediation indicators) are due to the total mass of reactive organic carbon;
3)   estimate the temporal stability of hydrogeologic and geochemical conditions which may favor microbial transformations in background, source and downgradient zones during the first year of characterization and monitoring;
4)   derive initial estimates of net microbial transformations of contaminant-related organic matter over time which may be built into the long-term monitoring network design.

The first three objectives establish the environment of major contamination and the conditions under which bioremediation may occur. The latter two objectives are vitally important since the evaluation of the progress of intrinsic bioremediation processes depends on distinguishing compound "losses" due to dilution, sorption and chemical reactions from microbial transformations. This approach has been suggested emphatically by Wilson [10] and was recently developed into a technical U.S. Air Force (USAF) protocol by Wiedemeier, *et al.* [11].

The latter reference focuses directly on the implementation of intrinsic remediation for dissolved fuel contamination in ground water. The general approach is shown in Figure 1 which has been modified from the original work. The USAF Protocol [11] has as its goals the collection of data necessary to support:

1) Documented loss of contaminants of the field scale,
2) The use of chemical analytical data in mass balance calculations, and
3) Laboratory microcosm studies using aquifer samples collected from the site.

These data, if collected in three dimensions for an extended period time should be sufficient to successfully implement intrinsic remediation [12]. The data collected in the initial site characterization effort (Figure 1) support the development of a site-specific conceptual model. This model is a three dimensional representation of the ground water flow and transport fields based on geologic, hydrologic, climatologic, and geochemical data for a site. The conceptual model, in turn, can be tested, refined and used to determine the suitability of intrinsic remediation as a risk-management strategy. The validity of the conceptual model as a decision tool depends on the complexity of the actual hydrogeologic setting and contaminant distributions relative to the completeness of the characterization database. The USAF Protocol is quite comprehensive in identifying important parameters, inputs and procedures for data collection and analysis. The major categories of necessary data are listed in Table 2 from the USAF Protocol [11]. Detective monitoring datasets available prior to in-depth site characterization are more likely to contain contaminant-related information rather than the three-dimensional aquifer property, hydrogeologic or geochemical data needed to formulate a conceptual model. A recognition of the variability inherent in these parameter distributions is critical to site-characterization efforts.

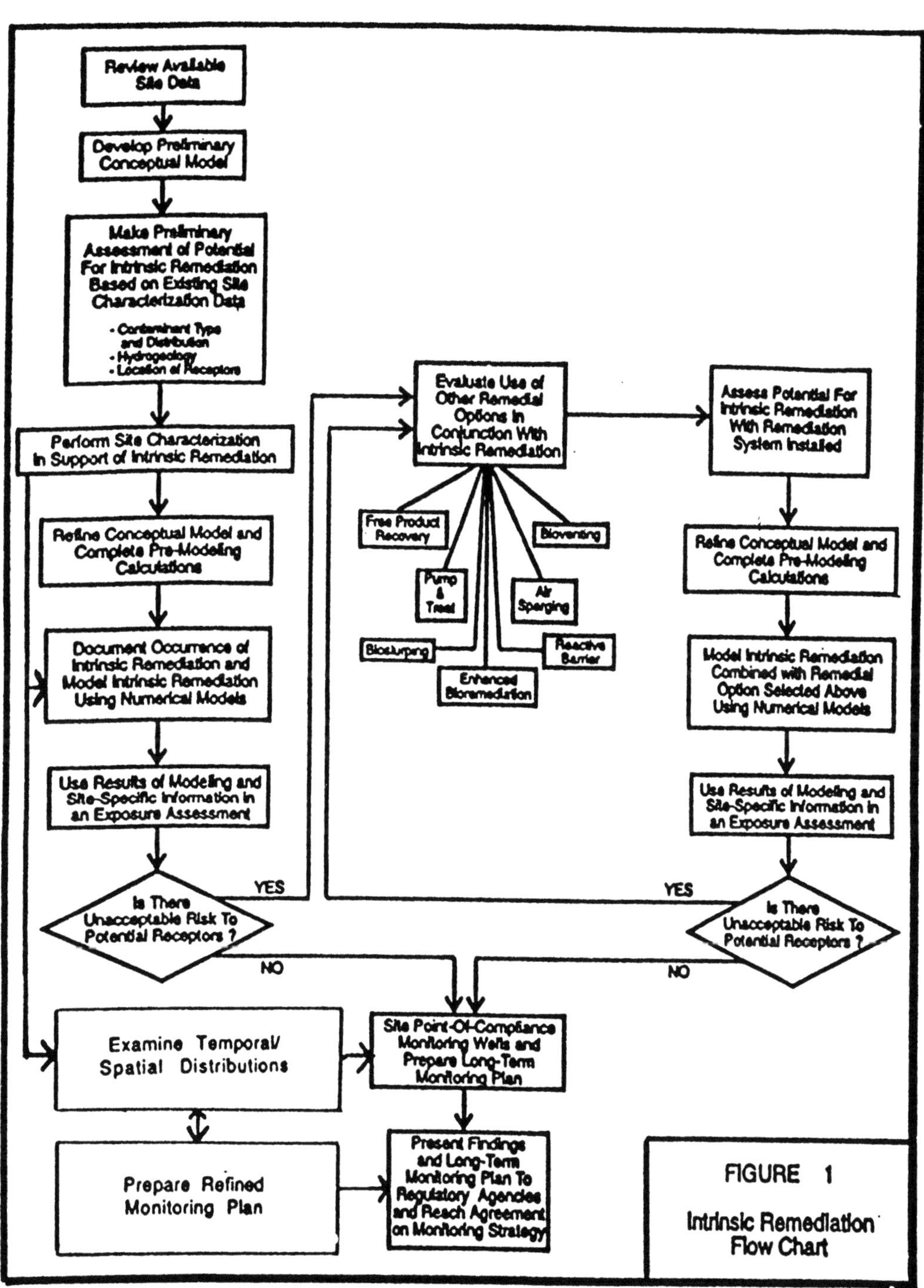
Review Available Site Data
Develop Preliminary Conceptual Model
Make Preliminary Assessment of Potential For Intrinsic Remediation Based on Existing Site Characterization Data
- Contaminant Type and Distribution
- Hydrogeology
- Location of Receptors
Perform Site Characterization In Support of Intrinsic Remediation
Refine Conceptual Model and Complete Pre-Modeling Calculations
Document Occurrence of Intrinsic Remediation and Model Intrinsic Remediation Using Numerical Models
Use Results of Modeling and Site-Specific Information in an Exposure Assessment
Is There Unacceptable Risk To Potential Receptors ?
YES
NO
Evaluate Use of Other Remedial Options In Conjunction With Intrinsic Remediation
Free Product Recovery
Bioventing
Pump & Treat
Air Sparging
Bioslurping
Reactive Barrier
Enhanced Bioremediation
Assess Potential For Intrinsic Remediation With Remediation System Installed
Refine Conceptual Model and Complete Pre-Modeling Calculations
Model Intrinsic Remediation Combined with Remedial Option Selected Above Using Numerical Models
Use Results of Modeling and Site-Specific Information in an Exposure Assessment
Is There Unacceptable Risk To Potential Receptors ?
YES
NO
Examine Temporal/ Spatial Distributions
Site Point-Of-Compliance Monitoring Wells and Prepare Long-Term Monitoring Plan
Prepare Refined Monitoring Plan
Present Findings and Long-Term Monitoring Plan To Regulatory Agencies and Reach Agreement on Monitoring Strategy
FIGURE 1
Intrinsic Remediation Flow Chart

FRACTIONATION AND SPATIAL EXTENT OF CONTAMINATION

1. Extent and type of soil and ground water contamination
2. Location and extent of contaminant source area(s)
   (i.e., areas containing free- or residual-phase product)
3. The potential for a continuing source due to leaking tanks or
   pipelines

HYDROGEOLOGIC AND GEOCHEMICAL FRAMEWORK

4. Ground water geochemical parameter distributions (Table 3)
5. Regional hydrogeology including:
   - Drinking water aquifers and
   - Regional confining units.
6. Local and site-specific hydrogeology, including:
   - Local drinking water aquifers;
   - Location of industrial, agricultural, and domestic water
   wells;
   - Patterns of aquifer use;
   - Lithology;
   - Site stratigraphy, including identification of transmissive
   and nontransmissive units;
   - Grain-size distribution (sand vs. silt vs. clay);
   - Aquifer hydraulic conductivity determination and
   estimates from grain-size distributions;
   - Ground water hydraulic information;
   - Preferential flow paths;
   - Location and type of surface water bodies; and
   - Areas of local ground water recharge and discharge.
7. Definition of potential exposure pathways and receptors.

TABLE 2
Site Specific Parameters to be Determined during Site Characterization
(modified from Reference 11)

3.1   SAMPLING IN SPACE

The initial site characterization phase should be designed to provide spatially dense coverage of critical data over volumes corresponding to ten to one-hundred year travel times along ground water flow paths. The "volume-averaged" values of the contaminants, hydrogeologic and geochemical parameters within zones along the flow path(s) should be derived from large enough datasets to permit estimation of statistical properties (e.g., mean, median, correlation distance, variance, etc.). Specifically, this means that the datasets for derived mass loadings of contaminants, aquifer properties, and geochemical constituents (Table 3) derived from spatial averages of data points must include approximately 30 or more data points [13,14,15]. Indeed, this minimum dataset size strictly applies to points in a plane. Two major decisions which must be made with regard to how spatially averaged masses of contaminants, electron donors (e.g, organic carbon, Fe^{2+}, $S^=$, NH_3, etc.) and electron acceptors (e.g, O_2, NO_3^-, NO_2^-, Fe and Mn oxides, $SO_4^=$, etc.) are to be estimated.

The first question deals with identification of the media in which the bulk of the constitutent's mass resides. For aquifer properties (e.g., grain size, laboratory estimates of hydraulic conductivity, etc.) the answer is simple. In this case, the solids are clearly the media of interest. For constituents particularly VOC's which are sparingly water soluble, the bulk of the contaminant mass may in fact reside in the solids though both solids and water samples must be collected carefully.

The second question pertains to the depth interval over which "planar" data points may be averaged. With fuel-related aromatic contaminants the depth interval above and below the capillary fringe/water table interface typically exhibits order of magnitude solid-associated concentration differences. This type of situation is typified by the BTEX data shown in Figure 2 for a fire training area at the recently decommissioned Wurtsmith AFB near Oscoda, MI. In this case the bulk of the contaminant mass along the axis of a dissolved BTEX plume with concentrations less than 1000 µg/L resides in aquifer solids below the watertable. In this situation, averaging data points over depths of > 0.5m could easily lead to order of magnitude errors in estimated masses for a site. Continuous coring of subsurface soids and close interval (i.e., < 1m) sampling of water should be considered in many VOC investigations. In order to approach this level of depth detail in sampling, the use of "push" technologies and/or multilevel sampling devices present very useful tools for site characterization.

CONTAMINATION AREA	APPARENT/ GEOCHEMICAL REDOX ZONE	CONTAMINANT MIXTURE	INORGANIC CONSTITUENTS	INTRINSIC CONSTITUENTS
SOURCE	REDUCING ANOXIC	FUELS CHLORINATED SOLVENTS	O_2, CO_2, H_2S; pH Fe^{2+}, $HS^-/S^=$, NO_2^-, NH_3, ALKALINITY	ORGANIC CARBONS, CH_4 ORGANIC ACIDS PHENOLS AS ABOVE AND: CHLORINATED METABOLITES ETHYLENE, ETHANE
DOWNGRADIENT	TRANSITIONAL/ SUBOXIC	FUELS CHLORINATED SOLVENTS	O_2, CO_2, H_2S; pH, Fe^{2+} ALKALINITY, NO_2^-, NO_3^-, NH_3, $HS^-/S^=$	ORGANIC CARBON, CH_4 ORGANIC ACIDS PHENOLS AS ABOVE AND: CHLORINATED METABOLITES ETHYLENE, ETHANE
UPGRADIENT/FAR-FIELD DOWNGRADIENT	OXIC	FUELS CHLORINATED SOLVENTS	O_2, CO_2, H_2S ALKALINITY, Fe^{2+}, NO_3^-, NO_2^-, NH_3	ORGANIC CARBON, CH_4 ORGANIC ACIDS PHENOLS AS ABOVE AND: CHLORINATED METABOLITES ETHYLENE, ETHANE

TABLE 3

Target Constituents for Site Characterization in Support of Intrinsic Bioremediation

Fire Training Area - BTEX Plume (µg/kg)
Distance From Water Table (feet)
5.0
2.5
0.0
-2.5
-5.0
-7.5
-10.0
-12.5
-15.0
X - Distance Along Transect (feet)
0
100
200
300
400
500
600
700
800
900
1000
1100
2500
2000
1500
1000
500
2500
3000
3500
4000
4500
5000
5500
1000
2000

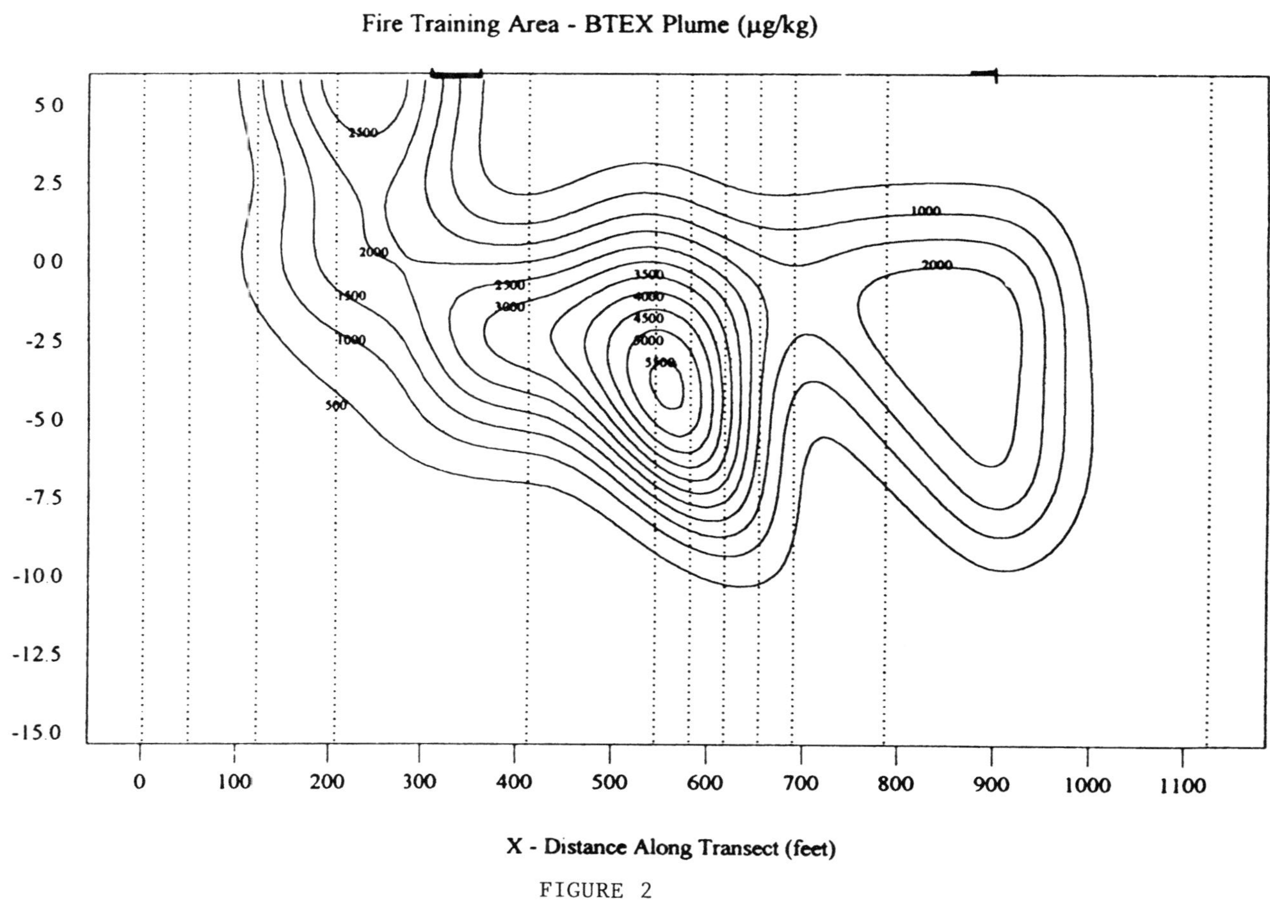

FIGURE 2

The approach to site characterization for chlorinated hydrocarbons is significantly more difficult. There are very few models of site characterization for these contaminants which have estimated mass loadings in specific media. Many of the previously referenced methods may work satisfactorily. However, free-phase detection, assessment and quantitation may be expected to be more a matter of luck and exhaustive sampling rather than intuition based on experience.

3.3   SAMPLING OVER TIME

VOC compounds (e.g., aromatic hydrocarbons, chlorinated solvents) are among the target contaminants which have been considered as constituents of concern in remedial investigations.. Their aqueous solubility and demonstrated association with aquifer solids requires sampling of these media during the site characterization phase. This suggestion also applies to organic metabolites of complex organic mixtures (e.g., ethylene, vinyl chloride, aromatic acids, phenols). Aqueous plumes which develop subsequent to the release of these organic mixtures and by-product compounds have received the most attention in the past. The fact that the mass of these contaminants frequently resides in the solids strongly suggests that the solids should receive the most attention in the initial site characterization effort. This should also be the case for the physical, geochemical, and microbial determinations.

Initially, conventional nested monitoring wells with screened lengths of 1 meter or more will be useful for estimating the spatial extent of the dissolved plume, for delineating apparent geochemical zones and to provide water level and aquifer property (e.g., slug and pump test derived hydraulic conductivity estimates). Semi-annual to annual sampling of wells, particularly multilevels appropriately designed and completed, should be quite useful over the course of the long term monitoring program. In this vein, their use should track the downgradient progress of risk-associated target compounds and permit testing predictions of intrinsic bioremediation effects on risk reduction.

However, proof of the effects of the net removal of specific solid-associated contaminants due to intrinsic bioremediation will depend on solid sampling and analysis at annual intervals or greater. This is because solid-associated concentration may be expected to change slowly. Unless biotransformation can be shown to be a major loss mechanism for contaminants mainly in solids over extended periods of time it will remain an area of research rather than practice.

Since very few contamination situations have been monitored intensively for periods exceeding several years, it is difficult to define specific sampling frequencies for the range of hydrogeologic and contaminant combinations which may be encountered. Suffice it to say that the adoption and future refinement of technically defensible protocols which have been developed recently will improve intrinsic remediation approaches to risk management in subsurface contamination situations.

Acknowledgements

The author would like to express his gratitude to the following individuals who aided in the preparation of the manuscript including: Dr. Gary Robbins, Todd H. Wiedemeier, Dr. John T. Wilson, Dr. Fran Kramer, Ms. Rebecca Mullin, C. Till, R. LaCasse, M. Henry and M. Lee.

REFERENCES

1.   Scalf, M.R., J.F. McNabb, W.J. Dunlop, R.L. Cosby, and J.S. Fryberger: Manual of Groundwater Sampling Procedures, National Water Well Association, (1981).

2.   Barcelona, M.J., J.P. Gibb, J.A. Helfrich and E.E. Garske: Practical Guide for Ground-Water Sampling. Illinois State Water Survey, SWS Contract Report 374, Ada, OK: U.S. Environmental Protection Agency (1985a).

3.   U.S. Environmental Protection Agency, Office of Waste Programs Enforcement, Office of Solid Waste and Emergency Response, RCRA Technical Enforcement Guidance Document, OSWER-9950.1, U.S. Government Printing Office, Washington, D.C., (1986), 208 pp. Appendices.

4.   Eklund, B.: Detection of Hydrocarbons in Ground Water by Analysis of Shallow Soil Gas/Vapor, API Publication No. 4394, Washington, D.C., (1985).

5.   Plumb, R.H.: A comparison of ground-water monitoring data from CERCLA and RCRA sites, Ground Wat. Mon. Rev., 7 (1987), 94-100.

6.   Robbins, G.A.: Influence of using purged and partially penetrating wells on contaminant detection, mapping and modeling, Ground Wat., 27 (1989), 155-162.

7.   U.S. Environmental Protection Agency, Office of Solid Waste, RCRA Ground Water Monitoring: Draft Technical Guidance Document, U.S. Environmental Protection Agency, Washington, D.C., (1992a) EPA/530-R-93-001.

8.   Barcelona, M.J., T. R. Holm, M.R. Schock and G.E. George: Spatial and temporal gradients in aquifer oxidation-reduction conditions, Wat. Res. Res., 25 (1989), 998-1003.

9.   U.S. Environmental Protection Agency, Proceedings of the Ground Water Sampling Workshop, Dallas, TX, December 8-10, 1993. U.S. EPA-R.S. Kerr Laboratory, Ada, OK, EPA Office of Solid Waste, Washington, D.C., (1994).

10.   Wilson, J.T. 1993: Testing Bioremediation in the Field, p. 160-184 in In-Situ Bioremediation - When Does It Work?, Committee on In-Situ Bioremediation, Water Science and Technology Board National Research Council, National Academy Press, Washington, D.C. 207 pp.

11.   Personal communication. T.H. Wiedemeier - T.H. Wiedemeier, D.C. Downey, J.T. Wilson, D.H. Kampbell, R.N. Miller, J.E. Hansen, Draft Technical Protocol for Implementing the Intrinsic Remediation (Natural Attenuation) with Long-Term Monitoring Option for Dissolved-Phase Fuel Contamination in Ground Water. Air Force Center for Environmental Excellence, Brooks AFB, San Antonio, TX, March 1994.

12.   National Research Council, *In*-Situ Bioremediation -When Does It Work?, National Academy Press, Washington, D.C. (1993), 207 pp.

13.   Journel, A.G.: Geostatistics: models and tools for the earth sciences, Math. Geol., 18 (1986), 119-140.

14.   Hoeksema, R.J. and P.K. Kitanidis: Analysis of the spatial structure of properties of selected aquifers, Wat. Res. Res., 21 (1985), 563-572.

15. Gilbert, R.O. and J.C Simpson.: Kriging for estimating spatial patterns of contaminants: potential and problems," Environ. Monit. and Assess., 5 (1985), 113-135.

DESIGN OF GROUND WATER MONITORING QUALITY NETWORKS

J.C. Tracy and T.J. Van Lent

South Dakota State University, Brookings, SD, USA

and

M.A. Mariño

University of California at Davis, Davis, CA, USA

ABSTRACT

A brief review of approaches to the design of subsurface water quality monitoring networks is presented. A theoretical discussion about methods to estimate parameters related to information based design approaches is presented. A hypothetical problem is then constructed to analyze the effect that different estimation procedures have on extending the design of a monitoring network. Results of this analysis indicate that although different parametric estimation procedures can produce significantly different parameter estimates, the resulting monitoring network design will be relatively unaffected.

1. INTRODUCTION

1.1 Background

Recent concerns over the effect that human activities have on the environment has led to an increased interest in understanding the fate of a pollutant once it enters a natural system. A variety of studies have been undertaken that examine the movement and degradation of contaminants through atmospheric, surface and subsurface systems. Some of these studies have resulted in new chemical monitoring techniques that help produce better descriptions of how a contaminant progresses through the environment. Other studies have developed better mathematical and statistical simulation techniques for predicting how a contaminant will behave in a given environment. The integration of the results of these advanced monitoring methods and simulation techniques can result in a powerful tool to aid in mitigating the effect of a contaminant on the surrounding environment, especially if employed in a symbiotic fashion. This is particularly true when applied to understanding the fate of contaminants in a subsurface environment, in which the transport of contaminants is more difficult to visualize, the medium tends to be highly heterogeneous, and it is very difficult to obtain spatially averaged soil or water quality samples without highly intrusive sampling methods.

The inability to obtain soil or water quality samples that are representative of large spatial areas is particularly troublesome, since a true measure of the mass or distribution of a contaminant within the area cannot be directly obtained. Rather, measurements of the contaminant concentration must be made at discrete locations within the subsurface medium, then interpolated over the entire problem domain so that an estimate of the contaminant distribution and total contaminant mass can be obtained. The group of discrete measurement locations can be referred to as a monitoring network, with the reliability of predictions of a contaminant's distribution highly dependent on the network's configuration.

Previous studies have undertaken efforts to produce systematic methods for designing networks for monitoring soil and subsurface water quality, so that accurate depictions of a contaminant's distribution can be obtained. A thorough review of approaches to designing ground water quality monitoring networks was presented in [1]. The general objectives of monitoring network design can be classified in one of two areas, these being designs based on contaminant detection, and designs based on describing a contaminant plume. In general, the approaches to designing monitoring networks for the purpose of detecting subsurface water contamination can be classified as being based on one or more of three general areas, classified as [1]: (1) hydrogeologic based approaches; (2) simulation based approaches; and (3) optimization based approaches.

1.2 Hydrogeologic Approaches

The hydrogeologic approach to monitoring network design is based on

judgements made by a ground water engineer or hydrogeologist without the use
of advanced mathematical or statistical modeling techniques. The main goal of
the hydrogeologic approach is to detect contamination of a subsurface
environment as early as possible [2]. The configuration of the monitoring
sites and sampling frequency is based on how the hydrogeologist perceives a
contaminant might move through a subsurface environment in relation to
potential contamination sources and subsurface water uses in the area. An
example of a hydrogeologic approach can be found in the Resource Conservation
and Recovery Act (RCRA) guidelines for ground water monitoring [3] that require
a minimum of three down gradient and one up gradient sampling sites near a
contamination source.

1.3 Simulation Approaches

Simulation based approaches to monitoring network design can be conducted
using either a deterministic or stochastic approach to predict the shape of a
contaminant plume for a potential contamination scenario. The effective use
of simulation based approaches requires a thorough understanding of the
physical characteristics of a subsurface medium as well as the identification
of sources and types of contamination. Simulations of potential contamination
scenarios are developed so that contaminant migration paths can be predicted.
Monitoring wells can be then be placed in the most probable locations to detect
contamination, thereby providing an early contaminant detection system.
Examples of simulation approaches to subsurface monitoring networks can be
found in [4][5].

1.4 Optimization Approaches

The optimization approach to monitoring network design is developed as
a mathematical programming problem. A quantifiable objective function is
developed that can be maximized with respect to the placement of a number of
monitoring sites, subject to certain physical or regulatory problem
constraints. The objective function and constraints are typically developed
as a function of the risks of not detecting contaminants and the cost of the
installation and sampling of monitoring sites. Examples of optimization
approaches for designing ground water quality monitoring networks can be found
in [6][7][8].

1.5 Information Approaches

The hydrogeologic, simulation, and optimization approaches can be used
individually or in tandem to develop a ground water monitoring network to
detect contamination that threatens nearby water supplies. However, none of
these approaches can be used to design a monitoring network to assess the level
of contamination in a subsurface environment once contamination is discovered.
Rather, information-based approaches must be employed that attempt to minimize
the uncertainty of contaminant predictions over a specified area when designing
a monitoring network. The minimization of the prediction uncertainty is akin
to maximizing the information on a contaminant's distribution throughout an

area and has been proposed as a network design methodology for many problems related to hydrogeologic phenomena [9][10][11][12].

The design of a monitoring network based on an information maximization approach requires the use of geostatistics for estimating the distribution of a contaminant's concentration throughout a subsurface environment. Geostatistics is a collection of statistical techniques applied to the estimation of spatially variable random functions [13][14]. The essential idea behind geostatistics is to assume that the concentration of a contaminant in a subsurface environment varies in space in a random fashion, but has some pattern to the randomness. Geostatistics allows for the prediction of a contaminant's concentration using all of the available measurements in a consistent and mathematically rigorous manner. However, the use of geostatistics involves the estimation of parameters related not only to the mean predictive behavior of a contaminant, referred to as the mean model parameters, but also parameters related to the correlation of contaminant concentrations, referred to as covariance model parameters. Much work has been presented on systematic methods to estimate parameters related to describing the mean behavior of hydrogeologic phenomena [15][16][17]. These studies have produced methods to estimate mean model parameters using statistical estimation techniques, but have largely relied on highly subjective nonparametric methods for estimating the parameters related to the covariance model, or have neglected the effect of the covariance model entirely in the model development process. Many times the use of nonparametric methods is dictated by the availability of information gained from an existing network. Nonetheless, when at all possible more objective parameter estimation approaches should be employed for estimating a covariance model's parameters.

Several parameter estimation techniques are available for use in estimating a covariance model's parameters, with the possibility that different methods will produce different parameter estimates. This could pose a problem when determining which parameters should be used in an information maximization approach to modify a monitoring network design. However, this will only become a problem if the resulting parameter estimates produce significantly different descriptions of the predicted contaminant concentration and error variance. The purpose of this paper is to examine the effect that different parametric estimation techniques have on the estimates of the covariance model parameters, and how they subsequently affect the predicted contaminant concentration and error variance distributions.

2. ESTIMATION OF COVARIANCE PARAMETERS

2.1 Geostatistical Approach

The geostatistical approach to the design of a soil or water quality monitoring network begins by assuming that the contaminant concentration is a stochastic process, such that

$$C(x) = m(x) + \epsilon(x) \tag{1}$$

where $C(x)$ is the concentration at some location defined by the spatial location vector, x; $m(x)$ is the mean concentration; and $\epsilon(x)$ is a zero-mean random component.

The characteristics of the random variable C can be quite complicated. In general C is said to be strictly stationary if

$$m(x) = m \tag{2}$$

$$E[(C(x) - m(x))(C(x') - m(x'))] = R(h) \tag{3}$$

where $h = |x - x'| =$ the separation distance. That is, the mean is constant and the two-point covariance is only a function of the separation distance. Any function for which the above is not true is termed a non-stationary function. We will concern ourselves with only one special type of non-stationary function: that with stationary increments. An increment of the observations is constructed as

$$z_1 = \sum_{i=1}^{n} \lambda_{i1} C_i \tag{4}$$

The properties of the increment z_1 depend upon those of C. Stationary increments are developed by choosing the λ_{i1} such that z_1 is stationary.

In this analysis, we will consider some types of nonstationary functions of a special class. Matheron [13] examined nonstationary functions in which C is a polynomial of order k. He called these functions intrinsic random functions of order k (IRF-k). For an IRF-k, z is a generalized increment of order k when λ satisfies the following condition (for a two-dimensional problem)

$$\sum_{i=1}^{n} \lambda_{i1} x_{i1}^{P_1} x_{i2}^{P_2} = 0 \quad for \quad l = 1, 2, \ldots k \tag{5}$$

for all nonnegative integers P_1, P_2 such that $P_1 + P_2 \leq k$. Thus, the conditions for k = 0, 1, 2 are

$$\sum_{i=1}^{n} \lambda_{i1} = 0; \quad for \quad k = 0 \tag{6}$$

$$\sum_{i=1}^{n} \lambda_{i1} = 0; \quad \sum_{i=1}^{n} \lambda_{i2} x_{i1} = 0; \quad \sum_{i=1}^{n} \lambda_{i3} x_{i2} = 0; \quad for \quad k = 1 \tag{7}$$

$$\sum_{i=1}^{n} \lambda_{i1} = 0; \quad \sum_{i=1}^{n} \lambda_{i2} x_{i1} = 0; \quad \sum_{i=1}^{n} \lambda_{i3} x_{i2} = 0; \quad \sum_{i=1}^{n} \lambda_{i4} x_{i1} x_{i2} = 0;$$

$$\sum_{i=1}^{n} \lambda_{i5} x_{i1}^2 = 0; \quad \sum_{i=1}^{n} \lambda_{i6} x_{i2}^2 = 0; \quad for \quad k = 2 \tag{8}$$

For an IRF-k, one is allowed to take the mean as a polynomial of order k

$$m(\boldsymbol{x}) = \sum_{i=1}^{p} b_i f_i(\boldsymbol{x}) \tag{9}$$

where p is the number of monomials $f_i(x)$

$$f_i(\boldsymbol{x}) = x_{i1}^{P_1} x_{i2}^{P_2}; \quad P_1 + P_2 \le k \tag{10}$$

Several forms of covariance models have been examined for use in geostatistical approaches. Generalized covariance models of polynomial, exponential, and Gaussian forms have been used and the model chosen for a specific study should be based on the function that best describes the covariance as supported by the data measured by the monitoring network. In general though, it can be said that the covariance model can be described as a function of the separation distance between points, h, and the covariance model parameter set, θ, such that

$$K(h) = g(\boldsymbol{\theta}, h) \tag{11}$$

where h = the separation distance between two locations and g() can be either a linear or nonlinear function in both θ and h.

One of the more important uses for the generalized covariance models is for estimation. That is, suppose one wishes to estimate the value of C at some location x_0, in which there are no observations. One builds an increment of the observations as

$$C^*(\boldsymbol{x}_0) = \sum_{i=1}^{n} \lambda_{i0} C(\boldsymbol{x}_i) \tag{12}$$

where $C^*(x_0)$ is the estimate of the concentration at x_0, $C(x_i)$ = the observed concentrations at points x_i, i = 1, 2, ..., n, and λ_{i0} is a weighting factor for the concentration observed at monitoring location i for the estimate at location x_0.

If one calculates the variance of the above estimator, the result is

$$Var[C^*(\boldsymbol{x}_0)] = \sum_{i=1}^{n} \sum_{j=1}^{n} \lambda_{io}\lambda_{jo}K(h_{ij}) \tag{13}$$

Often the generalized covariance calculated between the points x_i and x_j is denoted as K_{ij} and the separation distance as h_{ij} for convenience. The best linear unbiased estimate of the concentration at location x_0 can then be found by minimizing Eq. 13 with respect to the weighting factors, λ_{io}, subject to the constraints given in Eq. 5. This can be accomplished using Lagrange multipliers, μ_{ko}, and minimizing Eq. 13 with respect to the weighting factors and the Lagrange multipliers, resulting in the following set of linear equations

$$\sum_{j=1}^{n} \lambda_{jo}K(h_{ij}) + \sum_{k=1}^{p} \mu_{ko}f_k(\boldsymbol{x}_i) = K(h_{ij}) \quad \forall \quad i = 1 \ to \ n \tag{14}$$

$$\sum_{j=1}^{n} \lambda_{jo}f_k(\boldsymbol{x}_j) = f_k(\boldsymbol{x}_0) \quad \forall \quad k = 1 \ to \ p \tag{15}$$

The above system of simultaneous, linear algebraic equations can be written concisely as

$$Aw_0 = a_0 \tag{16}$$

Equation 16 is known as the "kriging system" and has important implications for the design of a monitoring network. By definition, this method will minimize the error in estimating the value of $C(x_0)$. Moreover, we can use the solution to calculate the value of the estimation error at any point within the problem domain as

$$Var[C^*(\boldsymbol{x}_0)-C(\boldsymbol{x}_0)] = K(|\boldsymbol{x}_0-\boldsymbol{x}_0|) - \sum_{j=1}^{n} \lambda_{jo}K(|\boldsymbol{x}_j-\boldsymbol{x}_0|) - \sum_{k=1}^{p} \mu_{ko}f_k(\boldsymbol{x}_0) \tag{17}$$

or in matrix notation as

$$K = K_{00} - w_0^T a_0 \tag{18}$$

The relevant property of Eq. 18 that makes it useful in network design is that one need only know the locations of the observations, x_i, the actual observation at x_i are never used. That is, we can calculate the estimation error variance at any point x_0 if we know the locations of every monitoring site in the network. This information can then be used in the design of additional monitoring sites by placing new monitoring sites that result in the greatest reduction in the overall error variance, thereby providing the maximum gain in information over the problem domain.

2.2 Parametric Estimation

Parametric estimation is the term used to describe the estimation of model parameters from the observed data. The usual approach is to assume a functional form for the covariance model, and the use the observations to find the unknown parameters. The most common approach in parametric methods is to work with stationary increments of the data rather than the data itself. This results in some notational as well as computational simplicity. Assuming that the random function describing the contaminant concentration is described by a polynomial mean and a zero-mean random component,

$$C(\boldsymbol{x}_i) = m(\boldsymbol{x}_i) + z(\boldsymbol{x}_i) \tag{19}$$

This formulation can be interpreted in the notation typically found in linear regression as

$$\boldsymbol{C} = X\boldsymbol{b} + \boldsymbol{\epsilon} \tag{20}$$

where $\boldsymbol{b}$ are the regression coefficients describing the mean, ϵ is the regression residual vector, and X is a matrix of spatial coordinates. For example, if we describe the process as containing a linear (planar) mean

$$C_i^* = b_0 + x_{i1}b_1 + x_{i2}b_2 \tag{21}$$

then the matrix X will become

$$X = \begin{bmatrix} 1 & x_{11} & x_{12} \\ 1 & x_{21} & x_{22} \\ & \cdots\cdots & \\ 1 & x_{n1} & x_{n2} \end{bmatrix} \tag{22}$$

Starks and Fang [18] used a regression approach to filter the polynomial mean. The least squares estimate of the parameters are

$$\boldsymbol{b} = (X^T X)^{-1} X^T \boldsymbol{C} \tag{23}$$

So that the residuals become

$$z = C - C^* = [I - X(X^TX)^{-1}X^T] C = TC \tag{24}$$

where T is now called the transformation matrix. One can easily show that the rank of X is p, the number of conditions imposed by the intrinsic function. Kitanidis [19] demonstrates that this will reduce the degrees of freedom in T, as p increments will be linearly dependent on the rest. Therefore, p rows of T can be dropped and will not affect the covariance calculations. Therefore, the data increments can be constructed as

$$z = \Lambda C \tag{25}$$

where Λ is a N x n matrix of rank N which is derived from T by dropping p rows. Note that this transformation does not actually require the computation of the mean parameters, only the functional form need be known to build the matrix X. Since only the functional form is required, there is no inherent restriction on what the form of the mean, save that it he linear in the parameters. The zero-mean increments, z, are now stationary increments of the observations, and we will use these for most of the parameter estimation procedures.

2.3 Restricted Maximum Likelihood Estimation

One of the most popular and efficient methods that is available to estimate a set of parameters is the maximum likelihood estimation method, often times represented by the acronym, MLE. The general philosophy behind maximum likelihood estimation is that parameters are estimated by maximizing the probability that the true parameter values are obtained. This can be accomplished by minimizing the negative log-likelihood function of a set of observed contaminant concentrations with respect to the mean and covariance parameters. However, since we are using the data increments, as opposed to the concentrations, only the covariance parameters that define the covariance matrix, K, need be estimated. As described earlier, the data increments can be formed using Eq. 25, resulting in zero mean increments, with a covariance matrix defined as

$$Q = \Lambda K \Lambda^T \tag{26}$$

where Q is the covariance matrix of the data increments. The negative log-likelihood function for the data increments can then be written as

$$L(z|Q) = \frac{n}{2}\ln(2\pi) + \frac{1}{2}|Q| + \frac{1}{2}z^TQ^{-1}z \tag{27}$$

where L is the negative log-likelihood function. Equation 27 can be minimized with respect to the covariance model parameter set by taking its derivative with respect to the parameters and setting them equal to zero, resulting in [19]

$$L_{\theta_j} = \frac{\partial L}{\partial \theta_j} = \frac{1}{2} Tr \left(Q^{-1} \frac{\partial Q}{\partial \theta_j} \right) - \frac{1}{2} \mathbf{z}^T Q^{-1} \frac{\partial Q}{\partial \theta_j} Q^{-1} \mathbf{z} \qquad (28)$$

where L_{θ_j} = the derivative of the negative log-likelihood function with respect to parameter θ_j. The resulting set of equations is nonlinear and must be solved in an iterative fashion. A variety of methods exist that can be used to solve this type of problem. However, it has been demonstrated [20] that for the problem being examined in this paper, the Gauss-Newton solution method provides superior computational efficiency over other alternatives. Using the Gauss-Newton solution method the second derivative of the negative log-likelihood function, referred to as the Hessian matrix, is approximated as [18]

$$M_{jk} = \frac{\partial L}{\partial \theta_j \partial \theta_k} = \frac{1}{2} Tr \left(Q^{-1} \frac{\partial Q}{\partial \theta_j} Q^{-1} \frac{\partial Q}{\partial \theta_k} \right) \qquad (29)$$

where M_{jk} = the approximate Hessian matrix, often referred to as the Fischer information matrix, and the subscripts j and k refer to the appropriate model parameter. The estimation then proceeds by taking an initial guess at the best parameter values, then computing the gradient vector, Eq. 28, and the approximate Hessian, Eq. 29, using these guesses, then updating the parameter estimates as

$$\boldsymbol{\theta} = \boldsymbol{\theta}^0 - M^{-1} \boldsymbol{L_\theta} \qquad (30)$$

where θ^0 = the previous guess of the parameter values. The new estimate then becomes the previous guess, and this procedure continues until there is little change in the parameter values and the gradient vector is approximately zero, which indicates that a local optimal solution has been obtained.

2.4 Minimum Variance Unbiased Quadratic Estimation

Kitanidis [19] described minimum variance unbiased quadratic estimation (MVUQE), and proves that an iterative application of MVUQE estimation will yield results equivalent to a restricted maximum likelihood approach. However, there is some opportunity to reduce the computational effort as compared to RMLE, so MVUQE should be pursued as a potential estimation technique.

The method begins by assuming that the unknown parameters are quadratic functions of the data, such that

$$\theta_j^* = \mathbf{z}^T F_j \mathbf{z} \qquad (31)$$

where F_j is a matrix selected so that the parameter estimates are unbiased and of minimum variance. These conditions can be expressed as

$$\theta_j = E(\theta_j^*) \tag{32}$$

$$\min \quad E[(\theta_j^* - \theta_j)^2] \tag{33}$$

A system of linear, simultaneous algebraic equations results [19], where

$$\sum_{l=1} Tr(Q_0^{-1} Q_l Q_0^{-1} Q_i) \lambda_l = 4\delta(l-i) \quad for \quad i = 1, 2, \ldots m \tag{34}$$

in which Q_0 is the best prior estimate of Q. The kriging weights λ_j are then used to calculate the matrix F_j as

$$F_j = \frac{1}{4} \sum_{i=1}^{m} \lambda_l Q_0^{-1} Q_l Q_0^{-1} \tag{35}$$

Lastly, the parameter estimates are found using Eq. 31.

The application of the MVUQE procedure is iterative. One begins with the *a priori* estimate of the covariance Q, using estimates of the new set of parameters, θ. These new estimates become the best prior estimates for forming Q_0, and the procedure is repeated until convergence is achieved.

2.5 Minimum Norm Estimation

The minimum norm approach potentially represents a significant reduction in computational effort when compared to restricted maximum likelihood estimation and minimum quadratic unbiased estimation. The minimum norm procedure [19] is predicated on the criterion of minimizing the norm of the difference between TKT and TCC^TT, where T is the project matrix defined by Eq. 24 and $K = E[CC^T]$. However, the method is only cost effective if the covariance is a linear combination of the unknown parameters, where K can be written as

$$K = \sum_{i=1}^{m} K_i \theta_i \tag{36}$$

where K_i are the individual matrix components that are independent of θ. The above is true for all generalized polynomial covariances, but not true for the stationary exponential and Gaussian variograms.

The minimum norm is based upon the criterion

$$\min_\theta Tr[(TKT - TCC^TT)(TKT - TCC^TT)] \tag{37}$$

The trace is interpreted as the Euclidean norm of the matrix of the difference between TKT and TCC^TT. In that sense, it is a minimum norm procedure. Taking the derivative with respect to the covariance parameter set, θ, and setting the result equal to zero will yield

$$2Tr[TK_iT(KT - CC^T)] = 0 \tag{38}$$

Rearranging and substituting Eq. 36 into Eq. 38 yields

$$\sum_{j=1}^{m} Tr[TK_iTK_j]\theta_j = C^TK_iTC \quad for \ i = 1, 2, \ldots m \tag{39}$$

The above represents an m x m system of linear, algebraic equations that can be written as

$$A\theta = e_1 \tag{40}$$

$$A_{ij} = Tr[TK_iTK_j] \tag{41}$$

$$e_1 = C^TTK_iTC \tag{42}$$

The primary advantage over the restricted maximum likelihood estimation and minimum quadratic unbiased estimation procedures is that there is no need to iterate and invert large matrices if K is linear in the covariance parameter set.

3.   HYPOTHETICAL PROBLEM

3.1 Problem Description

The hypothetical problem used in this paper was developed by generating a two-dimensional contaminant concentration distribution assuming a given mass of contaminant, M, was suddenly released into a homogeneous medium at a time of t = 0 days. The analytical solution for the contaminant concentration can be derived as function of time, and spatial coordinates as

$$C = \frac{M}{2\pi t\sqrt{D_xD_y}} [\exp\{ - \frac{x^2}{4D_xt} - \frac{y^2}{4D_yt}\}] \tag{43}$$

where C = the contaminant concentration, D_x and D_y = the dispersion coefficients in the x and y direction, respectively, and t = time. For this problem, the initial contaminant mass was set to 1,000 g per meter of medium thickness, the dispersion coefficients were $D_x = D_y = 0.0001$ m^2/day, and the measurement were taken at t = 2,000 days. Equation 43 was computed for a 25 by 25 grid in the x and y dimensions at even increments for $0 < x < 250$ meters and $0 < y < 250$ meters. A random error term consisting of white noise with a gaussian correlation term represented as

$$\epsilon = 0.25\delta_{ij} - 0.25e^{\left(-\frac{h_{ij}}{8}\right)^2} \qquad (44)$$

was added to each grid point, with the resulting values taken to be the "true" concentrations. A sampling network of 40 sites was then generated to represent an existing monitoring network. Both the "true" concentration distribution and the location of the sampling sites are shown in Figure 1. As seen in Figure 1 the shape of the contaminant distribution is somewhat like a plume, with a peak concentration near the center point, at $x = 125$ m and $y = 125$ m, with a moderately noisy surface. As also can be seen in the figure, the monitoring network was generated with gages near the higher contaminant concentrations, which tends to be the pattern that typically results when a monitoring system is initially paced with little site information.

3.2 Parameter Estimation

Since the shape of the concentration distribution is representative of a plume, a quadratic mean model was assumed, resulting in a mean model with 6 parameters. The variogram of the residuals was then assumed to follow a linear form, such that

$$K(h_{ij}) = \theta_1\delta_{ij} + \theta_2 h_{ij} \qquad (45)$$

Estimation of the parameters θ_1, and θ_2 then proceeded using the RMLE, MVUQE and minimum norm (MNE) methods discussed earlier. The initial estimates using the minimum norm procedure produced a negative estimate for θ_1. This represents a situation where an estimate would produce negative variance, which is physically impossible. Thus, for the minimum norm, θ_1 was set to zero, and the θ_2 estimated, resulting in the value shown in Table 1. The RMLE and MVUQE procedures did not produce negative values for θ_1 and hence their estimates could be used directly. Table 1 lists the results of all three estimation procedure and Figure 2 presents the resulting variograms for each estimation method. As can be seen in both the table and figure, the RMLE and MVUQE methods result in nearly identical parameter estimates. As discussed earlier, it has been proven theoretically [18] that these methods should produce identical estimates, hence these results are not surprising. However, it can also be seen in Table 1 and Figure 2 that the MNE for θ_2 varies quite a bit from the RMLE and MVUQE estimates. These differences in parameter estimates appear to be quite significant when viewing the variograms in Figure 2. However, the real measure of the effect of the estimates on future network designs can only be judged by viewing the estimated concentration and error variance distributions.

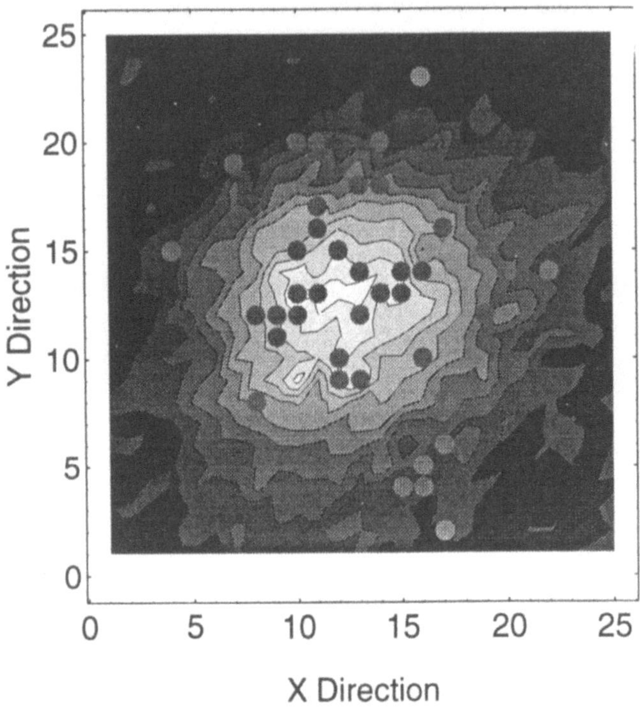

Figure 1.   True Concentration Distribution and Monitoring Site Locations.
Distances are in meters x 10^{-1}. Concentration contours are from 0 mg/l
(darkest) to 80 mg/l (lightest) in uniform increments.

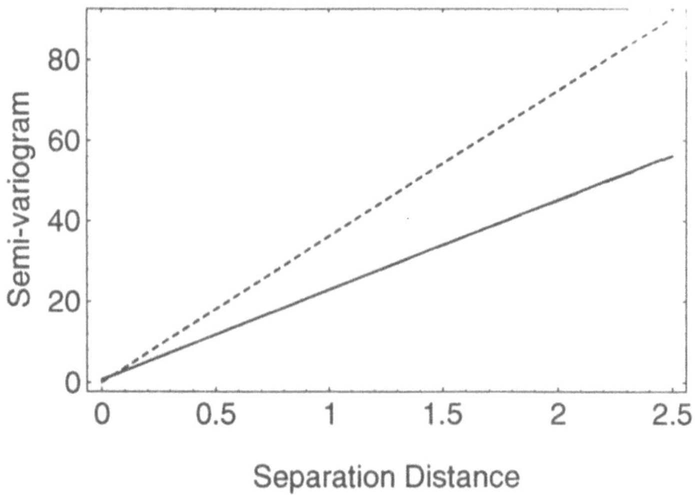

Figure 2.   RMLE (solid), MVUQE (long dash) and MNE (short dash) variograms.
Variogram values are in (mg/l)^{2} and distances are in meter x 10^{-1}.

Figures 3, 4 and 5 represent the estimated contaminant concentrations and error variances using the best linear unbiased estimation approach discussed in Section 2.1. As can be seen in Figures 3 and 4 the estimated concentration and error variance distributions for the RMLE and MVUQE methods results in identical estimates for the contaminant concentration and error variance distributions. This result was anticipated, since the parameter estimates were nearly identical. It can also be seen in Figure 5 that the predicted contaminant concentration using the MNE method is nearly identical to those developed using the RMLE and MVUQE methods. Also, as can be seen in Figure 5, although the magnitude of the error variance distribution is larger the MNE, as compared to the RMLE and MVUQE, estimates the shape of the error variance distributions are quite similar.

Table 1. Covariance Parameter Estimates.

Method (1)	θ_1 (2)	θ_2 (3)
RMLE	0.640	-22.36
MVUQE	0.688	-22.70
MNE	- - -	-36.23

These results demonstrate that although the estimated parameters appear to vary quite a bit depending on the parametric estimation procedure employed, the resulting prediction of the contaminant concentration and error variance distribution is quite similar for each parameter set. Ultimately, the selection of new monitoring sites will be determined by minimizing the uncertainty of the plume description. This involves the selection of new sites by predicting the effect that a new site would have on reducing the overall error variance. As seen in Figure 3, 4 and 5 the similarity in the shape of the error variance distribution indicates that new wells would be placed in the same general location, which implies that the estimation method would have very little effect on the ultimate selection of additional monitoring sites.

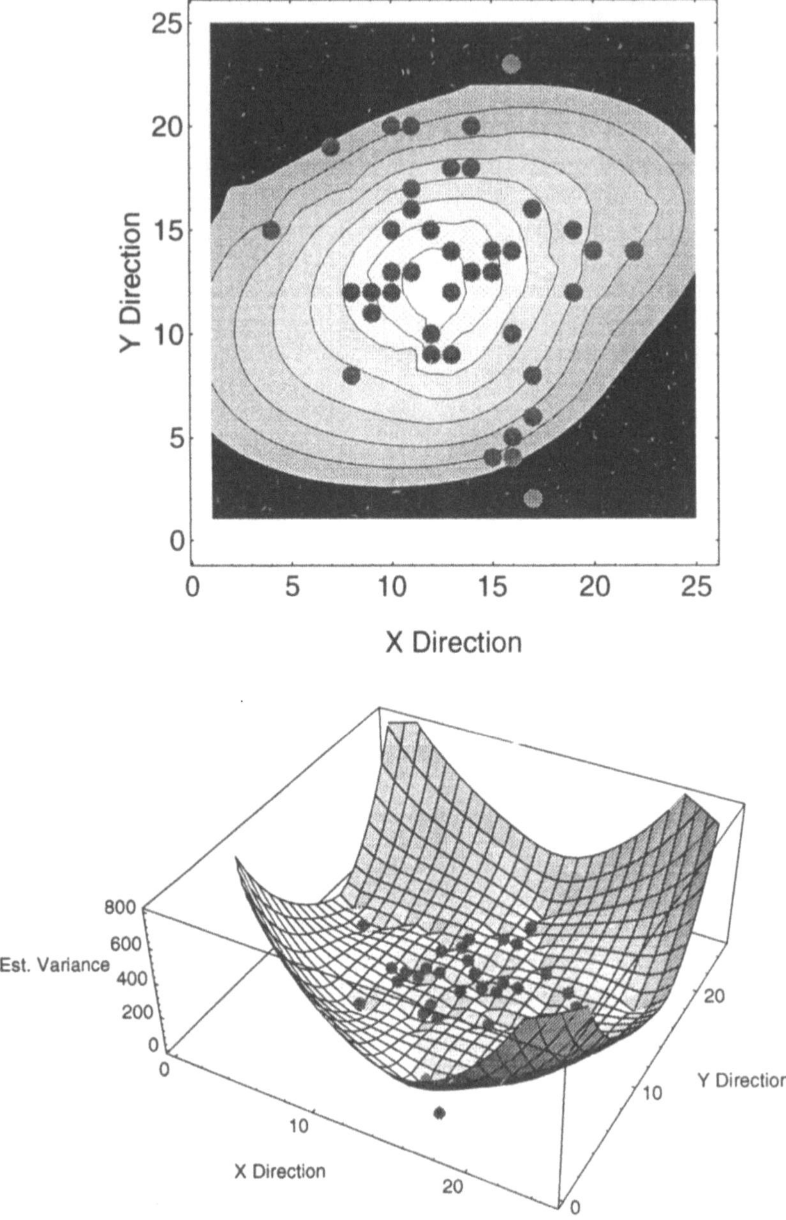

Figure 3. Estimated Concentration (Top) and Variance (Bottom) for RMLE. Concentration contours are from 0 mg/l (darkest) to 80 mg/l (lightest) in uniform increments. Estimated variance is in units of $(mg/l)^2$. Distances are in meter x 10^{-1}.

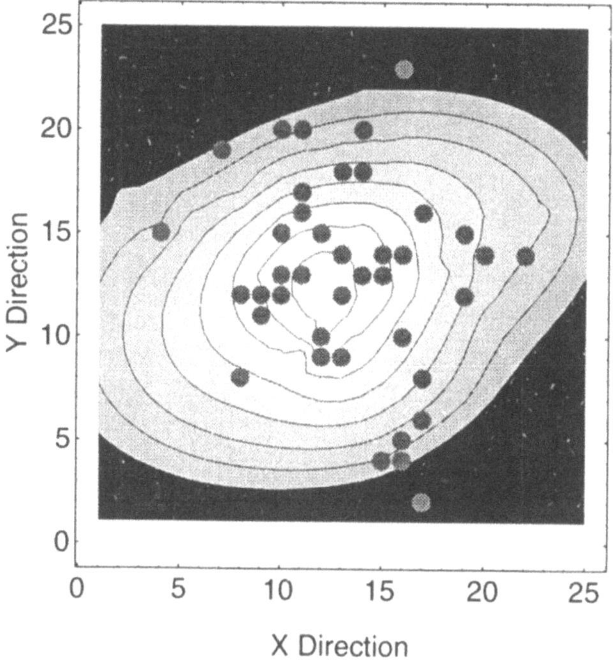

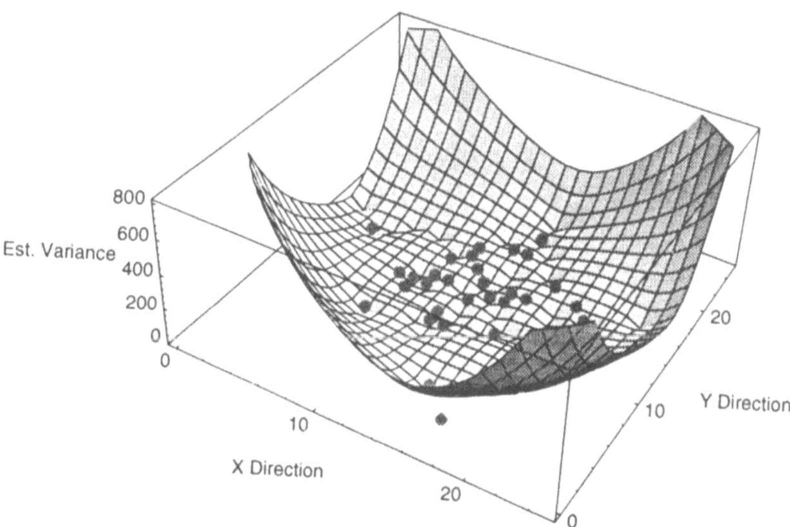

Figure 4. Estimated Concentration (Top) and Variance (Bottom) for MVUQE. Concentration contours are from 0 mg/l (darkest) to 80 mg/l (lightest) in uniform increments. Estimated variance is in units of $(mg/l)^2$. Distances are in meter x 10^{-1}.

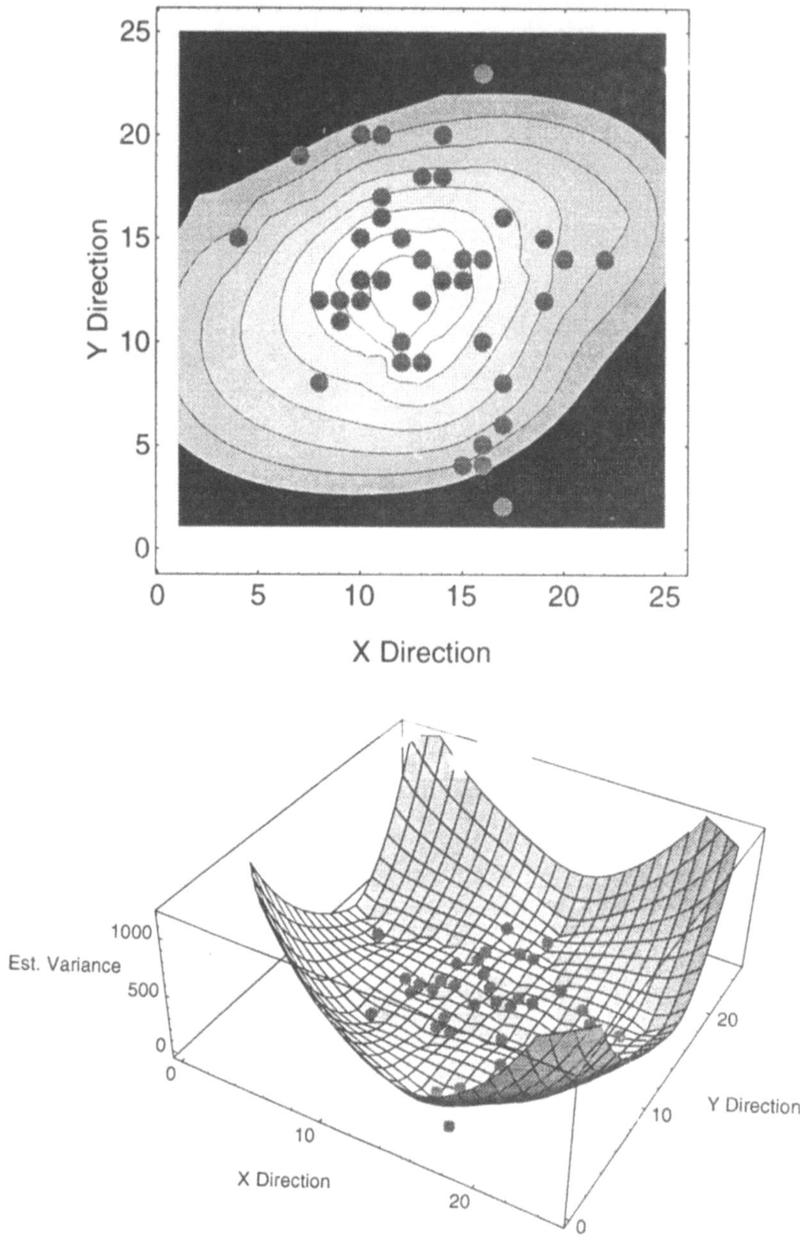

Figure 5. Estimated Concentration (Top) and Variance (Bottom) for MNE.
Concentration contours are from 0 mg/l (darkest) to 80 mg/l (lightest) in
uniform increments. Estimated variance is in units of $(mg/l)^2$. Distances are
in meter x 10^{-1}.

4. SUMMARY AND CONCLUSIONS

The results of this hypothetical study suggest that the selection of a parametric estimation method for estimating covariance model parameters can result in what appear to be significantly different parameter estimates. However, the use of these parameters in the extension of a monitoring network will most likely not result in significantly different designs. Thus, the selection of a parametric estimation procedure should be based on the robustness and computational efficiency of the procedure related to the problem being addressed.

ACKNOWLEDGEMENTS

This research was partially supported by the National Park Service and the South Florida Water Management District under contract CA4250-4-9015, the University of California Water Resources Center under Project UCAL-WRC-W-789, and the Ecotoxicology Program of the University of California Toxic Substances Research and Teaching Program.

REFERENCES

1.   Loaiciga, H. A., R. J. Charbeneau, L. G. Everett, G. E. Fogg, B. F. Hobbs, and S. Rouhani: Review of ground-water quality monitoring network design." J. Hydraul. Engr., 118 (1992), 11-37.
2.   Everett, L. G.: Ground Water Monitoring, General Electric Company, Schenectady, New York 1980.
3.   U.S. Environmental Protection Agency: Ground Water Monitoring Technical Enforcement Guidance Document, Office of Solid Waste and Emergency Response, Washington, D.C. 1986.
4.   Gilbert, R. O.: Statistical Methods for Environmental Pollution Monitoring, Van-Nostrand-Reinhold Co., New York, NY 1987.
5.   Ahfeld, D. P. and G. F. Pinder: A Ground Water Monitoring Network Design Algorithm, Report 87-WR-4, Department of Civil Engineering and Operations Research, Princeton, New Jersey 1988.
6.   Hsueh, Y. W., and R. Rajagopal: Modeling ground water quality decisions, Ground Water Monitoring Review, Fall, 121-134 (1988).
7.   Knopman, D. S., and C. J. Voss: Discrimination among one-dimensional models of solute transport in porous media: Implications for sampling design, Water Resour. Res., 24 (1988), 1859-1876.
8.   Hudak, P. F., H. A. Loaiciga, and M. A. Mariño: Regional Scale Ground Water Quality Monitoring: Methods and Case Studies, California Water Resources Center, University of California, Contribution 203 1993.
9.   Aboufirassi, M., and M. A. Mariño: Krigin of water levels in the Souss Aquifer, Morocco, Math. Geol., 15 (1983), 537-551.

10. Aboufirassi, M. and M. A. Mariño: Cokriging of aquifer transmissivities from field measurements of transmissivity and specific capacity, Math. Geol., 16 (1994), 19-35.
11. Rouhani, S.: Variance reduction analysis, Water Resour. Res., 21 (1985), 837-845.
12. Ben-Jemaa, F., M. A. Mariño, and H. A. Loaiciga: Multivariate geostatistical design of ground water monitoring networks, J. of Water Resour. Plan. and Manag., ASCE, 120 (1994), 505-522.
13. Matheron, G.: The intrinsic random function and their applications, Adv. Appl. Prob., 5 (1973), 539-568.
14. Journel, A. and Huijbregts, C.: Mining Geostatistics, Academic Press, New York, NY 1978.
15. Sadeghipour, J. and W. W-G. Yeh: Parameter identification of ground water aquifer models: A generalized least squares approach, Water Resour. Res., 20 (1984), 971-979.
16. Carrera, J. and S. P. Neuman: Estimation of aquifer parameters under transient and steady-state conditions: 1. Maximum likelihood method incorporating prior information, Water Resour. Res., 22 (1986), 199-210.
17. Loaiciga, H. A. and M. A. Mariño: The inverse problem for confined aquifer flow identification and estimation with extensions, Water Resour. Res., 23 (1987), 92-104.
18. Starks, T. and J. Fang: The effect of drift on the experimental semivariogram, J. Int. Assoc. Math. Geol., 14 (1982), 309-319.
19. Kitanidis, P.: Statistical estimation of polynomial generalized covariance functions and hydrologic applications, Water Resour. Res., 19 (1983), 909-921.
20. Kitanidis, P. and W. Lane: Maximum likelihood parameter estimation of hydrologic spatial processes by the Gauss-Newton method, J. Hydrol., 79 (1985), 53-71.

A SYSTEMATIC APPROACH TO DESIGNING A MULTIPHASE UNSATURATED ZONE MONITORING NETWORK

S.J. Cullen

University of California, Santa Barbara, CA, USA

and

J.H. Kramer

Condor Earth Technologies, Sonora, CA, USA

and

R.T. Ogg

EG&G, Golden, CO, USA

ABSTRACT

A systematic approach is presented for the design of a multiphase vadose zone monitoring system recognizing that, as in groundwater monitoring system design, complete subsurface coverage is not practical. The approach includes identification and prioritization of vulnerable areas, selection of cost-effective indirect monitoring methods which will provide early warning of contaminant migration, selection of direct monitoring methods for diagnostic confirmation, identification of background monitoring locations, and identification of an appropriate temporal monitoring plan. An example of a monitoring system designed for a solid waste landfill is presented and utilized to illustrate the approach and provide details of system implementation. The example design described incorporates the use of neutron moisture probes deployed in horizontal access tubes beneath the leachate recovery collection system of the landfill. Early warning of gaseous phase contaminant migration is monitored utilizing whole-air active soil gas sampling points deployed in gravel-filled trenches beneath the subgrade. Diagnostic confirmation of contaminant migration is provided utilizing pore-liquid samplers and conservative tracers. A discussion of background monitoring point location is also presented.

1   Hydrologic Specialist, Vadose Zone Monitoring Laboratory, Institute for Crustal Studies, University of California, Santa Barbara, 93106-1100, and Principal Scientist, Geraghty & Miller, 5425 Hollister, Suite 100, Santa Barbara, California, 93111, United States of America, (805) 964-2399 Telephone, (805) 967-9722 Fax.

2   Senior Hydrogeologist, Condor Earth Technologies, Sonora, CA, USA 95251; (209) 532-0361 Telephone

3   Senior Hydrogeologist, EG&G, Rocky Flats Plant, Golden, CO, USA.

OVERVIEW OF SYSTEMATIC APPROACH

While the subsurface environment varies widely at a regional and even local level, a systematic approach can be followed which will result in the design of the most appropriate and feasible monitoring network for unsaturated regions of the vadose zone at any given hazardous waste site. Everett et al. (1984) stated that, "A vadose zone monitoring program for a waste disposal site includes premonitoring activities followed by an active monitoring program". Other authors refer to premonitoring activities as site characterization (Sara, 1991) and have identified characterization tasks which must be accomplished before selecting locations and depths of groundwater monitoring wells. Everett et al. (1984) identified alternative techniques which can be used to characterize a site in preparation for development of a vadose zone monitoring program.

Once site characterization is complete, the essential elements required to develop a vadose zone monitoring strategy at a site include the following:

1) Initially establishing the feasibility of monitoring unsaturated regions of the vadose zone,
2) Identifying the goals of the monitoring effort ,
3) Interpreting the chemical and physical characteristics of the contaminants of concern and the likely mode of transport,
4) Interpreting characterization data to:
   a) construct a conceptual model of the subsurface,
   b) identify likely contaminant transport pathways in light of facility engineering design features,
   c) determine the likely effect of the unsaturated zone stratigraphy and lithology on retarding or attenuating contaminant migration,
5) Designing the unsaturated zone monitoring network.

Groundwater monitoring networks often do not provide complete coverage for detection of contaminants. Similarly, currently available technologies will not provide complete coverage for unsaturated regions of the vadose zone in a cost-effective manner. For example, comprehensive suction-lysimeter network coverage cannot be reasonably achieved because it requires a large number of sampler installations which is prohibitively expensive. Model simulations for a 1 ft^2 leak source in homogeneous medium-textured soil (Bumb et al., 1988) resulted in a lysimeter spacing of approximately 15 feet. Using this spacing, complete coverage of a forty acre site would require 6,000 to 35,000 samplers depending on soil physical properties. Aside from installation costs, the annual costs for chemical analyses to satisfy a quarterly monitoring order would be U.S. $20 million (in 1993 dollars). Lysimeters provide the best means for collecting pore-liquid samples and confirming chemical contamination. However, lysimetry can be effectively complemented by the use of other more economical indirect sampling methods to infer the status of pore-liquids, and consequently to detect contaminant movement. Monitoring programs using intelligently combined, selected techniques focused at vulnerable areas beneath waster disposal facilities can maximize the likelihood of leachate release detection at reasonable cost.

The conceptual design approach recommended herein will result in a reasonably comprehensive unsaturated zone monitoring system which balances the need to monitor areas most likely to transmit fugitive waste fluids (liquids or gases) and financial constraints. It is founded on a risk-based evaluation of the site and landfill design in which critical areas judged most vulnerable to leachate escape are prioritized and monitored.

DESIGNING THE UNSATURATED ZONE MONITORING NETWORK

The critical process of designing an unsaturated zone monitoring network includes the following steps:

1) Identification and prioritization of critical areas most vulnerable to contaminant migration;

2) Selection of indirect monitoring methods which provide reasonably comprehensive coverage and cost effective, early warning of contaminant migration;

3) Selection of direct monitoring methods which provide diagnostic confirmation of the presence and migration of contaminants;

4) Identification of background monitoring points which will provide hydrogeologic monitoring data representative of preexisting site conditions;

5) Identification of a cost-efficient, temporal monitoring plan which will provide early warning of contaminant migration in unsaturated regions of the vadose zone.

Identification of Vulnerable Areas

The design of an unsaturated zone monitoring system for is dependent upon an evaluation of the site hydrogeologic characterization data and the waste disposal facility engineering design to identify areas vulnerable to contaminant migration. Vulnerable areas should be identified and prioritized from most to least vulnerable. Landfills and surface impoundments pose the greatest challenge in terms of designing an unsaturated zone monitoring system. Both are similar in that, once constructed, the subsurface directly beneath both of these types of facilities is less accessible with conventional drilling equipment and installation of monitoring points after the fact becomes quite difficult. Figure 1 illustrates a schematic design of a waste landfill with a compacted earthen liner and leachate collection and recovery system (LCRS). The EPA (1988) details the provisions of the minimum technology requirement for liners and caps at hazardous waste facilities.

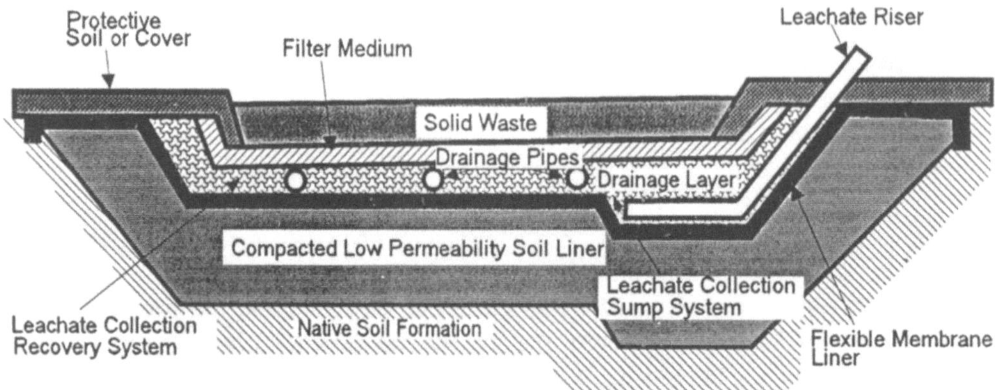

Figure 1. Schematic of a composite liner system at a landfill (adapted from USEPA, 1988).

Presented in order of decreasing vulnerability, critical areas at a landfill which should be identified include unsaturated regions of the vadose zone beneath:

1)   unlined, uncapped portions of facilities,
2)   lined but uncapped portions of facilities,
3)   leachate sumps,
4)   leachate recovery collection lines,
5)   intermodule seams,
6)   subsurface preferred flow pathways.

Unlined, Uncapped Portions of Facilities.

Uncapped portions of hazardous waste landfills are unprotected from infiltration from above. Infiltrating water from precipitation will generate leachate within the waste pile which can escape the unlined bottom of the facility. The presence, extent, and nature of suspected contamination in this area must be investigated and monitoring positions established in potential vadose zone flow pathways. Generally, unlined and uncapped portions of facilities require the most intensive monitoring networks.

Lined but Uncapped Portions of Facilities.

Uncapped portions of facilities are also unprotected from production of leachate due to precipitation which is not intercepted and diverted from the facility. The produced leachate will pond on top of the liner and be channeled into the leachate collection and recovery system (LCRS). When the landfill is properly designed and constructed, this may not represent a vulnerable location if it can be assured that there are no low spots in the LCRS where ponded leachate can develop a hydraulic head. Risk of a leak is greater because of the added leachate produced by infiltration. Depend on the cap design, leachate percolation can be reduced by decreasing infiltration and increasing surface runoff due to compaction, or by storing soil water and removing it at a later time by evapotranspirative pumping. Increased infiltration through uncapped portions of an active site can result in greater hydraulic head developing at the liner, LCRS and sumps.

Leachate Sumps.

Leachate sumps are potentially the highest risk areas in lined landfills because they are most likely to pond leachate, resulting in a hydraulic head which can drive contaminants through the compacted earthen liner. In some designs, leachate sumps may not be designed as double-lined liquid retaining storage areas, nor with flexible membrane liners. If no accommodation is provided for preventing and disposing of leachate build-up in sumps, the monitoring concerns are increased. The schematic diagram shown in Figure 1 shows sidewalls of a leachate collection sump as compacted, low permeability soil. Compacting sidewalls and insuring leak-proof joints presents some logistical problems for compaction equipment operators. Assuming that compacted earthen liners can be uniformly and reliably compacted to a 10^7 cm/sec permeability specification and have a porosity of 35%, it can be conservatively calculated, using Darcy's Law and the assumption of piston (Green-Ampt) flow, that a continuous one foot hydraulic head (hydraulic gradient assumed to be .3 feet per foot of liner) would be sufficient to drive liquid through three feet of earthen liner in approximately 10 to 30 years. Breaches in the sumps could allow immediate leachate escape.

Leachate Recovery Collection Lines.

As depicted in Figure 1, landfills are typically graded to drain to a central trough containing a leachate collection line or l ines. These lines are linear design elements more likely to leak than other areas of the lined landfill because these are areas where leachate flow is concentrated by design. Liquid is collected and available for episodic, saturated flow through breeches, or to eventually saturate and migrate through the compacted clay liner. In most designs, the liner is not overbuilt (thickened) along these designed drainages.

Intermodule Seams.

In an effort to distribute the cost of construction over the life of a landfill, some landfill designs provide for construction of the facility in cells or modules. Typically, the perimeter of a module is used as a haul road until the module is filled. When the next adjoining module is built, the liners from old and new modules are overlapped at the site of the haul road. However, operational practices observed at sites have resulted, at times, in filling over the haul road with refuse. In order to join the two liners, refuse from the old module must then be cut back to expose the liner edge to be coupled with the new module liner under construction. Time lags between construction of the module liners (typically measured in years), variability of material available for construction at different times, and fill practice make modules joints problematic and the seams potential vertical breeches in the liner. Since refuse is filled over the ridge formed at the joint between module liners, seams will be exposed to leachate and are more likely than adjoining areas to fail. The fact that they are at leachate drainage divides diminishes the probability of long continuous leaks. However, it does not eliminate the possibility for sudden flushing events such as might occur in response to high intensity rainfall events. Furthermore, during the period of time before the newest module is placed into operation, the seams are often located at the bottom of fill slopes which could generate a considerable concentration of runoff at times during the filling process. Landfill module construction designs and module seam designs which prevent the accumulation and ponding of leachate and water as the result of precipitation and runoff reduce the vulnerability of these to contaminant release.

Subsurface Preferred Flow Pathways.
Constructed fill placed in topographic lows, unless compacted carefully and meticulously, will most likely be at lower dry bulk densities than naturally occurring geologic material. When these areas are also underlain by clay rich deposits, they can become saturated flow pathways for leachate running along perching horizons. The source of that leachate can be from directly above or from any point topographically upgradient. Pre-existing drainages, often identifiable from pre-development topographic maps, also provide appropriate conditions for this type of subsurface flow. Other potential preferred pathways might be coarse-textured layers overlying fine textured layers identifiable on geologic cross sections constructed from borehole logs collected during the site characterization phase.

IDENTIFYING APPROPRIATE MONITORING METHODS
In order to illustrate the process of identifying appropriate monitoring methodologies, an example landfill is presented in Figure 2.

To spread the cost of construction over a longer period of time, the example landfill is being constructed in modules. Part of the landfill was constructed during a period of time before regulations in the U.S. required earthen liners. Module I (Figure 2) is not underlain by a compacted earthen liner nor overlain by an earthen cap. Module II (Figure 2) is lined but has no cap. Existing modules are being built to current landfill specifications and future adjoining modules are also planned.

Existing landfilled modules can present unique monitoring problems in that portions (for example, Module I in Figure 2) are not underlain by a compacted earthen liner nor overlain by a compacted earthen cap; likewise, Module II (Figure 2) is not capped. As such, they are susceptible to penetration of precipitation water into the existing landfilled cells which could result in landfill leachate and gases migrating from beneath the cells, particularly for unlined portions.

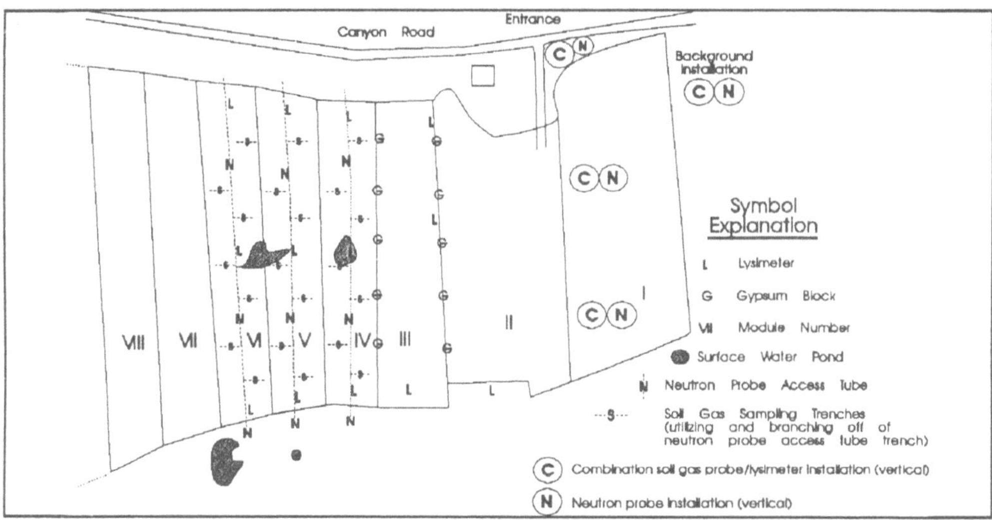

Figure 2. Plan view of example landfill illustrating conceptual design of an unsaturated zone monitoring network (Cullen and Kramer, 1992).

Drilling through a landfill to install monitoring instruments is a viable option, but it is often avoided because of health and safety considerations. Additionally, drilling through liners can compromise the integrity of low permeability layers and create flow pathways into the underlying vadose zone.

Unsaturated zone monitoring installations can be installed in areas in and adjacent to an existing unlined landfill (Modules I & II in Figure 2). When landfilled refuse (as in Cell I) has been placed on uncompacted material brought to grade, it is likely that migration of potential contaminants moving as a saturated or near-saturated wetting front in the subsurface would be in the direction of the topographic surface drainage which existed prior to grading and landfilling operations. When fill material is placed such that its depth is greatest in the topographic low-lying areas during grading operations, compaction at the time of grading will typically be least in these deeper areas. The lower density also tends to promote these relic drainage surfaces as soil gas migration pathways under dry subsurface conditions, particularly in the presence of advective or temperature gradients.

In the example, monitoring installation locations are placed on the downslope (southwest) side of cell I to identify the impact of the unlined portion of the facility on the monitoring program designed for the downslope, lined facility. Downslope installations (Module III) can be positioned using historical topographic data and constructed in the pre-existing natural drainage ways. Unsaturated-zone monitoring installations should also be placed to identify whether lateral migration of contaminants has occurred (near Canyon Road in Figure 2). Approximate locations of the unsaturated-zone monitoring installations in the example are identified on Figure 2.

Soil water content or matric potential can be measured during site characterization drilling operations in order to identify horizons of sufficient water content and matric potential which would act as contaminant migration barriers or facilitate pore-liquid sampling using suction lysimetry. Continuous soil core sampling in the fill and sediments beneath the cell can be conducted at the time of site characterization to identify the extent and degree of any existing contamination. Field screening of samples using portable gas chromatographic techniques can be used to identify the depth of penetration of volatile organic contaminants and reduce unnecessary laboratory costs associated with "non-detects". Samples suspected of being contaminated based on the field screening can be sent to a chemical laboratory.

Using Combination Installations to Monitor Contaminant Vertical Distributions
To take advantage of the most favorable aspects of both direct (lysimeters) and indirect (neutron probe) pore-liquid monitoring techniques, lysimeters and neutron access tubes can be installed in the same borehole. Combined installations can be utilized to maximize the benefit of each borehole drilled into the vadose zone. Soil gas monitoring probes can also be installed at upper levels (intersecting the most permeable strata) in the borehole to provide early warning of landfill volatile gas emissions.

Figure 3 is a schematic illustration of a vadose zone monitoring installation with neutron access tube, lysimeter and soil gas monitoring probes. Detailed geologic boring logs and "as-built" monitoring installation completion diagrams should be recorded during the drilling and monitoring station installation. A neutron probe access tube of 2-inch I.D., flush-threaded, aluminum extending the full length of the borehole is most frequently recommended. Larger size access tubes may be needed when tight curves which cannot be negotiated by the logging tool are likely to exist.

Relatively inexpensive neutron data can be used to substantiate the assumption of no leachate migration below the site, or if fluctuations occur, to identify potential preferred flow pathway strata for future lysimeter placement. Neutron logging can be used to approximate infiltration depths and rates within or beneath the waste pile and provide the data required for subsequent remedial and feasibility investigations. Neutron access tubes can provide access for moisture logs to document the presence or absence of fluctuating moisture conditions in the soils above the lysimeter installations. Cullen et al. (1991) have successfully installed and monitored combination installations at depths down to 300 feet below grade in deep unsaturated zones in California.

The goal of the combined installation is to place a lysimeter below the depth of any present contamination. When it is known that contamination is not present, the lysimeter should be placed at the first potential perching zone beneath the facility. The expected result is that if the lysimeter is able to obtain a soil pore-liquid sample, migration of particular contaminant compounds can be documented. Under dry subsurface conditions, the neutron probe can detect moisture level changes which could be associated with liquid phase migration. Such changes can also be documented along the entire extent of the access tube. Neutron logging, an indirect

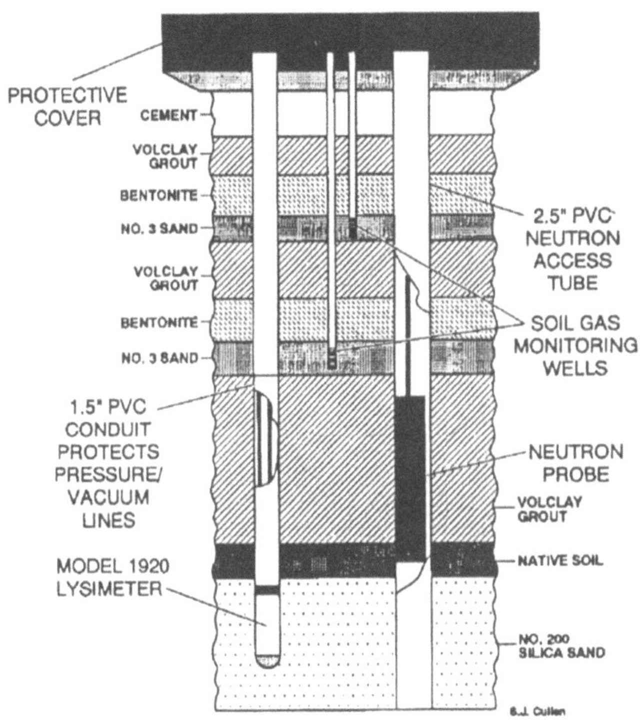

Figure 3. Schematic of combination vadose zone monitoring installation: soil gas/ neutron probe/ lysimeter (Cullen et al., 1991).

technique enables ongoing quantitative vadose zone monitoring even when subsurface conditions arc too dry to directly collect soil pore-liquid samples using suction lysimeters. Access borings should be backfilled in a manner such that a vertical conduit for contaminant migration is not created.

Direct Pore-Liquid Sampling Beneath Sumps

The second vulnerable areas at landfills are located beneath the leachate sumps. Installation of suction lysimeters beneath these sumps is recommended to detect leachate migration which may occur.

The depth at which lysimeters should be installed will be determined by the physical properties of materials encountered at depth in each boring. An important criteria is that if a low permeability silt or clay bed is encountered near the bottom of a boring, the lysimeter should be installed at that bed to intercept any leachate migrating along the bedding interface.

The lysimeters most frequently used for these types of installations are pressure/vacuum ceramic cup lysimeters. Numerous comparative studies indicate that these lysimeters deliver the best monitoring results, and generally function better than lysimeters with other porous membrane materials when the sampling range is 0 to 60 kilopascals. The pressure/vacuum units can be used at depths down to approximately 50 feet (15.2 m). A schematic illustration of a typical completed installation of a pressure/vacuum lysimeter is shown in Figure 4.

Prior to initiating field work, lysimeters should be decontaminated in order to remove any residual chemical contaminants which may have sorbed to the ceramic cups during manufacture. The decontamination procedure involves passing 1 N hydrochloric acid through each ceramic cup and then repeating the process twice using 500 ml of distilled water. Lysimeters should also be pressure- and vacuum-tested to ensure that they are functioning properly. The instruments are fragile and hairline cracks in the ceramic cups can prevent proper functioning.

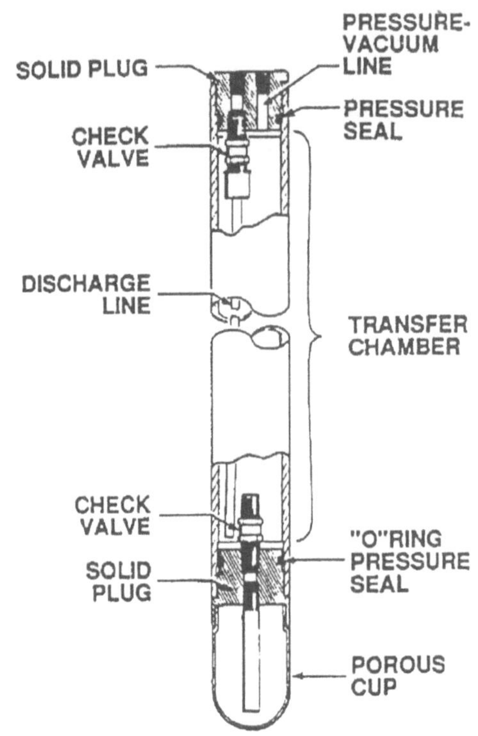

Figure 4. Cross-sectional detail of a pressure/vacuum lysimeter (courtesy Soilmoisture Equipment Corp).

Prior to installation, lysimeters should be vacuum tested and fitted with borehole centralizers. Lysimeter access tubes are threaded through small 1-1/2 inch (38 mm) diameter PVC conduit in order to protect the tubes during and after installation, and to provide a means to convey the lysimeters to the installation depth. Each borehole should be accurately depth-sounded and each lysimeter casing assembly measured to ensure precise depth of placement. It is important to ensure that the ceramic cup end of the lysimeter not be crushed by the weight of the PVC casing during or after installation. Lysimeters should be installed at the appropriate depth in a slurry of 99% pure, 200 mesh silica sand to prevent ceramic pore clogging.

Indirect Monitoring Using Tracers at Leachate Collection Sumps

One disadvantage with pore-liquid sampling is the high cost of repetitive analyses for a large number of chemicals of concern. Many landfill contaminant parameters, such as chlorine, electrical conductivity, nitrates and sulfates are commonly found at high background concentrations, particularly in arid soils. This can be compounded at some sites by the presence of nearby, offsite sources of contaminants. For example one facility observed was bounded by a feedlot directly upslope and a liquid waste discharge area for a vineyard immediately downslope.

A specific indicator parameter diagnostic of leachate escape is desirable for two reasons. First, detection of such a parameter would provide conclusive proof of leachate escape without the interpretation and uncertainty required if commonly present background constituents are used. Secondly, the analytical costs for ongoing monitoring could be substantially reduced. A commonly used conservative groundwater tracer is bromide. In a recent review, Leap (1992) concluded that bromide may actually be a better tracer than tritium in a widely varying flow environment. Bromide has been used to prove hydraulic connection between landfill leachates and springs (Murray et al. 1981). An interesting aspect of bromide which enhances its early alert warning characteristics for vadose zone monitoring, is anionic exclusion which causes it to move ahead of the advective front in clay rich materials. Bromide is nontoxic, rare in the natural environment and relatively inexpensive to analyze (by ion chromatography, or by specific ion electrode in the field). Risers at landfill leachate sumps are suitable locations for introduction of small appropriate quantities of bromide solution. Alternatively, bromide can be codispersed in the landfill leachate as a broader approach to using tracers. Knowledge of the tracer constituency in the waste is required to avoid confusion in interpreting sampling results.

Neutron Probe Monitoring Beneath Leachate Collection Drain Lines: An Indirect Technique

Monitoring by neutron probe in horizontal access tubes has several advantages discussed at length by Kramer et al., 1991, 1992; Franklin et al., 1992; Unruh, 1990; and Brose and Shatz, 1986. Its primary advantage is increased coverage at low cost, but it has the disadvantage of not discriminating between leachate and natural water flow or identifying steady state flow. The natural false positive problem is addressed by placing access tubes as close to liners and sumps as possible, without placing the sensor such that its radius of influence intersects the leachate source(s) themselves. This minimizes the natural transport pathways to the detection system. The steady-state issue is addressed by collecting background information before the landfill accepts wastes. Steady state flow from a slow continuous leachate leak would become apparent at its outset when compared to background. Use of indirect techniques require that indicated leaks subsequently be confirmed chemically with a direct sampling technique.

The neutron probe monitoring design considered here is horizontal neutron access tubes beneath leachate collection lines emplaced approximately two feet below the liner in shallow trenches dug prior to liner construction. This design completely covers the third high risk area of concern: the leachate collection pipeline along the axis of each module. Figure 5 shows a landfill liner cross section with typical trench design to facilitate installation of the neutron access probe, lysimeter access tubes, and soil gas sampling lines. Figure 6 illustrates cross-sectional detail of the neutron probe access tube trench with the emplacement of pressure/vacuum lysimeters beneath. Figures 7 and 8 illustrate cross-sectional detail of the termination of the neutron probe access tube and lysimeter/soil gas monitoring tubes at the landfill margin.

Indirect monitoring with electrical resistance blocks

Intermodule seams, the seams in an earthen liner which will occur at the joints between modules represent potential contaminant migration pathways and leachate leak points. Monitoring points should be sited in order to assess the performance of the engineering design and operational practices. In the example, gypsum blocks, equally spaced along the upslope (northeast) margin of module III, are installed to monitor for pore-liquids emanating from the landfill at this vulnerable point. Additional gypsum blocks are similarly installed at the downslope (southwest) margin of module III. The approximate location of the gypsum block installations are shown on Figure 2.

Gypsum Blocks.

Gypsum blocks are a specific design of what are more generically known as electrical resistance blocks. Electrical resistance blocks measure the electrical potential, using an AC half bridge, between two wires spaced at a known distance apart. This distance is typically on the order of centimeters. The wires are imbedded in a porous matrix through which soil pore-liquids can freely enter and leave. The water in the porous matrix of the block achieves an equilibrium with the surrounding soil water. When soils (and thus the electrical resistance

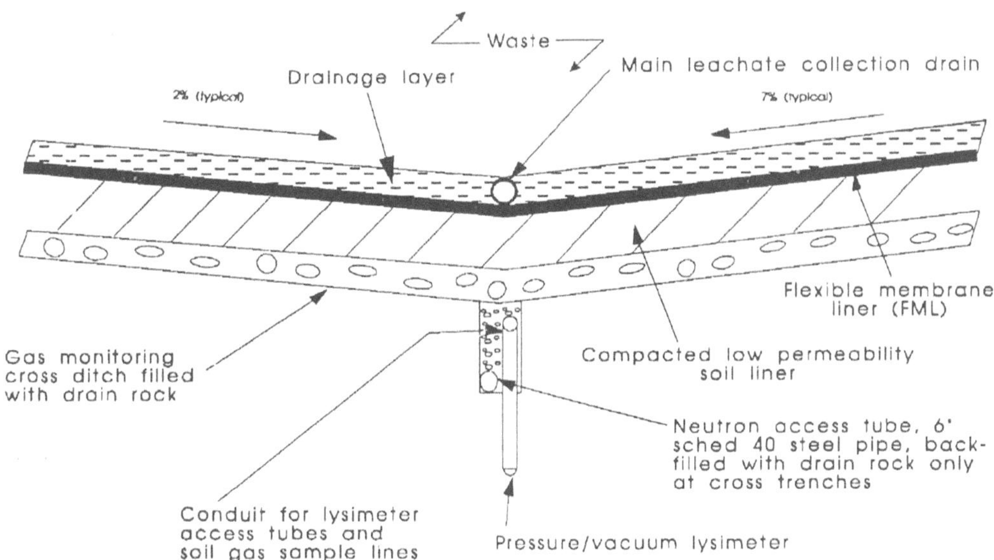

Figure 5. Landfill liner and monitoring cross section, end view (Cullen and Kramer, 1992).

block) are dry, the resistance to electrical current flow is high and can be read on a resistance meter. Conversely, when the soil and block become wet, a low resistance is measured on the meter.

Native and landfill-originated salts can affect the electrolytic concentration of pore water and consequently the electrical resistance of the pore-liquid solution. To avoid this problem, the most effective electrical resistance blocks are constructed of gypsum (calcium sulfate), one of the most soluble of naturally occurring soil salts. Because of its high solubility, the gypsum in gypsum blocks is continuously dissolving into soil solution creating a background condition of electrolytes which buffers the effect of native or landfill-generated soil salts on the measurement.

This high solubility of gypsum also limits the life and accuracy of gypsum blocks. The accuracy of gypsum blocks is generally considered to be approximately +/- 15%. They should not be considered as a highly accurate method for the measurement of soil suction. However, they are quite acceptable as a means of detecting the arrival of a wetting front in an otherwise dry soil.

Gypsum blocks are recommended only as part of a short term unsaturated zone monitoring program. They can be effectively used to provide an inexpensive means of quality assurance. In the example, they are used to insure the integrity of the compacted earthen liner joint between modules.

When monitoring beneath intermodule seam joints, lysimeters are also recommended for installation at the seam between modules to enable collection of pore-liquid samples for chemical analysis in the event of a wetting front detection by the gypsum blocks. These kinds of installations can be used as a dual function monitoring point to detect vertical migration of leachate from the module seam joint, and to monitor lateral leachate migration from beneath upgradient modules. Approximate location of the lysimeter installations in the example are depicted in Figure 2.

Direct Monitoring of Soil Gases Beneath Landfills

Soil gas migration, particularly under advective, temperature and density-driven gradients and/or low soil water potential gradients, can precede pore-liquid migration in unsaturated regions of the vadose zone. When soil is dry beneath a facility, unsaturated hydraulic conductivities typically are very small (on the order of 10^{-7} cm/sec and smaller). Air-filled pore space is increased and conditions conducive to vapor transport by advection or diffusion are maximized. A soil gas monitoring system is an effective means of detecting volatile organic compound (VOC) contaminant migration which can occur in the vapor phase under unsaturated soil conditions.

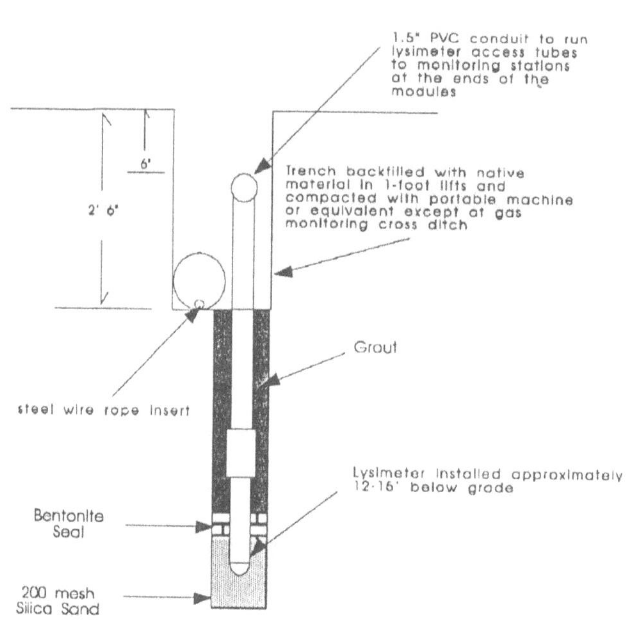

Figure 6. Detail of neutron probe access trench with lysimeter/ soil gas conduit (Cullen and Kramer, 1992).

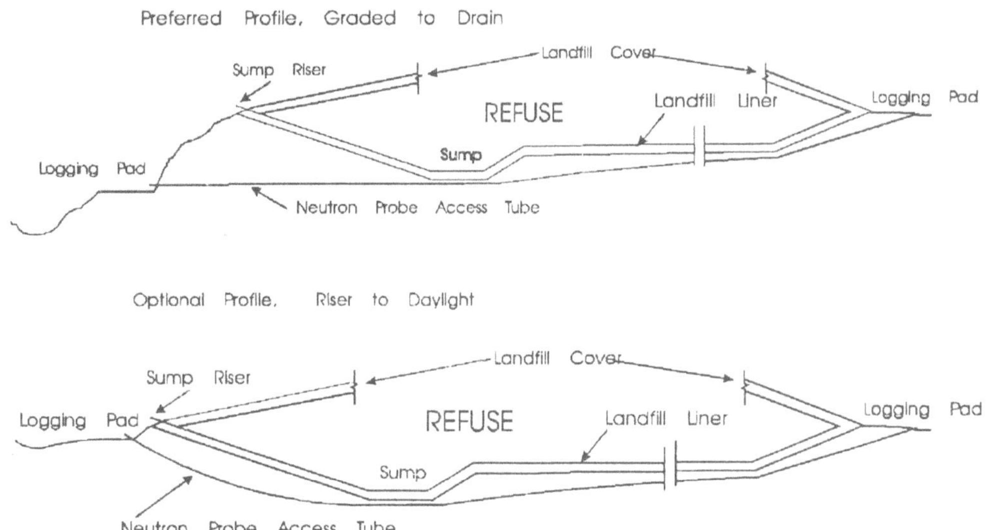

Figure 7. Cross section through typical landfill module showing neutron probe access tube position (Kramer, 1994).

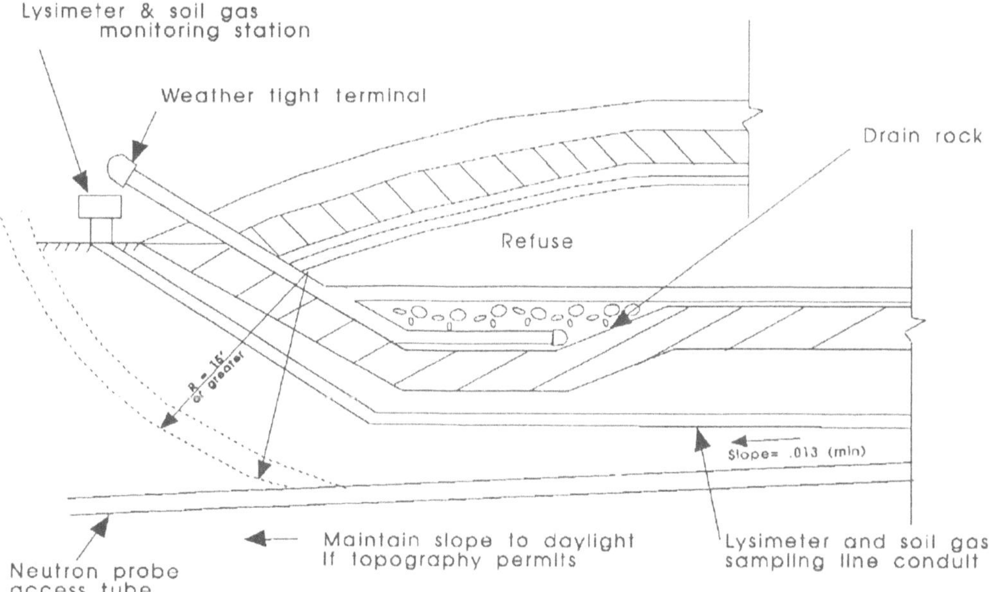

Figure 8. Termination of neutron probe access tube, lysimeter access tubes, and soil gas sampling lines at landfill margin (Kramer, 1994).

A soil gas monitoring design is presented for the example waste landfill. In the example, soil gas sampling points consist of shallow trenches which are dug concurrently and perpendicular to the trench to be used for the neutron probe access. The trenches are spaced along opposite sides of, and perpendicular to, the module centerline approximately 125 feet (39 m) apart (Figure 2). Each trench is excavated six inches (15 cm) wide, six inches (15 cm) deep below grade and backfilled with drain rock. Each trench is approximately 50 feet (15 m) in length perpendicular to the neutron access tube trench (Figure 5).

In the example, four of the sample lines are run to the northwest margin of the landfill module IV and three run to southeast margin of the landfill module. One main conduit, 1000 feet (303 m) in total length (1" [25.4 mm] inside diameter Schedule 80 PVC), is buried below the upper most lift (15 cm deep) in the backfill of the neutron probe access tube trench. **(Note: In all cases where protective conduit is used, no glues or adhesive should ever be used to join sections. These adhesives may contain VOCs which can result in false positives during monitoring activities).** At each soil gas trench, a reducing tee is connected to a smaller (.5" [1.3 cm] PVC, Schedule 80 for example) conduit consisting of five flush-threaded 10-foot sections. Three alternate 10-foot (3 m) sections are screened .5 mm. Going away from the tee, the conduit can be described as 10 feet (3 m) screened, 10 feet (3 m) unscreened, 10 feet (3 m) screened , 10 feet (3 m) unscreened, 10 feet (3 m) screened for a total of 50 feet (15 m) in length. A fitting is installed on the inside of the reducing tee to connect to 1/8" (3.2 mm) polyethylene tubing which is threaded inside of the 1" (25.4 mm) conduit for protection and run back to the margin of the landfill for sampling access.

Analysis of soil gas samples can be cost-effectively made in the field using field-portable gas chromatographs or organic vapor totalizers. The method is similar to the method recommended for soil sample screening. If VOCs are detected and confirmed using portable soil gas analysis instrumentation, a follow-up sampling event should be immediately scheduled with a whole air sample taken for laboratory analysis and speciation to provide final confirmation of the contaminant.

Direct Monitoring of Subsurface Preferred Flow Pathways
In the case of a catastrophic leak caused by a breach in the landfill liner, saturated pore-liquid migration is likely to result. Subsurface preferential flow of this nature will likely follow the topographic surface which existed prior to filling activities.

In order to intercept and detect these migration events, suction lysimeters are recommended. Depending on the final location and elevation of the modules as the result of grading activities, lysimeters can be located along the length of and into the bottom of the trench used to install the neutron probe access tube (Figure 2). The lysimeters should be installed where permeable layers overlay finer-textured, less permeable material. Final depth of placement should be determined when the actual installation boring is conducted. Cross-sectional detail of the installation of a lysimeter in the bottom of the neutron probe access tube trench is shown in Figure 8. In the example, one lysimeter is installed beneath each module approximately 50 feet southeast of the northwest margin of the landfill to detect potential offsite leachate migration. A lysimeter is also installed along the historic surface drainage that bisects the length of the landfill modules.

Background Instrumentation
Careful consideration should be given to the locations of background monitoring points. Monitoring data collected at these points will be compared to data collected from the "active" areas. The resulting comparison and statistical analysis will form the basis of decision-making regarding a suspected release from a facility.

In groundwater monitoring system design, background sampling points are typically selected in an upslope or upgradient position with respect to groundwater flow. In any case, the goal is to place background instrumentation in locations where samples are representative of the nonactive area of the site. The direction of water flow in unsaturated regions of the vadose zone, however, does not necessarily coincide with the direction of saturated groundwater flow and procedures for locating background groundwater monitoring wells are not applicable to unsaturated zone applications.

Problems in locating appropriate background monitoring points are not uncommon. Based on site characterization data, uncontaminated areas of the site should be identified which are as similar as possible to the active site with respect to site stratigraphy, lithology, and geomorphic position. Background data should be collected following protocols which mimic those used in the active area beneath the active site.

Since background sampling points are often near property boundaries, background data can be influenced by offsite conditions. In the example, the site was bounded by two feedlots that were located upslope and a vineyard applying reclaimed wastewater was located just downslope. All three nearby enterprises represent potential sources of contaminants which could compromise the goal of background sampling at the subject site. Background sampling points should be located away from areas towards which natural surface and subsurface drainage patterns will channel water during precipitation events. Areas towards which the engineered design of the facility will channel leachate or create ponding should also be avoided.

Whenever possible sampling should be conducted in the active area prior to initiation of activities at a waste disposal site in order to provide a baseline comparison of the active versus background areas. Adequate background consists of data collected from all the phases monitored in the active portions (i.e. soil cores, pore-liquid, and soil gases). Background data should be collected for the indirect as well as direct sampling techniques.

CONCLUSION

As with groundwater monitoring system design, complete coverage is often not a realistic goal for monitoring the vadose zone. Early warning of contaminant migration can be reliably accomplished by implementing a design approach which addresses waste disposal site locations most vulnerable to contaminant release and migration. Since contaminants can migrate in the liquid or gaseous phase beneath waste disposal sites, an effective monitoring system must have the capability of detecting multiphase migration. A realistic monitoring system should also minimize the cost of periodic monitoring. A monitoring system which combines indirect monitoring techniques (such as the neutron moisture probe, conservative tracer sampling, and electrical resistance blocks) and direct sampling techniques (such as pore-liquid sampling and soil gas sampling) can provide early warning of contaminant migration as well as diagnostic confirmation of contaminant species without the large expense typically attendant to monitoring systems which rely solely on regular direct sampling and laboratory analysis. While vadose zone monitoring can provide early detection of contaminant migration for the purpose of preventing eventual groundwater contamination, it should not be considered a replacement for groundwater monitoring. The water quality status in an aquifer can only be determined by direct groundwater sampling.

REFERENCES

Brose, Richard J. and Richard W. Shatz, 1986. Neutron Monitoring in the Unsaturated Zone. *In*, First National Outdoor Action Conference on Aquifer Restoration, Ground Water Monitoring and Geophysical Methods. NWWA, Dublin, OH, p. 455-467.

Bumb, A., C. McKee, R.B. Evans, and L.A. Eccles, 1988. Design of Lysimeter Leak Detector Networks for Surface Impoundments and Landfills. Groundwater Monitoring Review, Spring.

Cullen, Stephen J., and J.H. Kramer, 1992. Conceptual Design of a Vadose Zone Monitoring System for A Modular Solid Waste Landfill, Monterey County, California. Report to Woodward-Clyde Consultants by Stephen J. Cullen and Associates, P.O. Box 1825, Santa Ynez, CA, USA 93460.

Cullen, Stephen J., W.F. Allmon, and B.K. Keller, 1991. China Grade Sanitary Landfill: Vadose Zone Monitoring Program, Report to County of Kern. Department of Public Works,

Bakersfield, CA.

EPA, 1988. Guide to Technical Resources for the Design of Land Disposal Facilities. Technology Transfer, EPA/625/6-88/018. Center for Environmental Research Information, Risk Reduction Engineering Laboratory, U.S. Environmental Protection Agency, Cincinnati, OH 45268.

Franklin, James, J., Mark E. Unruh and Vince Suryasasmita, 1992. Neutron Probe Monitoring in the Unsaturated Zone Case Histories from Several Sites Comparing Problems and Utility of Horizontal and Vertical Access Tube Installations. *In*, Proceedings of the Sixth National Outdoor Action Conference, NGWA, Dublin, OH.

Kramer, John H., 1994. Vadose Zone Monitoring Strategies Employing Horizontal Neutron Moisture Logging, Ph.D. Dissertation, University of California.

Kramer, John H., Lorne G. Everett, and S.J. Cullen, 1991. Innovative Vadose Zone Monitoring at a Landfill Using the Neutron Probe. *In*, Proceedings of the Fifth National Outdoor Action Conference of Aquifer Restoration, Ground Water Monitoring and Geophysical Methods. NWWA, Dublin, OH.

Kramer, John H., Stephen J. Cullen, and L.G. Everett, 1992. Vadose Zone Monitoring with the Neutron Moisture Probe, Groundwater Monitoring Review, v.12, No.3, pp 177-187.

Leap, T.H. 1992. Apparent Relative Retardation of Tritium and Bromide in Dolomite, Ground Water, v.30, no.4, pp 549-558.

Murray, J.P., J.V. Rouse, and A. B. Carpenter, 1981. Ground Water Contamination by Sanitary Landfill Leachate and Domestic Wastewater in Carbonate Terrain: Principal Source Diagnosis, Chemical Transport Characteristics and Design Implications, Water Research, v.15, pp 745-757.

Sara, Martin N., 1991. Ground-Water Monitoring System Design. *In* Nielsen, David M. (ed.), Practical Handbook of Ground-water Monitoring, Lewis Publishers, Chelsea, MI.

Unruh, Mark,E., Christopher Corey, and John M. Robertson, 1990. Vadose Zone Monitoring by Fast Neutron Thermalization (Neutron Probe): a 2-year Case Study. *In*: Ground Water Management, Number 2, NWWA, Dublin, Ohio, p.1303-1317. 431-444.

CITATION:

Cullen, Stephen J., John H. Kramer, and R.T. Ogg, 1994. "A Systematic Approach to Designing a Multiphase Unsaturated Zone Monitoring Network". *In* G. Gambolati (ed.), <u>Proceedings of the International Symposium on Advanced Methods for Groundwater Pollution Control, May, 1994, Udine, Italy</u>. Published by the International Center for Mechanical Sciences, Udine, Italy (undergoing review).

APPLICATIONS OF TIME-DOMAIN REFLECTOMETRY TECHNIQUES TO FIELD-SCALE TRACER TESTING IN UNSATURATED SEDIMENTS

D.L. Rudolph and P. Ferré
University of Waterloo, Waterloo, Ontario, Canada

ABSTRACT

Field scale tracer testing in partially saturated sediments has proven to be extremely challenging due, in part, to limitations of available monitoring instrumentation. Recent developments and applications of time-domain reflectometry (TDR) techniques to measure volumetric water content and solute mass flux rapidly and non-destructively have provided new flexibility in tracer testing design. Installation of networks of thin TDR probes in the subsurface have very little influence on natural flow conditions and can permit detailed spatial and temporal monitoring of transient water content distributions and solute migration in three dimensions. The TDR techniques are described in this paper along with two field-scale tracer tests to illustrate the application of the TDR methods to evaluating water flow and solute transport behaviour in the unsaturated zone.

1. INTRODUCTION

The movement and fate of dissolved contaminants in the vadose zone has been an area of increasing research activity over the last few years. Sources of contamination ranging from surface - applied agricultural chemicals, septic tile beds and organic compounds from tank leaks and spills are significantly influenced by solute transport behaviour in the partially saturated zone above the water table. Field investigations designed to study solute transport phenomena in the vadose zone are extremely challenging due, in part, to the non-linear characteristics of the main physical parameters and the difficulty in collecting sufficient temporal and spatial data at the field scale.

Controlled field-scale tracer testing has proven to be a useful approach in understanding transport behaviour under both saturated and unsaturated conditions. Large-scale tests in saturated media such as those reported by [1, 2] have contributed significantly to the understanding of contaminant migration in spatially-variable sediments. These long-term experiments rely on the natural groundwater gradient to transport carefully injected tracer mixtures through a dense network of monitoring wells. Samples extracted from these wells permit transient tracking of the tracer cloud which can subsequently to be used to determine field-scale values of the flow and transport parameters that control contaminant migration.

This type of field tracer testing has been attempted in partially saturated sediment, generally on a smaller scale, by many researchers. A common experimental design has included the installation of tensiometers, solution samplers and neutron access tubes in a field test plot where a controlled irrigation system has been constructed and a solute tracer can be released. These experiments have clearly indicated a significant degree of complexity in solute transport phenomena due to the spatial variability of natural sediments. In addition, field-scale tests have illustrated the requirement for dense temporal and spatial data to appropriately evaluate the transport phenomena [3, 4, and 5].

In comparison with tracer testing under saturated conditions, field tracer tests in unsaturated media present several additional challenges. Due to the irregularity of infiltration events, a natural, stable flow gradient is rarely encountered in the vadose zone and must be artificially established through controlled irrigation. In addition, not only are the controlling physical parameters variable in space but they are also a function of varying water content within the flow domain. This increases the complexity of the measurements required.

A significant limitation to implementing controlled tracer tests in the vadose zone relates to restrictions on the monitoring instruments themselves. Mapping solute distributions with solution samples requires an extensive instrument network which can be destructive of in situ conditions, labour-intensive to sample and often unacceptably slow to monitor in higher permeable sediments. Also, analysis of the samples can be extremely costly. Water content distributions monitored with a neutron probe can also be restrictive due to the requirement for numerous access tubes which are destructive. Neutron probe measurement can be restrictively time-consuming if a highly transient wetting front is being monitored.

In an attempt to overcome some of these limitations, recent field-scale tracer tests in unsaturated sediments have made use of emerging technologies for data collection [6, 7, 8]. Of particular interest has been the application of time-domain reflectometry methods (TDR) to the measurement of soil water content and solute distribution. These techniques are essentially non-destructive, permit dense spatial and temporal monitoring and can be used to measure in situ soil water content [9] and vertical solute mass flux [10]. The TDR techniques permit the design of field-scale tracer tests in unsaturated sediments that allow transient solute transport behaviour to be observed in much more detail even in heterogeneous porous media.

In this paper, two tracer tests are described and briefly evaluated to demonstrate the application of the TDR techniques. The first test was conducted at the U.S. Geological Survey's Toxic-Substances Hydrology research site on Cape Cod, Massachusetts and was designed to observe the transient nature of soil water and solute movement under controlled irrigation conditions in course sands and gravels. The second experiment was conducted at the Waterloo Groundwater Research Station at Canadian Forces Base Borden, Ontario. This experiment focused on evaluating the affectiveness of soil flushing as an enhanced remediation technique to remove solute contamination trapped in the vadose zone after standard pump and treat procedures were applied. The results from these experiments will be discussed in terms of the general application and limitations of the TDR methods to field-scale tracer testing and some recent advancements in TDR technology that appear promising will be described.

2. APPLICATIONS OF TIME DOMAIN REFLECTOMETRY

The application of time domain reflectometry (TDR) techniques to the estimation of soil water content has been gaining popularity since its introduction in the early 1980's [9]. The TDR method involves the measurement of the rate at which an electromagnetic wave propagates through soil. A step pulse is generated by the TDR device (a Tektronix Model 1502B cable tester is commonly used) and transmitted along a pair of parallel metal rods. With the rods placed in the soil, the transmitted electromagnetic pulse propagates as a plane wave along the rods which function as a wave guide [11]. The wave velocity is controlled by the dielectric characteristics of the porous medium between the rods. The cable tester records the time for the plane wave to enter the soil and return back by reflection from the ends of the rods. The travel time can then be used to determine the dielectric constant (K) of the soil.

The dielectric constant of the soil is a function of soil material K_s and the water K_w. Since a large contrast exists between K_s ($\approx$ 4) and K_w ($\approx$ 80), as the water content in the soil increases, the bulk dielectric constant K_b progressively increases. The measured K_b from the TDR method can then be used to estimate volumetric water content (Θ) through, for example, the relationship developed by [9],

$$\Theta = -0.053 + 2.92 \times 10^{-2}(K_b) - 5.5 \times 10^{-4}(K_b)^2 + 4.3 \times 10^{-6}(K_b)^3$$

(1)

This relationship was shown to be relatively insensitive to the physical characteristics of the soil.

The metal rods used as wave guides are typically less than 0.5 cm in diameter and are generally installed in the field in a vertical orientation. Typical rod lengths range between 20 cm and 250 cm. These limits are controlled by the sensitivity and transmission strength of the cable tester. The TDR method estimates the K_b of the material contained along the entire length of the wave guides and as such represents a length weighted average. In order to measure a vertical water content profile, sets of wave guides of various lengths are installed in close proximity at a point of interest. Soil water content is determined for each wave guide pair and the average water content over a given depth increment is determined by differencing the values from adjacent rod pairs. The water content of the non-overlapping portion of different length rod pairs is determined by

$$\Theta_n = \frac{\Theta_l L_l - \Theta_s L_s}{L_l - L_s}$$

(2)

where L is the interval length and n, s and l represent the non-overlapping portion, short rod length and long rod length respectively [12].

Several aspects of the TDR method for the determination of soil water content are particularly advantageous. Due to the narrow diameter of the wave guides, the instrumentation is not very destructive to the in situ sediment. Numerous sets of TDR rods or probes can be installed in a given soil volume without interfering with the hydraulic behaviour of the soil. Consequently, a dense network of spatial data can be collected. In addition, an individual measurement with the TDR device can be made in seconds by reading data directly from the cable tester screen or by storing the data in a portable computer. As such, the network of TDR probes can be monitored rapidly so that transient water content data can be obtained on a detailed level temporally.

The TDR method can also be used to monitor the movement of an electrically conductive tracer pulse introduced to the soil water flow system as a conservative solute [13]. The bulk electrical conductivity or related impedance load (R_L) of the soil can be measured by the TDR method and is a function of the total mass of tracer between the TDR probes. The change in R_L over time is linearly related to solute mass flux past the TDR probes. This permits a direct measurement of the travel-time density functions or solute breakthrough curves at each point where the TDR probes have been installed [10].

The electromagnetic signal recorded by the TDR device (cable tester) becomes weakened or attenuated as the R_L of the soil decreases. When the water content remains constant between a given set of TDR probes (i.e. flow system is at steady-state) the total mass of solute ($M_{T,L}$) contained within the probes between ground surface and the ends of the probes is given by [10]

$$M_{T,L} = (\frac{L\,\Theta}{\propto})\left[\frac{1}{R_{L(t_o)}} - \frac{1}{R_{L(t_i)}}\right] \tag{3}$$

where L is the probe length, Θ is average volumetric water content between the probes, $\propto$ is an empirical constant related to soil and measurement device properties, t_i refers to time before the solute pulse is applied and t_o is a given time after application of the pulse but before any solute moves past the end of the probes.

The relative solute mass ($M_{R,L}$) at any given time (t) after release of the pulse can be estimated by [10]

$$M_{R,L}(t) = \left[\frac{R_L(t)^{-1} - R_L(t_i)^{-1}}{R_L(t_o)^{-1} - R_L(t_i)^{-1}}\right] \tag{4}$$

or the solute flux concentration $C_{F,L}(t)$ can be determined by [10]

$$C_{F,L}(t) = \frac{\partial(R_L(t)^{-1})}{\partial t}\left[\frac{1}{R_L(t_o)} - \frac{1}{R_L(t_i)}\right]^{-1}\frac{M_{T,L}}{q} \tag{5}$$

where q is the steady-state infiltration rate.

Through the impedance measurements with the TDR method, (4) can be used to construct solute breakthrough curves for each set of TDR probes. By fitting the solution to the convective-dispersive equation for relative mass remaining [14] to the breakthrough curves, estimates of effective solute velocities and dispersion coefficients can be obtained. Due to the non-destructive nature of the TDR instruments and the ease with which the measurements are made, detailed information on solute transport properties can be obtained in situ without the use of labour-intensive solution sampling. As with any geophysical technique, however, the physical information required is inferred or derived from a different set of material properties measured by the geophysical device and care must be taken during the interpretation process.

The TDR techniques provide a method for rapid and non-destructive measurement of soil water content and solute distribution in partially saturated sediments. Through appropriate implementation these techniques can prove extremely valuable in field-scale tracer tests. Two such tracer tests are described below with examples of the types of data sets that can be obtained employing the TDR methods.

3.0 INFILTRATION EXPERIMENTS AT CAPE COD, MASSACHUSETTS

The United States Geological Survey (U.S.G.S.) has been involved in studies of large-scale groundwater contamination problems on Cape Cod, Massachusetts since the mid-1980's. Many of these studies have been focused at the U.S.G.S. Toxic-Substances Hydrology research site located near Falmouth, on Cape Cod (Figure 1). The majority of the field investigations have focused on groundwater flow and solute transport in the saturated zone and have included a large-scale natural gradient tracer test [1] and investigations of the spatial variability of hydraulic properties in the aquifer [15].

The stratigraphy at this site consists of approximately 30 m of glacial outwash sand and gravel that overlie fine sand and silt and crystalline bedrock. The water table at the site is on average 5 m below ground surface and typically fluctuates about 1 m annually. This aquifer represents the only source of drinking water for the region [6].

Very little work had been done at the research site related to the investigation of water and solute movement in the vadose zone above this unconfined aquifer. Considering the number of surface sources of contamination in the vicinity, significant interest in the behaviour of contaminants in the vadose zone had developed. To this end, a series of field-scale tracer tests were designed and implemented at the research site to investigate the hydraulic and solute transport characteristics of the unsaturated sand and gravel sediments.

The main objectives of the tracer experiments were to evaluate the feasibility of using TDR methods to track soil water and solute tracers in controlled flow field conditions and to provide field-scale data to improve the interpretation of contaminant transport behaviour in the unsaturated sediments.

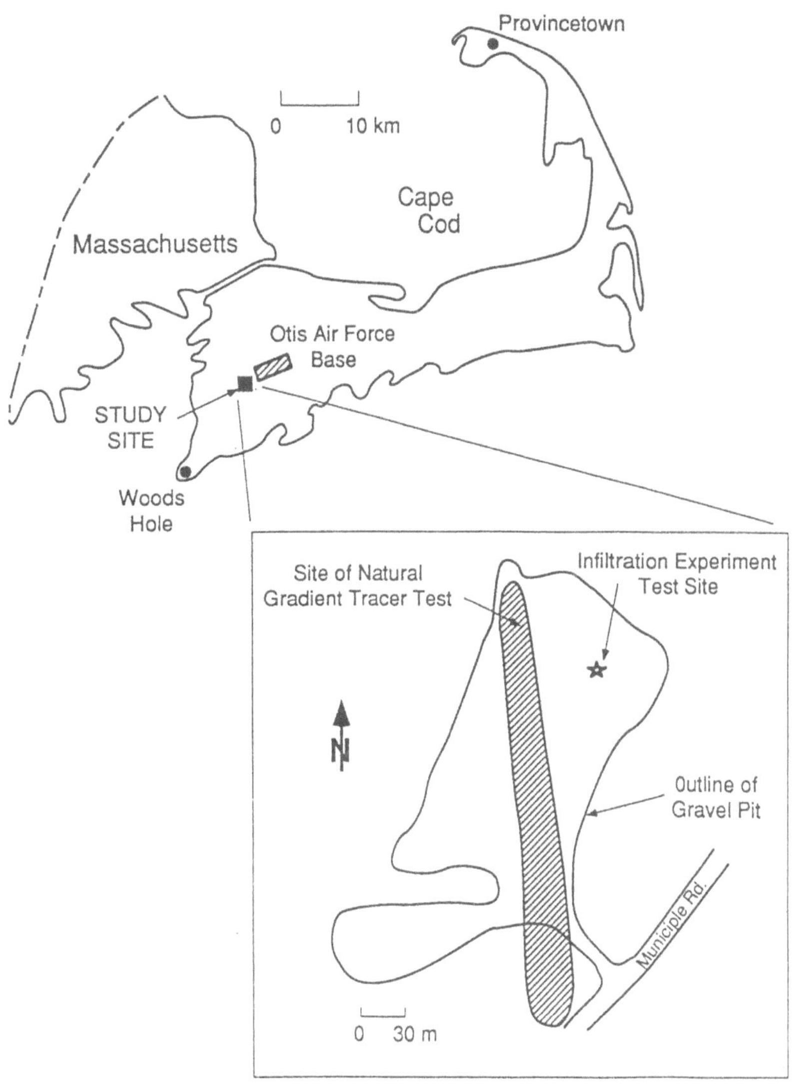

Fig. 1,   Location map of Cape Cod field site (after [16])

3.1 Experimental Design

The research site is located in an abandoned gravel pit and most of the ground surface was void of vegetation. A rectangular area 12 m long and 3 m wide was prepared by excavating 20 cm of the pit floor to expose a relatively undisturbed surface. Within this cleared area, a test plot approximately 10 m long and 2 m wide was selected as the focus of the field experiments. This site was located immediately adjacent to the area where the natural gradient test was conducted (Figure 1).

The general design of the experiment involved two parts. The controlled irrigation of the test plot at various flow rates along with the tracking of the transient water content distribution below the test plot was the first step. Once steady-state flow conditions were established at each infiltration rate, a solute pulse was released at ground surface and flushed through the monitored domain and transient spreading of the tracer cloud was tracked.

The subsurface instrumentation involved alternating rows of porous cup lysimeters and TDR probes separated by 12.5 cm (Figure 2a). The lysimeter rows consisted of four cups, each placed at equal spacings from depths of 50 cm to 200 cm for a total of 112 instruments (Figure 2b). The rows of TDR probes included 6 sets, each increasing in depth by 25 cm from 25 cm to 100 cm depth, and by 50 cm up to 200 cm total depth (Figure 2b). A total of 168 sets of TDR probes were installed. The entire instrumented plot was 7.0 m long and 0.5 m wide located in the centre of the overall test plot (Figure 2a). Irrigation water was applied to the ground surface through a series of overhead sprinklers that provided a uniform distribution over the 10 m x 2 m test plot. The solute pulse was applied as a liquid solution through hand sprinklers over the entire irrigated test plot.

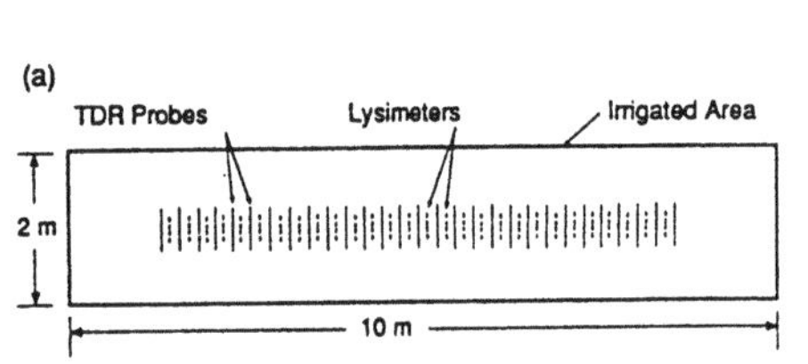

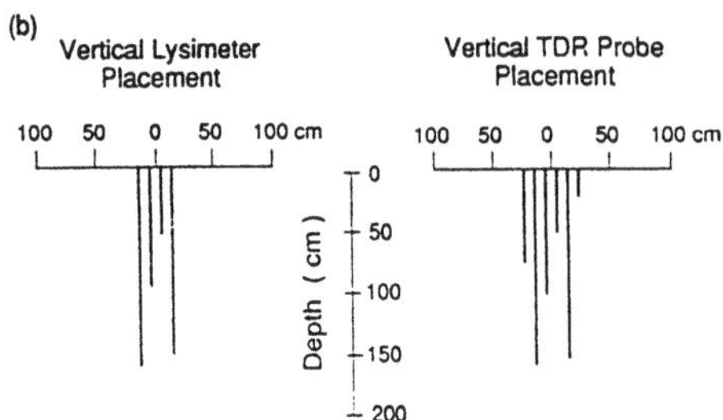

Fig. 2,   a) Plan-view of irrigated test plot and instrumentation, b) Vertical configuration of instruments (after [16]).

The lysimeters were connected to a common vacuum system so that the entire network could be sampled simultaneously. The TDR probes were connected through transmission cables to common access boxes to permit easy and rapid measurement with a cable tester. Infiltration experiments were conducted at irrigation rates ranging from 7.9cm/hour to 37cm/hour and the solute pulse applied was a NaCL solution. A more complete description of the experiments can be found in [6].

3.2 Results

A subset of the data collected during the experiments is presented here to illustrate the performance of the TDR methods [17]. One of the objectives of the experiment was to examine the spatial distribution of the infiltrating water front at discrete points in time along the test plot. The 168 TDR probes could be scanned in less than 8 minutes using two cable testers. The results of two sets of water content measurements are shown in Figure 3 for an infiltration rate of 24.8 cm/hour. The variable nature of the water content distribution is clearly evident as the wetting front proceeds down through the 2 m instrumented depth. In order to make this density of measurements with a neutron probe, 28 access tubes would be required and if 2 probes were used, approximately 2 to 2.5 hours would be required, assuming 1.5 mins. per measurement and physical movement of the probe. At these high infiltration rates, it would be impossible to capture the water content distribution that would be representative of a specific time period after the start of infiltration.

These types of data sets can be used to examine the statistical nature of the water content distribution as irrigation rates change and can provide a valuable data base for numerical simulation.

By monitoring the water content variations with the TDR probe network, one can determine when steady-state flow conditions are achieved. For the solute transport experiments, the conservative, electrically-conductive tracer was released into the steady-state flow field. The movement of the tracer was monitored by both the solution samplers and the TDR probes. For the case of an infiltration rate of 7.9 cm/hour, Figure 4 shows the breakthrough curve recorded by both techniques at a depth of 50 cm midway along the plot length. Both curves show very similar characteristics indicating the applicability of the TDR method.

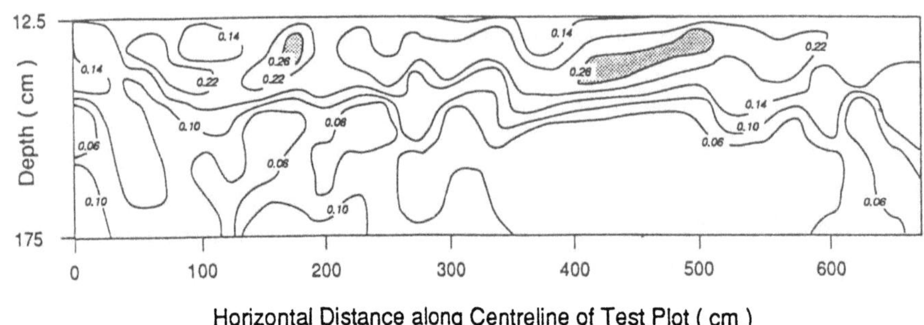

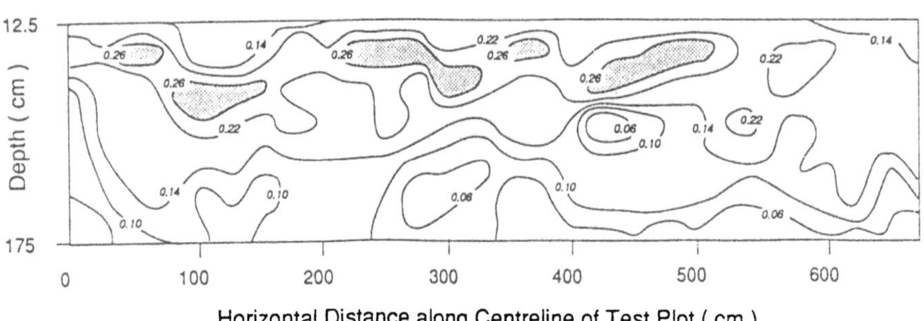

Fig. 3,   Distribution of volumetric water content along centreline of test plot at infiltration rate of 24.8 cm/hour, a) 30 mins. after start of infiltration, b) 60 mins. after start of infiltration (after [17]).

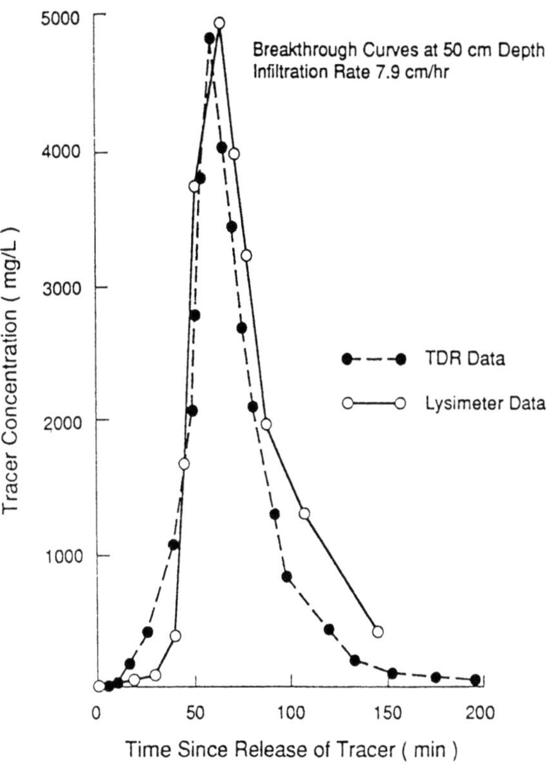

Fig. 4,   Tracer breakthrough curves at 50 cm depth at midpoint of test plot under 7.9 cm/hour infiltration rate as measured by solution samplers and TDR probes (after [17]).

If the breakthrough curves at each depth are averaged along the entire length of the test plot, field-scale behaviour can be examined as indicated in Figure 5. Here the plot-scale breakthrough curves at 50 cm and 100 cm are indicated based on the data from the TDR probes. Again scanning of the TDR network could be done in approximately 5 minutes and could easily be repeated as required during the solute transport experiment. An equivalent data set from the solution samplers would be significantly more labour-intensive to collect and would result in the need to analyze approximately 3,500 samples for each infiltration experiment.

The detailed spatial and temporal data that can be collected by the TDR method in this type of tracer test can permit estimates of solute transport parameters to be made at various scales and infiltration rates through appropriate curve fitting techniques. As an example, the dispersion coefficient (D) determined for the two field-scale breakthrough curves illustrated in Figure 5 were determined by matching the data to the analytical solution as described earlier and a clear increase in D was observed as the solute cloud moved deeper in the profile (D_{50} = 2.63 cm^2/min, D_{100}=4.73 cm^2/min). Again, this type of field data is useful in the development of predictive simulations of contaminant transport behaviour in unsaturated sediments.

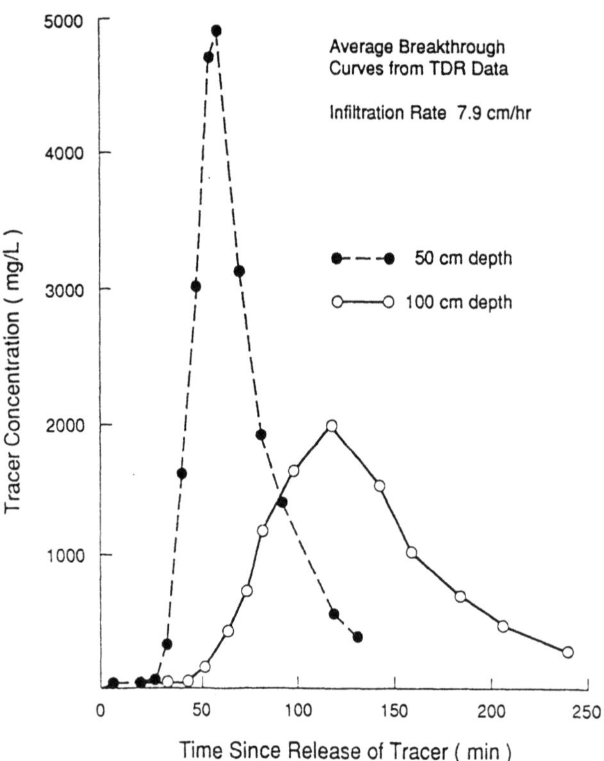

Fig. 5,   Field-averaged tracer breakthrough curves under 7.9 cm/hour infitlration rate at 50 cm depth and 100 cm depth (after [17]).

4.0 SOIL FLUSHING EXPERIMENTS AT BASE BORDEN, ONTARIO

The shallow soils at Base Borden are uniform, fine-grained progradational beach sands that have been widely studied and described [16]. University of Waterloo researchers have performed natural gradient tracer tests to investigate the spatial variability of both the flow and transport properties of the saturated Borden aquifer material [2]. Further investigations have also been conducted to study the movement of pure phase NAPLs through the unsaturated zone [18] in addition to the response of the unsaturated zone to operation of a pumping well [19]. The objectives of the experiment described below were to examine flow and solute transport continuously through the unsaturated zone and across the water table to assess the ability of a pumping well to remove dissolved contaminants from the subsurface with and without the use of soil flushing. A more complete description of the experiment can be found in [20].

4.1 Experimental Design

The surface of a 3 metre by 3 metre experimental site was excavated to provide a relatively undisturbed surface approximately 2 metres above the natural water table. A 7.5 cm diameter monitoring well was installed in the centre of the site. Six monitoring points were distributed around the well as shown in Figure 6. Each monitoring point included an array of monitoring probes designed to measure throughout the unsaturated and saturated zone while accommodating potential variations in the water table elevation. A multilevel piezometer with 10 screens between 1.15 and 3.40 metres depth and a centre stock for water table elevation measurement was located at each point. A DC resistivity probe with 26 electrodes allowed for 23 measurements of the bulk electrical conductivity between 1.05 and 3.25 metres depth. Five lysimeters were used to collect soil water samples from the unsaturated zone at each monitoring point. Finally, a multilevel TDR probe as described in [12] was used to measure both water content and bulk electrical conductivity at 10 cm intervals throughout the vadose zone.

The more complex design of the multilevel TDR probe requires further calibration than those presented for standard continuous rod probes. For the experimental conditions, the multilevel probe was calibrated to measure the water content profile beneath each measurement point. As described above, the bulk electrical conductivity can be measured independently by TDR. The bulk electrical conductivity is a function of both the water content and the pore water electrical conductivity as described by Archie's Law [21]:

$$\sigma_b = a\sigma_w \Theta^b \tag{6}$$

where σ_b is the bulk electrical conductivity, σ_w is the pore water electrical conductivity, $\ominus$ is the volumetric water content, and a and b are constants. Therefore, the bulk electrical conductivity measured by the multilevel TDR probe could be corrected for the independently measured water content to determine the pore water electrical conductivity. Finally, the pore water electrical conductivity can be calibrated to the concentration of an electrolytic solute used as a tracer.

The field experimental took place in three steps: source emplacement, operation of a pumping well in an attempt to extract the source followed by fresh water flushing from ground surface. Initially, a concentrated KBr solution was infiltrated on the ground surface through a drip line irrigation system for two hours. This was followed by a two hour fresh water infiltration after which the subsurface was allowed to drain. With a stable source now located in the unsaturated zone and spanning the water table, the central pumping well was operated to test the ability of a well to remove the salt mass from the subsurface. The movement of both water and salt were followed during the operation of the pumping well until no further changes were apparent with time. When no further mass could be removed by pumping alone, fresh water was applied at the surface through a separate drip line irrigation system for two days and the removal of the remaining salt mass was monitored.

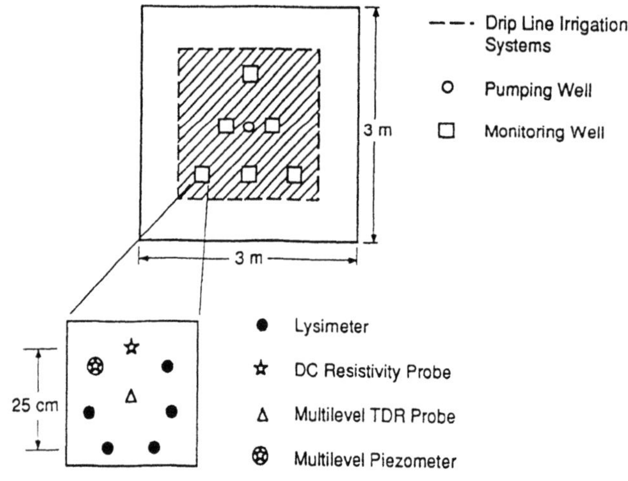

Fig. 6,   Instrumentation array at Base Borden test site.

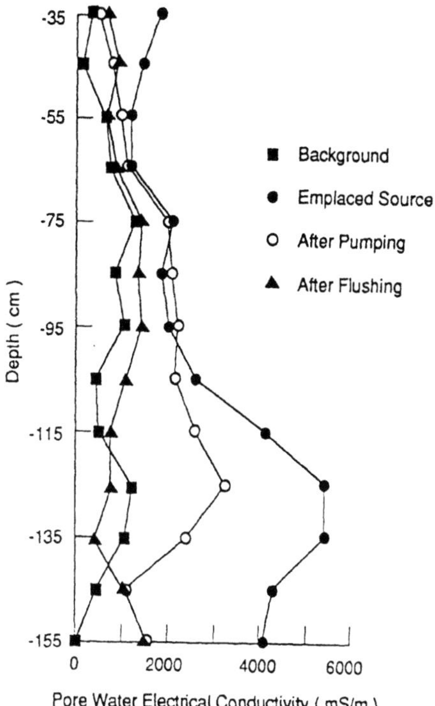

Fig. 7,   Vertical distribution of tracer after emplacement, pumping and soil flushing.

4.2 Experimental Results

A subset of the experimental results is shown to demonstrate the utility of TDR for transient water content and solute concentration monitoring. The profiles presented in Figure 7 show the pore water conductivity determined from one of the six multilevel TDR probes following complete drainage after the end of each experimental step. The data clearly show that pumping at the water table cannot completely remove the solute mass in the unsaturated zone. Subsequent soil flushing successfully remobilizes the trapped solute down to the water table where it can be removed by the pumping well.

Because the multilevel probe results require more complex interpretation, each TDR reading requires approximately 45 - 60 seconds, making it less applicable for three dimensional monitoring of rapid events such as infiltration into the Cape Cod sediments. However, the probe design, employing small diameter access tubes, causes as much disturbance as two lysimeters and the ensuing water content and electrical conductivity measurements are nondestructive. In addition, for the dry conditions following drainage, the lysimeters used in this experiment could not provide any samples. Similarly, the DC resistivity probes under dry conditions were affected by high contact resistance and did not produce useful measurements of bulk electrical conductivity of the soil. Therefore, although calibration of the multilevel TDR probe to estimate pore water conductivity to very low water contents is not exact, it produced the only relative measure of solute concentrations under low water content conditions.

Development of the multilevel TDR probe extends the ability of TDR to profile both the water content and solute concentrations, allowing for fully three dimensional, transient monitoring of flow and transport phenomena. In this experiment, only the results from the multilevel TDR probe were able to demonstrate the inability of pump and treat without flushing to remove contaminant mass from above the capillary fringe. Other applications of the multilevel probe include measurement of water content or salt concentration in highly heterogeneous environments. In addition to extending the ability to monitor field conditions, these data will allow for benchmarking of advanced three dimensional numerical models.

5.0 SUMMARY AND CONCLUSIONS

The implementation of transient, field-scale tracer tests in highly permeable, unsaturated sediments can be limited due to the destructive and time consuming nature of conventional monitoring instruments. Time-domain reflectometry (TDR) methods for measuring volumetric water content and solute mass flux rapidly, with minor disturbance and without the need for costly sample analysis, provide a useful approach to enhancing the scope of field tracer testing. Networks of TDR probes installed in dense spatial distributions can permit accurate and extensive tracking of a three-dimensional infiltration front or migrating solute cloud even under rapid flow conditions.

Experiments such as those conducted at the Cape Cod and Base Borden study sites illustrate some of the potential applications of these emerging techniques. Continued research focused on improvements and extensions of the TDR methods are beginning to provide even greater flexibility in their field applications. The increasing focus on enhanced remediation techniques in unsaturated media will require more extensive field-scale investigation and the TDR methods may prove to be a useful tool in these types of experiments.

ACKNOWLEDGEMENTS

We wish to acknowledge the significant scientific contributions to this research project of Drs. R.G. Kachanoski, M.A. Celia and R.W. Gillham along with Mr. D.R. LeBlanc and Mr. Peter Von Bertoldi. Graduate students form both Princeton University and the University of Waterloo provided technical support in many aspects of the data collection and synthesis, particularly Jonathon Stevens, Andy Mace, Xiuhui Xie and George Schneider. We gratefully acknowledge the support of the Marlborough, Massachusets office of the United States Geological Survey throughout the work conducted at Cape Cod. Funding for the projects were provided by the Waterloo Centre for Groundwater Research, Princeton University, University of Guelph and the Natural Sciences and Engineering Research Council of Canada.

REFERENCES:

1.   LeBlanc, D.R., P.S. Garabedian, K.M. Hess, L.W. Gelhar, R.D. Quadri, K.G. Stollenwerk, and W.W. Wood, Large-scale natural-gradient tracer test in sand and gravel, Cape Cod. Massachusetts: 1. Experimental design and observed tracer movement, Water Resour. Res., 27, 895-910, 1991.

2.   Mackay, D.M., D.L. Freyberg, P.V. Roberts, and J.A. Cherry, A natural gradient experiment on solute transport in a sand aquifer, 1. Approach and overview of plume movement, Water Resour. Res., 22, 2017-2029, 1986.

3.   Biggar, J.W., and D.R. Nielsen, Spatial variability of leaching characteristics of a field soil, Water Resour. Res., 12, 78-84, 1976.

4.   Bowman, R.S., and R.C. Rice, Transport of conservative tracers in a field under intermittent flood irrigation, Water Resour. Res., 22, 1531-1536, 1986.

5.   Van Wesenbeeck, I.J., and R.G. Kachanoski, Spatial scale dependence of in situ solute transport, Soil Sci. Soc. Am. J., 41, 10-13, 1991.

6.   LeBlanc, D.R., D.L. Rudolph, R.G. Kachanoski and M.A. Celia, Design and operation of an infiltration experiment in the unsaturated zone, Cape Cod, Massachusetts, USGS Toxic Substances Hydrology Program, Proceedings of the Technical Meeting, Monterey, California, USGS Water Resources Investigation Report, eds. G.E. Mallard and D.A. Aronson, 1992.

7.   Roth, K., W.A. Jury, H. Fluhler, and W. Attinger, Transport of chloride through an unsaturated field soil, Water Resour., Res., 27, 2533-2541, 1991.

8.   Wierenga, P.J., R.G. Hills, and D.B. Hudson, The Las Cruces trench site: Characterization, experimental results, and one-dimensional flow predictions, Water Resour. Res., 27, 2695-2705, 1991.

9.   Topp, G.C., J.L. Davis, and A.P. Annan, Electromagnetic determination of soil water content: Measurements in coaxial transmission lines, Water Resour. Res., 16, 574-582, 1980.

10. Kachanoski, R.G., E. Pringle, and A. Ward, Field measurement of solute travel times using time domain reflectometry, Soil Sci. Soc. of Am. J., 56, 4752, 1992.

11. Topp, G.C., and J.L. Davis, Measurement of soil water using time-domain reflectometry (TDR): A field evaluation, Soil Sci. Soc. Am. J., 49, 19-24, 1985.

12. Ferré, P.A., D.L. Rudolph and R.G. Kachanoski. A multi-level wave guide for profiling water content using time domain reflectometry. Time domain reflectometry in environmental infrastructure and mining applications, symposium proceedings, pp. 81-89, United States Bureau of Mines, Evanston, Illinois, 1994.

13. Dalton, F.N., W.N. Herkelrath, D.S.Rawlins, and J.D.Rhoads, Timedomain reflectometry. Simultaneous measurement of soil water content and electrical conductivity with a single probe, Science (Washington, D.C.) 224, 989-990, 1984.

14. Elrick, D.A., R.G. Kachanoski, E.A. Pringle and A.L. Ward, Parameter estimates of field solute transport models based on time-domain reflectometry measurements, Soil Sci. Soc. Am. J., 56, 1663-1666, 1992.

15. Hess, K.M., S.H. Wolf, and M.A. Celia, Large-scale natural-gradient tracer test in sand and gravel, Cape Cod, Massachusetts, 3. Hydraulic conductivity variability and calculated macrodispersivities, Water Resour. Res., 28, 2011-2027, 1992.

16. MacFarlane, D.S., J.A. Cherry, R.W. Gillham, and E.A. Sudicky, Migration of contaminants in groundwater at a landfill: A case study, 1. Groundwater flow and plume delineation, J. Hydrol., 63, 1-29, 1983.

17. Rudolph, D,L., R.G. Kachanoski, M.A. Celia, and D.R. LeBlanc, Infiltration and solute transport experiments in unsaturated sand and gravel, Cape Cod, Massachusetts: Experimental design and overview of results, in submission, Water Resour. Res., 1994.

18. Poulsen, M.M., and B.H. Kueper, A field experiment to study the behaviour of tetrachloroethylene in unsaturated porous media, Environ. Sci. and Tech., Vol. 26, No. 5, 889-895, 1992.

19. Nwankwor, G.I., R.W. Gillham, G. van der Kamp, and F. Akindunni, Unsaturated and saturated flow in response to pumping of an unconfined aquifer: Field evidence of delayed drainage, Ground Water, Vol. 30, No. 5, 690-700, 1992.

20. Xie, X., Solute transport and remediation in the interface zone: Mathematical Modelling and field investigations, Ph.D. thesis, Univ. of Waterloo, 1994.

21. Archie, G.E.. The electrical resistivity log as an aid in determining some reservoir characteristics. Trans. AIME 146, 54-62. 1942.

GROUNDWATER POLLUTION CONTROL IN FRACTURED AND KARSTIFIED ROCKS

M. Veselic

University of Ljubljana, Ljubljana, Slovenia

ABSTRACT

Concepts of seepage flow and mass transport were originally developed for soils and rocks with granular structure and intergranular porosity. Fractured and karstified rocks are formed of blocks of solid rock (or matrix), intersected by discontinuities that may be locally developed into (karst) channels. Due to the intergranular, fracture and channel porosities, they represent a much more heterogeneous porous media. Detailed research into the flow mechanisms within the individual fractures and within the fractured media as a whole showed that the flow and mass transport through such media are scale dependants. Observation and description of the intrinsic structural features of the rock masses are scale dependant and so is the study of the flow and transport features of the fractured and karstified rock masses. The scale of the intended flow and mass transport study will predetermine the selection of the flow model type and, consequently, the selection of the observation technique furnishing the requested structural and flow detail. For small scale groundwater pollution problems (such as waste repository studies) the network and channel models will be applied. With large scale regional groundwater pollution problems, the use of the advection-dispersion models must be coupled with the proper internal aquifer regionalization. For such problems, also the flow and transport mechanisms within the unsaturated zone of the fractured and karstified aquifers should be given attention.

1. INTRODUCTION

Groundwater pollution control in fractured and in karstified rocks is a difficult task, given the heterogeneous nature of these rocks as far as their porosity and hydraulic conductivity are considered. Conceptually, the regional flow pattern and characteristics can be deduced from the structural data and from some basic hydrologic and hydrogeologic data. Today, most of the pollution control in fractured and in karst aquifers is based on such considerations and, especially for karst aquifers, no specific mathematical groundwater modelling is generally applied.

To base the pollution control in fractured and in karstified rocks on the mathematical modelling of groundwater flow and mass transport should be:
1. the nature of the porosity of such media quantitatively adequately described on a local and on a regional scale, and
2. the nature of the flow and of mass transport in such media understood on a local and on a regional scale and formulated in adequate models.

An insight into the nature of these two crucial problems is tempted within this paper.

2. CONCEPT OF POROUS MEDIA

Concepts of seepage flow and mass transport were originally developed for soils and rocks with granular structure, representing a media with intergranular porosity.

Heterogeneity is inherent to the concept of porous media. The concepts of REV [1] and AEV [2] were therefore developed to overcome the problems of basic void-solid heterogeneity and of spatial variability of this media. Though at the very base of the REV concept there is a procedure of statistical averaging of the porous media - let it be a formal or an intuitive one - this concept allowed for a deterministic approach to the flow and mass transport through the porous media.

The volume, and therefore the size of the REV, should be selected such that the volumetric averages can be considered as satisfactory estimates of all the relevant statistical parameters of the void space configuration [2,3].

The above exposed REV concept, is illustrated in the Figure 1. The U_0 range defines the adequate range of the REV volume and therefore of its size.

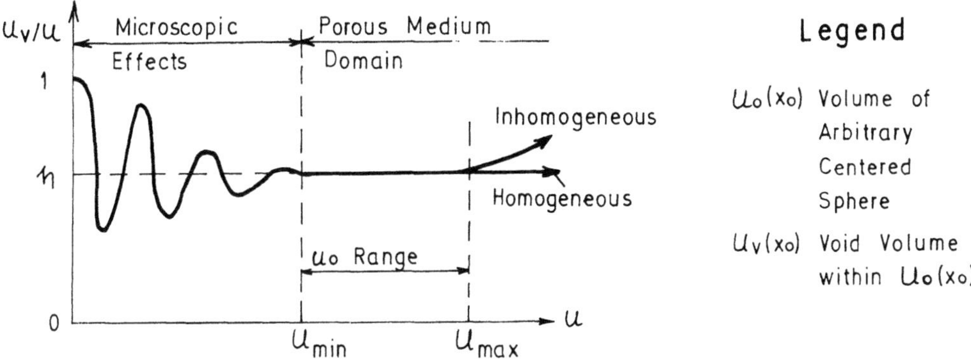

Figure 1 - Definition of Representative Elementary Volume

Practically, the questions on the minimum acceptable size of a REV and on the REV's relation to the natural environment arise and have to be clarified.

After Bachmat and Bear [3] the REV's size should be l≫d, where l the REV's diameter and d the mean hydraulic radius of the pores (equal to the reciprocal of the specific surface of the void space). Given the nature of the inequality sign this statement is not very conclusive. After Irmay [1], its size should be in the range $50d_e \leq l \leq 100d_e$, where d_e denotes the effective grain diameter. For sands and gravels, this leads to an l of the order of one cm to some dm.

While applying the REV concept to the natural environments, one has to bear in mind that a REV can be related to a homogeneous porous medium only. With natural aquifers, this may on a local scale and on a regional scale require the definition of several REV, relating to the portions of aquifer or rock volume that may be defined as both intrinsically homogeneous and characteristically different from each other.

3. FRACTURED AND KARSTIFIED ROCKS CHARACTERIZATION

Fractured and karstified rocks are made up of blocks of solid rock (or matrix), intersected by discontinuities that may be locally developed into (karst) channels. General aspect of the fractured rock is given in the Figure 2. General aspect of the karstified rock, with the karst (solution) channel development pattern following bedding planes and faults, can be seen from the Figure 3.

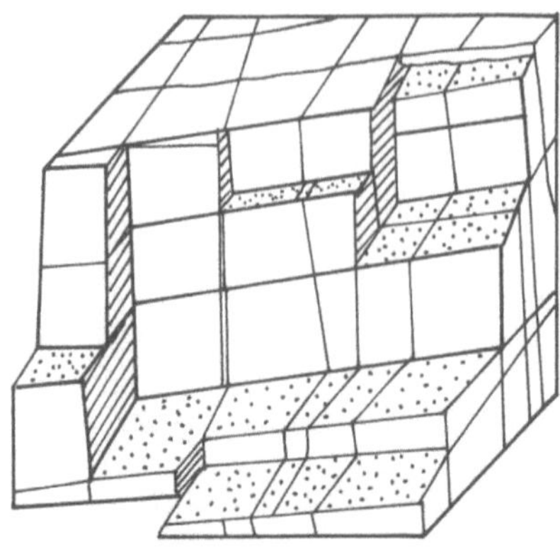

Figure 2 - View of a fractured rock mass

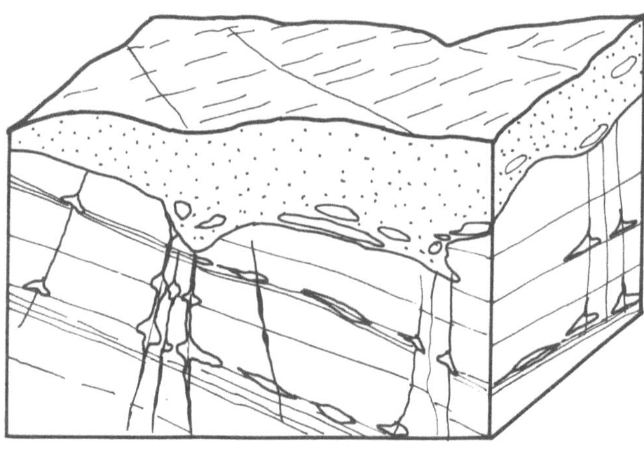

Figure 3 - Block diagram of a karstified rock
with the karstification pattern

International Society of Rock Mechanics (ISRM) Suggested Methods for
Quantitative Description of Discontinuities in Rock Masses [4] recommend the use
of term discontinuity instead of generic term for rock mass continuity breaks.
According to these recommendations, 'discontinuity' is the general term for any
mechanical discontinuity with zero or low tensile strength in a rock mass. The
term discontinuity is accepted here in the sense of these recommendations as
it was with other modern authors [5].

Discontinuities can be grouped by geometric and/or by generic criteria and several sets of discontinuities can be defined within a rock mass. This sets will usually be given some hierarchical order and individual sets ranked according to the important parameters. In groundwater hydraulics these will generally be the water flow influencing parameters or the flow itself.

Geometrical criteria denote size, dip and spacing of discontinuities and their total width, aperture (opening), wall rugosity and percentage of fill. By genetic criteria can the discontinuities be classified as bedding planes, stilioliths, cleavage or schistosity planes, fractures or joints and faults.

To characterize a rock mass hydraulically, Louis [6] proposed 11 parameters: type of structural element, orientation, continuity, aperture, nature of filling material, degree of void area, water inflows, decompression, spacing between elements, nature of fracture endings, roughness or angle of friction. ISRM Suggested Methods for Quantitative Description of Discontinuities in Rock Masses [4] recommend ten parameters: orientation, spacing, persistence, roughness, wall strength, aperture, filling, seepage, number of sets and block size [5]. Seven of these are identical with Louis [6] proposal, while two characterize only the fractured rock mass.

The real difficulty in fractured rock characterization, as already stated elsewhere [5] is not in the number of parameters involved but in the fact that only a few of these parameters can be determined with precision while most can only be estimated. The above statement was deduced from the engineering problems, where the data on fresh rock can be obtained from excavation or drilling. Unfortunately, with the regional groundwater flow problems, fresh rock exposures and on purpose drilling are less abundant. Therefore, the data on the morphology of the discontinuities and on their seepage flow are generally missing. They tend to be substituted by other structural data, obtainable by detailed mapping.

4. FRACTURED AND KARSTIFIED ROCKS AS POROUS MEDIA

It has already been stated that the fractured and karstified rocks are made of blocks of solid rock or matrix, intersected by discontinuities which may be locally developed into channels. Both matrix and discontinuities (and/or channels) have a certain porosity. In most fractured and karstified rocks the fracture and channel porosity are hydraulically preponderant. Yet, very often, the matrix porosity cannot be neglected.

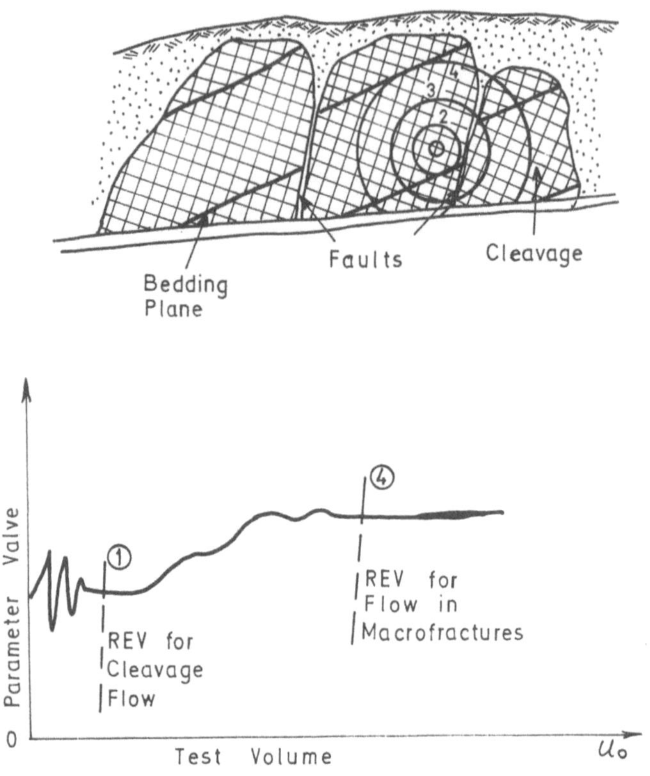

Figure 4 - Outcrop of a fractured dolomite and the corresponding
Representative Elementary Volumes

Heterogeneity of such media to the fluid flow is much more pronounced than in
the case of the media with an intergranular porosity, especially if different
sets of discontinuities are present. To overcome this problem, the concepts of
REV and AEV were extended to the fractured and karstified porous media. On the
Figure 4 is shown an outcrop of stratified dolomite with a pronounced cleavage
(paralelepipede in reality), intersected by faults. From the corresponding
diagram, analogous to that given in the Figure 1, it can be seen that a REV can
be defined for the cleavage domain and that a further REV can be defined for
a larger rock volume, when macrofractures (faults and bedding planes) influence
the rock parameter values. This diagram illustrates the double porosity
concept, often used to model fissured and especially karstified rocks [7]. As
seen here, clear difference can generally be observed between the porosity of
microfractures and that of macrofractures and channels, the former often being
assimilated to the matrix porosity.

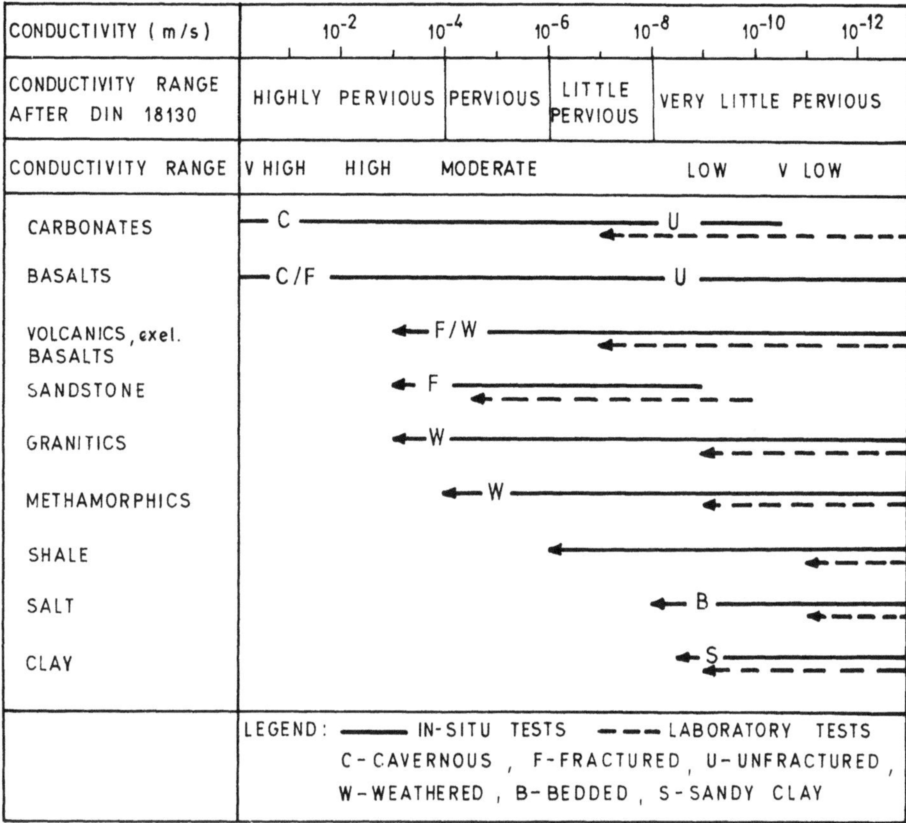

Figure 5 - Hydraulic conductivity of rocks
After Cheng-Haw-Lee & Farmer [5],
Johnson [8] and Hölting [9]

5. FLOW THROUGH FRACTURED AND KARSTIFIED ROCKS

A quick glimpse of the variability of the behaviour of fractured and karstified rocks with respect to the fluid flow is provided by the range of variability of hydraulic conductivity of these rocks, presented in the Figure 5. Clay, which is not a fractured rock, is added as a yardstick. Striking is the difference between the laboratory data, relating to what can be termed the matrix flow domain and between the in-situ tests data, relating to the fracture and channel flow domain.

Initial approach into modelling the fluid flow in fissured and karstified media was to assimilate it to a continuous porosity medium, with REV being the basis

for homogenization. The double porosity idea was added to better describe the
media's behaviour. Yet, obviously the heterogeneity of the rock masses, when
treated as fracture and channel porosity media, makes rather uncertain the
criteria for the REV definition. Often no REV may overcome the constraints
imposed by the limited size of the studied volume.

To model the fractured rocks in greater detail more sophisticated models were
developed. The first to mention is the fracture network flow model [5,7], with
four principal types: a) sheets model, b) match-stick model, c) cubes model and
d) disc type or Beacher model.

An ultimate conceptual development is the channel flow model [10,11,12]. This
idea can be illustrated by two possible types of channeling: a) channels at the
fracture intersections in a match-stick type fractured medium, and b) irregular
curvy channel network within fractures in a disc type fractured medium.

6. MASS TRANSPORT THROUGH FRACTURED AND KARSTIFIED ROCKS

Detailed research into the flow mechanisms within the single fractures and
within the fractured media as a whole showed that the flow and mass transport
through such porous media are scale dependant [5,10,11,12]. For very long flow
distances, orders of magnitude bigger than the adequate REV, the advection-
dispersion model seems appropriate. For long flow distances, equalling some
orders of REV, the network flow model is adequate. For short distance flow,
shorter than some orders of REV, the channel flow model seems the most
suitable. From the above statements it follows that the flow and transport
behaviour are relative scale and not absolute scale dependant. The appropriate
flow and transport model selection may be a question of the adequate
description of the intrinsic structural and flow features of the porous media.

7. POLLUTION CONTROL IN FRACTURED AND KARSTIFIED ROCKS

The observation and therefore the description of the intrinsic structural
features of the rock masses will always be scale dependant. The observation
techniques allowing for greatest detail are costly and not applicable on a
regional scale. Inversely, the observation techniques affordable on a regional
scale furnish no adequate data for the advanced fracture flow models.

The same is true for the study of the flow and transport features of the
fractured and karstified rock masses. The laboratory analyses may give valuable

data on the rock matrix porosity and permeability. The eventual laboratory data on fracture or channel porosity and permeability from large samples will generally lack representativeness. In-situ tests will, unfortunately, be scale dependant too. It is therefore vital to design the scale of such tests accordingly to the media's REV and accordingly to the scale of the intended flow and mass transport study. The scale of the intended flow and mass transport study will predetermine the selection of the flow model type and, consequently, the selection of the observation technique furnishing the requested structural and flow detail.

From the above discussion it follows, that, depending on the scope of the study, two basic approaches will continue to coexist in the study of the flow and mass transport phenomena in the fractured and karstified rocks. For small scale hydrotechnical and groundwater pollution problems (such as waste repository studies) the network and channel models will be applied. For regional groundwater flow problems, related to the aquifer pollution control, the advection-dispersion model will further be used.

With regional groundwater flow problems, the use of the advection-dispersion models should be backed by double porosity approach and coupled with proper internal aquifer regionalization, based on the respective REV and flow parameters definition. For the latter, appropriate structural and hydrogeological methods should be used. In principle, this approach differs little from that, applied to the aquifers with intergranular porosity.

To the regional groundwater flow and transport problems, the results of the small scale research on the flow and transport mechanisms of the fractured media should be implemented. This application may be twofold: either will the results of such studies be transposed to the studied aquifer or rock massive, or will the local small scale studies be used to allow for a better characterisation of the different parts of the studied aquifer or rock mass. Generally and on the conceptual basis, the results of such studies display the dependence of flow and mass transport concentration on the channel conductance standard deviation [10,11] - an effect long perceived qualitatively by karst hydrologists. But they allow for the insight into the diffusion and sorption mechanisms, which could not be but guessed yet.

Finally, but not unimportant to the groundwater pollution control in fractured and karstified rocks, the flow and transport mechanisms within the unsaturated zone of such rock masses should be given further attention. It is known that most of the pollution inactivation must be attributed to the unsaturated zone of the aquifers with the intergranular porosity. Unfortunately, the adequate data for the fractured and karstified aquifers are still very scarce.

REFERENCES

1.   Irmay S.: On the hydraulic conductivity of unsaturated soils, Trans.Am. Geophys.Union, 35(1954)398-406.
2.   Bear J. and A. Verruijt: Modelling Groundwater Flow and Pollution, D. Reidel Publishing Company, Dodrecht 1987.
3.   Bear J. and Y. Bachmat: Macroscopic modelling of transport phenomena in porous media, 2. Applications to mass, momentum and energy transport, Trans.Por.Med., 1(1986), 241-269.
4.   Brown E.T.(Ed.): Rock Characterization, Test and Monitoring, Pergamon Press, Oxford 1981.
5.   Cheng-Haw-Lee and I. Farmer: Fluid Flow in Discontinuous Rocks, Chapman & Hall, London 1993.
6.   Louis C.A.: Rock Hydraulics, in: Rock Mechanics (Ed. L. Muller), Springer Verlag, Vienna 1974.
7.   Reiss L.H.: The Reservoir Engineering Aspects of Fractured Formations, Editions Technip, Paris 1980.
8.   Johnson A.I.: Specific Yield - Compilation of specific yields for various materials, U.S.G.S.Wat.SupplyPap., 1662-1, Washington 1967.
9.   Hölting B.: Hydrogeologie, Ferdinand Enke Verlag, Stuttgart 1989.
10.   Moreno L. and I. Neretnieks: Fluid flow and solute transport in a network of channels, Journ.Contam.Hydrol., 14(1993)163-192.
11.   Abelin H., L. Birgersson, H. Widén, T. Ågren, L. Moreno and I. Neretnieks: Channelling experiments in crystalline fractured rocks, Journ.Contam.Hydrol., 15(1994)129-158.
12.   Cacas M.C., E. Ledoux, G. de Marsily, B. Tillie, A. Barbieu, P. Calmels, B. Gaillard, R. Margrita, E. Durand, B. Feuga and P. Peaudecerf: Flow and transport in fractured rocks: An in situ experiment in the Fanay-Augeres mine and its interpretation with a discrete fracture network model, Mem.22nd Congr.IAH, 22(1990)13-37.

HYDRODISPERSIVE PARAMETERS ESTIMATION IN GROUNDWATER CONTROL AND VALIDATION MODELS

S. Troisi

University of Calabria, Montalto Uffugo, Italy

ABSTRACT

The groundwater management is function of several elements, that are influenced by several sciences. The "role of knowledge" of a groundwater body is considered punctual over the domain (wells, springs, sampling points, etc.) as it is well known. This particular approach influences the conceptual model of the aquifer under the study and the experimental data to be used. Some data should help to describe hydrological behaviour of water body and the other should allow to analize and to predict the quantitative change of the aquifer during the time

This report, through some cases study, points out the centrality of the data in an ordinary groundwater management problem.

1.   INTRODUCTION

The principal aims of the groundwater management are the exploitation of the available water and the protection against pollution.

To obtain such aims the determination of the aquifer pararmeters and of the pollution sources have to be known. Such knowledge is closely joined to the availability of data and of mathematical model (code).

In the past the data and the code were examined separately determining a high development of numerical codes. Nevertheless, the few available data influenced the mathematical outputs too.

A right approach to the groundwater problems places the knowledge of the aquifer and, if present, the cause of the pollution above the mathematical code. This report points out the centrality of the available data by means of four case studies in which the principal aim is the knowledge of the hydrogeologic and hydrodispersive parameters of the aquifer at issue.

The first two cases concern two test sites for measurements of hydrodispersive parameters, the first relative to a porous medium and the second relative to a fissured medium. In these ones the hydrogeologic parameters were calculated by means of pumping tests, in steady and transient states. Moreover, in the test site relative to a fissured medium, were determined the hydrodispersive parameters by means of the channelling theory.

The third case study concerns a pollutant migration beneath a M.S.W. landfill. A geognosic sounding was carried out by means of rotary drilling. Undisturbed samples were gathered in order to characterise the soil laying beneath the old landfill. The results of geotechnical tests carried out on the samples taken at different depths showed diverse behaviours between the deep and the superficial clay layers.

The last one concerns the calibration of a numerical code in order to evaluate the behaviour of a system subjected to salt water intrusion. This numerical code was calibrated both on the phenomenon and on the aquifer at issue by means of a "history matching" between measured data (head and salt concentration data) and the numerical code outputs (head and salt concentration calculated).

2.   AN EXPERIMENTAL SITE FOR MEASUREMENTS OF HYDRODISPERSIVE PARAMETERS IN POROUS MEDIA

A test site for measurements of hydrodispersive parameters is located near the University of Calabria, Department of Difesa del Suolo, in the area of Montalto Uffugo Municipality[*] , [1].

From a geological point of view this site shows the typical characteristics of a newly formed valley, with alluvial, conglomeratical and sandy deposits. These

---

[*] This study has been made in the ambit of the ENEL-CNR Convention-Sub-project 3- research line 3.1- ENEL-UNIVERSITY OF CALABRIA Convention

deposits are little consolidated, easily breakable and characterised by a large permeability. The whole surface of the test site is about 2100 m^2 (35 m x 60 m) wide.

The flow direction turns out to be a very important element in performing the test field; infact as far as the study on hydrodynamic dispersion are concerned, wells need to be lined up with precision considering a certain direction which is parallel to the flow direction, so that it is possible to intercept the water marked by a tracer.

Moreover, in order to verify the entity of the "transversal dispersion", it was necessary to forecast other wells lined up on an orthogonal direction with respect to the flow direction. This need has been satisfied through the installation of five wells: a central one and other four wells installed on two different alignments which were perpendicular between each other. One of this wells alignment was placed along the direction of the flow.

During the period between April 1992 and March 1993 systematic measurements of the levels have made to verify the constancy of the flow direction and the average slope of the aquifer, that turned out to be 0,0055 m/m.

In order to perform experiments both on the superficial aquifer and on the deep one, the above scheme of five wells has been identically repeated for both aquifers. The whole ten wells have been realised, that is five couples of lined up wells, like the following scheme, so that each couple of wells have one for the deep aquifer and another for the superficial aquifer.

2.a. **Wells**

The necessary information, to define the geology of the site and the hydrogeologic characteristics of the aquifers in detail, have been obtained through the mentioned perforations.

The stratigraphic column showed basically the existence of a unique, strong, sandy layer, covered by an alluvial layer with the interposition of a weak clay layer. This condition made possible the formation of a superficial aquifer which essentially interests the alluvial layer and having the already mentioned clay bed. This layer also represents the top of the below deep aquifer.

2.b. **Pumping tests**

The pumping tests, in steady and transient states, were made in order to obtain the hydrodynamic characteristics of the confined aquifer. By pumping only from the central well, level variations have been spotted by means of a tele-acquisition system both in the central well and in the other four ones which worked as piezometers. The gathering of the hydrodynamic levels during the tests has been performed through a remote data collection system. The portable system for the acquisition of the levels is represented by a central unit consisting of a Personal Computer, by a card for the acquisition of the analogic data transmitted by sensors, and by the sensors consisting of pressure transducers for recording piezometric levels inside the wells. For the tests, five transducers have been installed, one for each well penetrating in the deep aquifer. Through the analogic card and the connected sensors the Personal Computer carries out the measurements at definite time intervals; the minimum time is 1 minute.

The flow rate has been determined by means of volumetrical method, by using both a counter connected in set to the pipe and a previously calibrated caisson system consisting of a flow deviation device, of a graduated metallic shaft with decimal vernier and of a precision chronometer for time measurements. During the test, the pumping waters have been drained in a ditch near the experimental field.

The steady state test has been performed for 5 different flow rate values; the total scheme is shown in figure 1.

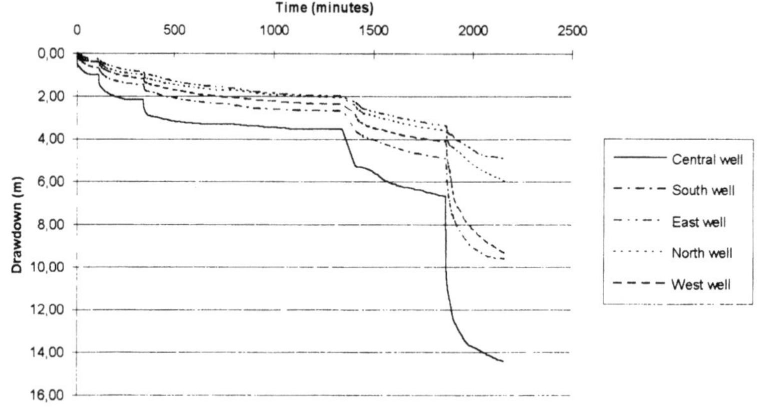

Fig. 1, Diagram of the steady state pumping tests.

In each pumping test the equilibrium state has been reached in times which proved to be shorter for lower discharge values, although drawdown did not turn out to be extremely fast. In figure 2 in the pumping well are shown.

The pattern of these curves points out that the aquifer is confined and, in particular, the curves demonstrate that Jacob's approximation hypothesis is acceptable.

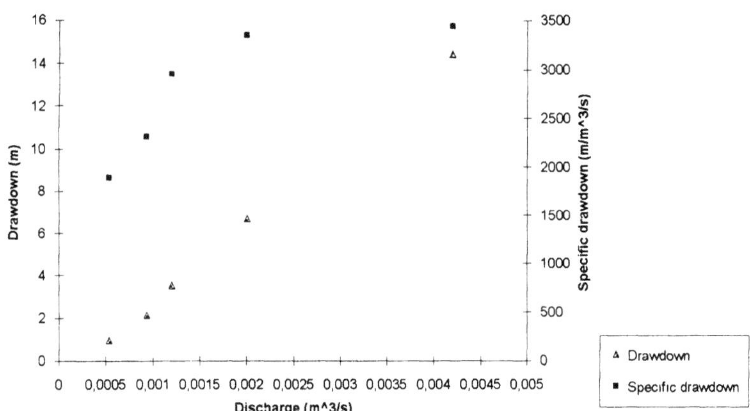

Fig. 2, Drawdown-discharge curve and specific drawdown-discharge curve during the steady state tests.

Once the pumping stopped, level variations during the upwelling, both in the central well and in each piezometer, were noticed for a timespan of about 12 days.

By observing the above mentioned curves it is possible to spot the low recharge of the aquifer. On the basis of the data collected during the tests, the efficiency of the central well has been determined by means of the Jacob's method. With reference to the pumping discharge values considered, the average efficiency is equal to 83%. Hydrodynamic parameters have been evaluated by means of various formulations of the steady state theory. The results obtained by Dupuit theory and the $\sigma - \vartheta$ method have been used to calculate transmissivity (table 1).

	Dupuit	$\sigma - \theta$
$Q\ (m^3/s)$	$T\ (m^2/s)$	$T\ (m^2/s)$
0,00053	$6.0*10^{-4}$	$4.6*10^{-4}$
0,00093	$4.8*10^{-4}$	$4.1*10^{-4}$
0,0012	$3.8*10^{-4}$	$3.5*10^{-4}$
0,002	$3.4*10^{-4}$	$3.4*10^{-4}$
0,0042	$3.3*10^{-4}$	$3.6*10^{-4}$

Table 1: Transmissivity values for the steady state test

The transient state test lasted 33.5 hours, during which a constant flow rate (3.2 l/s) has been pumped off. Discharge controls have often been performed by means of the above described measurement equipment, while the level variations, both in the central well and in the piezometers, have been controlled by means of the above mentioned method. At the end of the test, recovery of water levels occurred; this phase lasted about 13 days. In figure 3 the diagram of the transient state test in the pumping phase is shown.

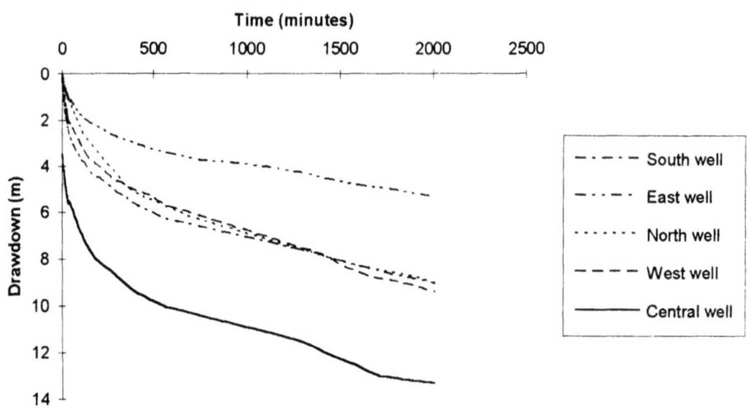

Fig. 3, Diagram of the transient state pumping test.

Hydrodynamic parameters have been evaluated by means of the Sheaham, Theis and Jacob methods (table 2).

Seaham		Theis		Jacob		
T (m^2/s)	S	T (m^2/s)	S	T (m^2/s)	S	Piezometer
$2.93*10^{-4}$	$1.60*10\text{-}3$	$3.82*10^{-4}$	$2.38*10^{-4}$	$1.86*10^{-4}$	$1.62*10^{-3}$	1
$1.84*10^{-4}$	$1.72*10^{-3}$	$2.68*10^{-4}$	$1.07*10^{-3}$	$1.95*10^{-4}$	$1.15*10^{-3}$	3
$3.50*10^{-4}$	$3.77*10^{-3}$	$4.69*10^{-4}$	$2.81*10^{-3}$	$1.75*10^{-4}$	$2.36*10^{-3}$	7
$4.50*10^{-4}$	$3.44*10^{-3}$	$5.89*10^{-4}$	$1.41*10^{-3}$	$3.19*10^{-4}$	$3.05*10^{-3}$	9

Table 2: Transmissivity values for the unsteady state test

	Theis
Piezometer	T (m^2/s)
1	$1.97*10^{-4}$
3	$2.10*10^{-4}$
7	$2.03*10^{-4}$
9	$3.79*10^{-4}$

Table 3: Transmissivity values for recovery phase in the unsteady state test

3. AN EXPERIMENTAL SITE FOR MEASUREMENTS OF HYDRODISPERSIVE PARAMETERS IN A FISSURED AQUIFER

An experimental field site, made up of five wells (one well located at centre and the others at the corners of a square, figure 4), has been carried out in a fissured aquifer to study hydrodispersive phenomena, [2,3,4].

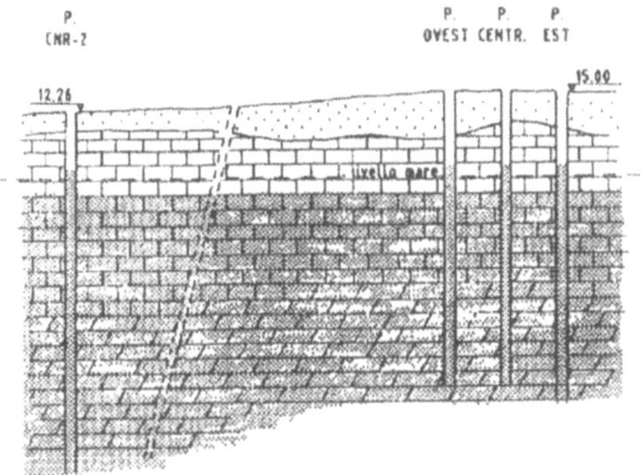

Fig. 4, Stratigraphy log of the test site (Di Fazio et Al., 1994).

The zone has been characterised as geological as hydrological peculiarities. All the wells have been drilled using reverse circulation rotary of water with a hammer drill. The coreshave shown the presence of three different layers from the top to the bottom of the wells. The first layer was made of quaternary calcarenite having a thickness of about 6 m, the second one was fissured and fractured limestone of about 25 m, the last one was dolomite of about 20 m, figure 5.

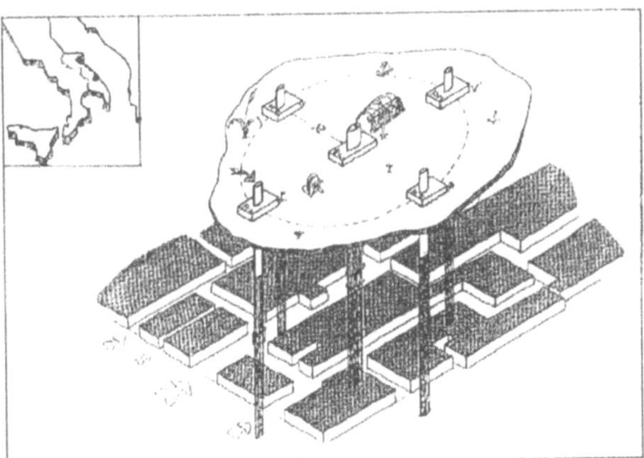

Fig. 5, Test site scheme (Di Fazio et Al., 1994).

Several pumping tests have been performed to evaluate the main hydraulic parameters in equilibrium as far as in non equilibrium conditions. Different methods have been used to calculate hydraulic conductivity, transmissivity and storavity and a comparison has been carried out to evaluate the differences in the results obtained. As a matter of fact these differences were acceptable.

Tracer tests have been performed using both single well and two wells techniques. The injection has been done along the whole depth of the well taking into account the possibility to find preferential flow paths. Small volume of water have been sampled from the wells for few months and breakthrough curves have been developed for different depths. A particular device has been set up to sample water without disturbing the natural flow of the groundwater system and to assure the representativity of the samples.

The breakthrough curves have been analysed and values of hydrodispersive parameters have been evaluated using the channelling theory, suggested by some Authors, for fractured system. Adsorption and matrix diffusion phenomena were not considered in this study.

Flow	Methods	Wells	Transmissivity $(10^{-2}\,\mathrm{m^2/s})$	Storage coefficient (10^{-3})
steady state	Dupuit	Central	$9.73 \pm 30\%$	
" "	Thiem	North	$8.93 \pm 12\%$	
" "	Thiem	West	$6.93 \pm 29\%$	
" "	Thiem	East	$7.20 \pm 15\%$	
" "	$\sigma - \theta$	Central	$8.75 \pm 18\%$	
unsteady state	Jacob	Central	9.8	6
" "	Jacob	Central	12.11	7
" "	Sheaham	Central	7.50	12.5
" "	$(h - x^2/t)$	South	7.94	5.7
" "	$(h - x^2/t)$	CNR2	10.56	12.2

Table 4: Values of the hydrogeological parameters evaluated in two flow cases comparing different methods

	Central Well $L_{ij} = 10$ m	North Well $L_{ij} = 20$ m	West Well $L_{ij} = 14.5$ m	East Well $L_{ij} = 13.8$ m
$t_{0.9}\text{-}t_{0.1}/t_{0.5}$	1.430	1.340	1.290	1.333
D_{ij} (m/g)	5.32	12.62	97.31	11.66
V_{ij} (m/g)	0.57	0.67	2.34	0.61

Table 5: Values of the hydrodispersive parameters evaluated by Channeling theory

4.   POLLUTANT MIGRATION BENEATH A M.S.W. LANDFILL - FIRST RESULTS

Among the risks of pollution coming from M.S.W. landfills, the risk of possible interaction between M.S.W. and the groundwater gives rise to very important problems. They are especially due to leachate produced by M.S.W., which has different characteristics depending on the substances present in the wastes and the age of the landfill.

It is common knowledge that the amount of leachate gathered on the bottom depends on both meteorological events and the way the landfill has been managed. The leachate tends to infiltrate underground and it can reach the groundwater in a time span which depends on the nature of the various layers of the soil.

In order to avoid this kind of pollution, it is necessary to isolate M.S.W. from the soils where they have been located by means of some barriers made of natural or artificial materials, or both, with a very low permeability. These barriers should be laid there during the realization of the landfill in order to prevent the leachate from infiltrating underground.

Synthetic protective liners are largely used as impermeable barriers, but they often cause considerable problems because of the characteristics of the polymers they contain, so they are subjected to phenomena such as diffusion, osmosis, or absorption. Another problem derives from significant spills which can take place as a consequence of lacerations, cuts and defective seams.

As far as natural materials are concerned, compact clay has been employed in the realization of the barriers. Though the material is characterised by low permeability, it is necessary to pay particular attention to phenomena deriving from the inevitable contact between clay and leachate. Particular reactions can be triggered by the chemio-physical characteristics of the clay. On the one hand, the reactions are positive because they generally induce an attenuation of the polluting power of the leachate. On the other, however, they can alter the structure of the clay, leading to possible negative variations in hydraulic conductivity [5,6].

Legal regulations in the field of M.S.W. indicate clay as the material of choice for the realization of impermeable barriers in landfills. Nevertheless, suitable analyses should be carried out in order to characterise the nature of leachate and clay to be employed. Moreover, the effectiveness of the barriers should be continuously monitored when the landfill is working and soon after in order to check the behaviour of the materials and to solve any possible problem.

Groundwater beneath, upstream and downstream the landfill should be monitored, as well as non-saturated soil and the impermeable layers between the soil and M.S.W. [7].

Thus, monitoring should be carried out in the very early realization stages so that sensors and suitable equipment can be provided in those sites which would be no longer accessible after the deposition of M.S.W. As a consequence of the great number of unauthorized landfills, it may happen that experts have to face risks of pollution deriving from the uncontrolled deposition of existing wastes. In this case, monitoring is certainly more complicated, especially in non-saturated layers laying beneath M.S.W. Soil reclamation interventions are even more difficult to perform.

If no artificial barrier is available, regulations in this field allow experts to use protective barriers made of only natural material such as compact clay having suitable thickness and characteristics.

It is advisable to chose the right site in the very early planning phases. Sites should be away from urban settlement and watercourses; moreover, they should be characterised by clayey soils, thus offering economic advantage and a more effective protection against pollution.

This is the case with a research program carried out by the Department of Soil Conservation - University of Calabria, in collaboration with the C.R.C. - Enel (Italian Electricity Board) in Brindisi. The aim is to study how leachate propagates in non-saturated soil layers above the groundwater, focusing attention particularly on interactions between leachate and clay.

With this in mind, we will monitor a new M.S.W. landfill which is not working yet, and a recently disused landfill close to the first one, both located on a sufficiently thick compact clay bank.

Being comprehensive, the research program will last for many years; this paragraph reports the results of a preliminary research carried out on the body of the old landfill, which was run without any control for many years. The aim of the preliminary study was the chemical and geotechnical characterisation of the underground, trying to assess the actual degree of pollution and evaluate the modifications induced by leachate on clays.

In the next phases of the study, groundwater in the entire area will be monitored through a series of wells which will function as piezometers and, at the same time, allow the characterisation of the aquifer by determining the corresponding hydrodispersive parameters. The wells will also supply information on water quality.

Moreover, a monitoring network will operate on the non-saturated layer of the new landfill, which is not working yet [8,9].

4.a. Description of the site.

The above mentioned M.S.W. disposal plant is located in Montalto Uffugo, North of Cosenza.

The old landfill, which had worked from 1986 to 1992, is now out of use. It covers a 8000 m^2 area and is about 1 km far from the town. No spring or other hydropotable outcrops are present in the area.

The landfill is located on a hill very close to a little stream.

The landfill serves about 12000 inhabitants; if the daily production is 0.84 kg/ab [10] total wastes amount to 26000 t.

Wastes have been buried in ditches between 3 and 5 m deep, and bailing is not very compact. When the landfill ceased to operate, it was covered with a 1 m-thick layer by employing the şame soil dug out before.

The plant is not equipped with any system for the disposal or recirculation of the leachate; wastes are directly stored on the soil, in contact with the clayey bottom of the landfill. No network operates to carry rainwater away, and it flows over the landfill and around it.

A new controlled M.S.W. landfill is being built close to the old one, at a 200 m distance from the above mentioned stream, in compliance with the law.

The new plant covers an area of 20.000 m^2 it will serve about 21000 inhabitants for a 5 year period. The landfill will be composed of 4 sectors, about 4 m deep, equipped with a system for leachate detention and recirculation.

The area is completely fenced and equipped with a network to convey and carry rainwater away.

Again, bottom impermeability is guaranteed by the mentioned clay bank, so no other protection barrier has been planned.

As far as geomorphological characterisation of the area is concerned, the landfill occupies an area of low hills, about 370 m a.s.l.. Slope is variable; the area where the landfill stands is quite flat; moving from west to east slope suddenly increases and becomes very steep; on the bottom of the escarpment, the above mentioned stream flows.

The area does not show signs of strong erosion; only west to the landfill, calanque formations are present, but they are not very significant.

The morphology of the area reproduces the geological characteristics of the soil. The landfill area lies on a formation of Pliocene grey marl clays, which are low permeable and easily subjected to erosion. The formation lies on bank of Pliocene sands and conglomerates with pale brown pebbles, locally intercalated with sandstones. The formation is often interdigitated with clay mentioned above. Sands and conglomerates have high permeability and they are resistant to erosion.

4.b. **Site stratigraphy**

A geognosic sounding was carried out by means of rotary drilling. Undisturbed samples were gathered in order to characterise the soil laying beneath the old landfill.

The drill hole was bored on the body of the landfill as to obtain information about the thickness of the waste layer.

The first 1 m layer is made of clay mixed with sand; this layer covers the wastes and it consists of the material resulting from the digging up of the tanks for waste storage.

Beneath this cover, a 2.50 m thick layer is present; when analysed, it showed that wastes are quite compact and show signs of combustion.

The bottom of the landfill lies at -3,50 M. The clay bank constitutes a 6.30 m thick layer which reaches a level of -9,80 m.

The characteristics of the clay change with increasing depth.

The layer between -3,50 m and -5,80 , is made of an almost plastic material, easy to be sampled and which shows signs of oxidation.

Down to -9,80 m, the clay becomes so compact that special barrels had to be used to obtain undisturbed samples.

The layer between -9,80 m and -18,30 m is made of silty sand and gravel.

Some very altered and degraded headers are present at depths between -11,50 m and 18,30 m.

Finally, the layer between -18.30 m and - 20.00 m is made of very dense sand with inglobation of more or less fine gravel. Logging was performed down to -20.000 m and no groundwater was found.

4.c. **Results of geotechnical tests and chemical.**

In the course of geognosic survey, five undisturbed samples were taken at the most significant layers, at depths of 4,60 m, corresponding to the bottom of the landfill, down to 11,00 m.

In order to characterise in a better way the soil layer examined, identification tests were performed on five samples, and the following elements were measured for each of them: weight per volume unit(γ), specific weight of the grains (G), natural water content (Wn) and Atterberg limits. The results are reported in table 6.

The samples were subjected to granulometric assay through sieving and sedimentation. As a result, the samples n. 1, n. 2, n. 3, n. 4, taken at depths between 4,60 m to 8,00 m, have been classified as "clay with sandy silt"; sample n. 5, taken at depths between 9,50 and 11,00 m, can be classified as "lightly coarse sand".

By examining plasticity index values (Ip) a remarkable gap between sample n. 4 and the rest of the samples emerged. Oedometric compression tests were, moreover, carried out on the first four clay samples, from 4,60 to 8,00 m deep, using a Geonor oedometer, up to

sample	depths m	γ (g/cm^3)	W_n %	W_l %	W_p %	I_p %	d<2 %	G (g/cm^3)
1	4.6-5.4	1.83	20.65	50.12	18.65	31.47	51	2.63
2	5.4-5.6	2.13	18.41	49.15	18.18	30.97	48	2.64
3	5.7-6.1	1.98	22.78	49.20	17.87	31.33	51	2.58
4	7.2-8.0	2.54	14.95	29.77	17.30	12.07	38	2.65
5	9.5-11.0	--	16.06	--	--	--	--	2.65

Table 6: Identification test results

a maximum load of 24,5 daN/cm. Under this load, the values of oedometric modulus (Eed) and permeability (k), were obtained and are reported in table 7.

Though the permeability values were similar in the four samples, as far as oedometric volume is concerned, a remarkable gap was reported between sample n. 4 and the first three, which averagely corresponds to about 1100 daN/cm .

In order to characterise the soil from a chemical viewpoint, automatic X-ray fluorescent emission spectrometry was carried out, by using the method of full correction of matrix effects.

Sample	Pressure failure (daN/cm)	Progressive index (mm)	Porosity	Oedometric modulus (daN/cm)	Permeability (cm/s)
1	24.5	3.160	0.4036	201.822	$0.4*10^{-8}$
2	24.5	1.860	0.3217	254.296	$0.5*10^{-8}$
3	24.5	3.401	0.3035	300.271	$0.5*10^{-8}$
4	24.5	0.481	0.1087	1469.810	$0.2*10^{-8}$

Table 7: Oedometric assay results

The analyses involved fractions of the undisturbed samples at the following depths: 4,60 m, 5,40 m, 8,00 m.

The main typical elements which constitute clays, as well as other substances present, were determined for each of them.

The 8,00 m sample was taken as chemically unaltered reference, and the other two samples were compared to this value.

Elements	Unaltered sample (white)		Sample A/white	Sample B/white
SiO	36.170	%tot	1.115	0.959
TiO	0.710	%tot	1.028	0.901
AlO	10.700	%tot	1.113	0.929
FeO	5.960	%tot	1.169	0.998
MnO	0.250	%tot	1.160	1.600
MgO	4.360	%tot	0.894	0.800
CaO	39.270	%tot	0.843	1.083
NaO	0.360	%tot	1.139	0.889
KO	2.080	%tot	1.072	0.918
PO	0.130	%tot	0.923	1.077
Rb	96000	ppm	1.167	0.896
Nb	4000	ppm	1.750	0.750
Zr	92000	ppm	1.098	0.761
Sr	699000	ppm	0.971	1.252
Y	23000	ppm	1.000	0.913
Ni	30000	ppm	1.300	1.033
Cr	77000	ppm	1.169	1.039
V	106000	ppm	1.236	1.094
Ba	202000	ppm	1.223	0.856
La	12000	ppm	1.250	2.083
Ce	·64000	ppm	1.281	1.109
Zn	87000	ppm	1.874	1.207

Table 8: Chemical analyses results on clay samples compared to the reference sample

The first ten elements reported in the table chemically characterise clay, and the values of the three samples slightly differ.

As far as the other elements are concerned, only Pb and Zn present values which are remarkably higher than those of the reference sample, both being more than twice the reference value in the more superficial sample and about 30% in the sample taken at depths of 5,40 m.

Geotechnical survey and chemical analyses were carried out respectively in the Geotechnical Laboratory and Earth Science Department of Calabria University.

The tests carried out on the samples taken at different depths showed diverse behaviours between the deeper clay layer and the more superficial layers.

As already said, plasticity index values, particularly indicate that surface layers present increased plasticity than deeper ones.

Moreover, the latter ones resulted to be easily compressed as well, as shown by the values of the oedometric modulus of the different samples.

Though the examined clay bank resulted to be homogenous from a granulometric viewpoint, the above elements show different characteristics between surface and deeper layers.

This difference could be due to changes in the structure of the clay as a consequence of the presence of leachate coming from the above waste mass.

The hypothesis is supported by the greater amount of Pb and Zn present in surface layers, being these elements generally present in the leachate produced by urban solid wastes.

Moreover, the concentration gradient reported in the 4,60 m and 5,40 m layers can be seen as a further support to the hypothesis.

Furthermore, as for the main elements (CaO, Na2O, etc...) a peculiar symmetry was reported between the two surface samples if compared with the reference; particularly the values of the 4,60 m sample are always higher, except for Mg an Ca.

The above hypotheses need further confirmation, which are expected in the future stages of the research, through other sounding aiming at characterising the groundwater beneath as well.

A complete mineralogical characterisation of the fine fraction will be performed and the possible structural variations of the clay will be examined as well.

5.   CALIBRATION AND VALIDATION

When a computer code was run in order to know the evolution of a natural system, it is necessary to know if the code provides an accurate solution to the partial differential equations with different boundary conditions. This is usually done by carrying out tests which compare numerical model results with those measured or with analytical solutions. However, numerical solutions greatly depend on spatial and temporal discretization, so even if there is a perfect agreement between data, this can only prove that the code can accurately solve partial differential equations, but it does not mean that it will happen under any and all circumstances.

The term "validation" indicates a process of verification which tries to ascertain that a model possesses a satisfactory degree of accuracy and certainty within its entire domain of applicability and over the entire spatial and temporal scales for which the model is intended to be used.

This definition has splitted the opinions of scientists and experts in two. The notion of validation is shared by those who accept the positive thought according to which "a theory is confirmed or refuted on the basis of critical experiments designed to verify the consequences of the theories", while those who share Karl Popper's thought which assumes that *"a theory can never be validated, only invalidated"* believe that this notion should be dismissed, being the notion of calibration itself enough.

Theory evolution, in every field, confirms Popper's thought. Many theories, which were considered "validated" by the scientific world, have been invalidated by new experiments which disproved them.

This is even clearer in hydrogeology where the estimated parameters are nonunique due to the phenomenon can never be directly observed and the fundamental processes are not completely known.

Finally, the term validation should be avoided in hydrogeological science because it gives a distorted view of model reliability. More meaningful terms, such as calibration and "history matching", should be used in order to avoid false confidence into model predictions.

In the calibration operation, different parameters, within reasonable ranges, are in turn employed, and the set giving the best agreement between computed values and measured data is then chosen [11].

The optimization of the calibration procedures is attempted through trial and error adjustments or through some automated inverse or parameter estimation procedure.

The model is considered "calibrated" when it reproduces historical sets of measured data within some acceptable level of coherence. A good match between data does not prove the validity of the model because the solution is nonunique and the model can include compensating errors. In other words, the model can adequately reproduce historical data but fail to predict future responses of the system under new boundary conditions or, more in general, new set of stresses. In the petroleum industry, calibration is called "history matching", that is the adjustment of computed data and measured ones. Once history matching is over, petroleum engineers, never predict system performance for more than two times the period of history sets. It is interesting to note that periodical updating of computed (or predicted) data and measured ones is very useful because "the longer the matched history period, the more reliable the predicted performance will be" [12].

5.a. Calibration of SUTRA code

All Reggio Calabria coastal aquifer is characterised by a strong saline intrusion which is mainly due to very high distribution of pumping wells on the entire basin above the aquifer.

The areas mainly subjected to intrusion are Calopinace and Valanidi valleys where the concentration chlorine in groundwater exceeds potability levels. The condition of Calopinace [13] is particularly worrying, because of the negative hydraulic gradient up to S. Giorgio wells field (1÷2 %) where salt concentration amounts to 600 mg/l. The well field supplies about 300 l/s to Reggio Calabria water system, and for this reason the situation is getting worse. Therefore a planning scheme has to be carried out to restore the water body and it made of:
- reclamation of water in order to push the saline wedge back to the sea;
- artificial recharge operations;
- reduce the abstraction from the wells;
- control the quality of the aquifer.

Being the intrusion so vast, the artificial recharge through injection wells was considered the better solution. Because of pumping well distribution all over the basin, the simulation of the intervention should have been carried out by means of a tridimensional computer code. A tridimensional analysis would simultaneously take into account density differences between salt and fresh water at the vertical level and the reciprocal influence among different wells at the horizontal level. However, tridimensional computer codes which simulate pollutant hydrodispersive phenomenon didn't yet give such good results to justify their employment. Moreover, being the most severe intrusive phenomenon is localized only in Calopinace and Valanidi fans which are hydraulically independent, the phenomenon could be considered bidimensional and, analysed in a cross section. For these reasons, the bidimensional code SUTRA [14] was employed. The area chosen to study the recharge was Calopinace valley.

5.b. **Schematization**

In order to schematize the cross section where intrusion had to be simulated, the geological section was reconstructed through geologic logs of the wells. The stratigraphies of the wells have been carried out in the wells field of S. Giorgio and Prumo. The phreatic and homogeneous aquifer mainly consists of alluvial material (sand and slightly compacted gravel).

Because of compaction, a 100% hydraulic conductivity anisotropy has been proposed ($K_h/K_v = 100$).

The hydrodynamics parameters chosen have been determined by Italpros, trough field measurements in the Calopinace valley:

$k_h = 7 \cdot 10^{-4}$ m/s

$\varepsilon = 0.25$

$S_w = 5 \div 11 \cdot 10^{-2}$.

The discretized area goes from the coast to 4 km inland.

The impermeable bottom, about 80 m b.s.l. near the coast, was considered horizontal, while the land plane was given a 1.5% slope.

Three abstraction sites were found in this area:

- two wells, for industrial supplies of Reggio Calabria town, located about 600 m from the coast with a 9 l/s pumping;

- S. Giorgio well field, for potable supplies, located about 1900 m from the coast with a 280 l/s overall discharge;

- Prumo well field, for potable supplies, located about 2800 m from the coast with a 40 l/s discharge.

The grid mesh is made of quadrangular finite elements with variable geometry. The side of the elements longitudinally varies as a function of the wells distribution (100⇒50 m), while it vertically varies according to aquifer thickness. In the cross section, thickness is constant and it has been set equal to 120 m. The number of elements used was 1380, while the number of nodes 1470. The phreatic aquifer generally gives rise to problems in water table simulation because of its variability within the discretized domain.

This problem is solved by setting specific yield different from zero above the initial water table. The influence of the unsaturated element on heads had also to be taken into account.

For this reason Van Genuchten [15] relationship between saturation coefficient and capillary pressure of the porous medium was assumed as valid.

Boundary conditions are described in the following figure 6

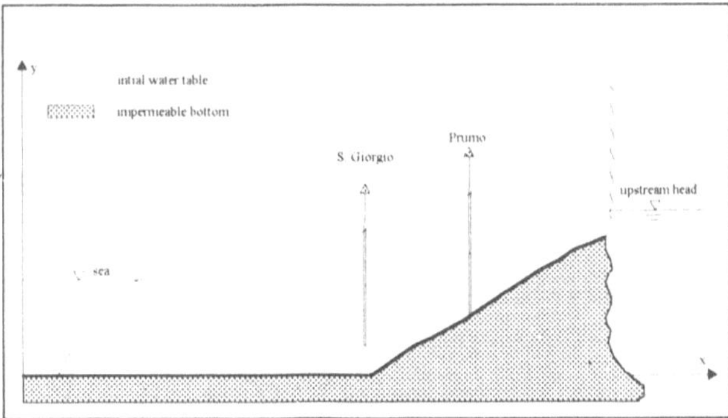

Fig. 6, Schematization of the simulation

5.c. Model calibration

Before starting the simulation in order to evaluate the responses of the system subjected to new stresses, we carried out model calibration to obtain a reliable match between measured and computed data by SUTRA. For this reason, we assumed different values of characteristic hydrodispersive parameters α_L, α_T and D_0 within a reasonable range [16].

In these conditions, the best approach to the problem is the inverse one, that is, knowing the initial and final conditions of the aquifer, the values of hydrodispersive parameters are those which minimize the differences between model output and the final condition of the aquifer. The distributions of concentration and pressure in 1981 (initial condition) and in 1988 (final condition) had been provided by Italpros survey. After a set of simulations carried out by means of Sutra computer code using different values of α_L, α_T and D_0, the following values were estimated:

α_L= 10 m
α_T=0.10 m
D_0=1.10 m^2/s

These values minimize the differences between pressure and salt concentration distributions in the model and those measured in situ.

The computed concentration distribution obtained through these values coincides with the 620 ppm measurement obtained in San Giorgio in 1988. Pressure distribution as well confirms the estimated parameters because a negative hydraulic gradient up to S. Giorgio was obtained as assumed by the above-mentioned Italpros study (in 1988 it was impossible to measure the static ground-water level in the wells which were pumping 24 hours a day).

Through calibration, the parameters α_L, α_T and D_0 express all the uncertainties due to hydrodynamic parameter estimation (k, T, ε, S_w,...), real abstraction history, hypotheses on boundary conditions and computer code. Pressure and concentration distributions in 1988 are here reported.

6. CONCLUSIONS

As we said previously, the best approach to the groundwater management is the global one. The application of such global approach, in general, involves the necessity to find a point of equilibrium between the demand of a better knowledge of the aquifer system and the solution of the actual problems.

Because of the limits of a mathematical approach, that influences also the experimental measurements of hydrogeologic parameters, it is necessary to be able to utilize the various knowledges , (the "Hydrogeologic information").

These ones seem not to be adequate to provide rules for the groundwater management, because they can't be utilized "immediately" in the running code.

Because of these hydrogeologic information out the problematic before reported the code outputs must be consistent with them.

The result of such comparison depends on the designer who must choose between different solutions obtained from the availability of data and code.

REFERENCES

- 1. Troisi, S., Fallico, C., Maiolo, M. and Coscarelli, R.: Campo prove per lo studio sperimentale di fenomeni idrodispersivi in acquiferi porosi., Dipartimento Difesa del Suolo - Università della Calabria, Montalto Uffugo, 1993.

- 2. Di Fazio A., Maggiore M., Masciopinto C., Troisi S., Vurro M.: Stazione per lo studio sperimentale di fenomeni idrodispersivi in acquiferi fessurati, Istituto di Ricerca sulle Acque - C.N.R., Quaderni , 88, Roma 1990.

- 3. Di Fazio A., Maggiore M., Masciopinto C., Troisi S., Vurro M.: Caratterizzazione idrogeologica di una stazione di misura in ambienti carsico, Acque sotterranee, 2 (1992), 17-26.

- 4. Di Fazio A., Masciopinto C., Troisi S., Vurro M.: Una interpretazione bidimensionale di prove di tracciamento in sistemi fratturati, in Atti del Pro. XXIV Convegno di Idraulica e Costruzioni Idrauliche, Napoli 1994.

- 5. Cancelli, A., Cossu, R. and Malpei, F.: Prove di laboratorio sull'impiego della bentonite come impermeabilizzante degli scarichi controllati, Rifiuti Solidi, 6. (1987).

- 6. Esposito, L. and Ferrante, A.P.: Sull'impiego delle argille costipate per la impermeabilizzazione dei fondi delle discariche controllate, Rivista Italiana di Geotecnica; anno XXIV, n. 2 (1990).

- 7. Galassi, S.: Migrazione degli inquinanti dalle discariche alle falde, Rifiuti Solidi, Vol 2, n. 1 (1988).

- 8. Gervasoni, S. and Piepoli, A.: Sistemi di monitoraggio delle discariche Rifiuti Solidi, Vol. 4, n. 4 (1990).

- 9. Carabelli, E., Gera, F.,Gruszka, A. T.,Piepoli, A. and Schneider, A.: Advanced procedures for landfill monitoring, in Proceedings Sardinia '91, 1991.

- 10. Frega, G., Fallico, C. and Maiolo, M.: Ricerche in corso e in programma per la valutazione qualitativa e quantitativa di rifiuti solidi nella regione calabrà, in Atti del IV Workshop "Clima ambiente e territorio nel Mezzogiorno", Lecce, 1991

- 11. Konikow, L. F. and Bredehoeft, J. D.: Ground-water models cannot be validated, Advances in Water Resources 15 (1992), 75-83.

- 12. Thomas, G. W.: Principles of hydrocarbon reservoir simulation, IHRDC Pbubl., Boston, 1982.

- 13. Italpros S.R.L.: Aggiornamento modello matematico fiumare della provincia di Reggio Calabria, Relazione Zona Sud, rapporto tecnico non pubblicato, 1989.

- 14. Voss, C.I.: A finite-element simulation model for satured-unsatured, fluid-density dependent groundwater flow with energy transport or chemical-reactive single-species solute transport, U.S. Geological Survey National Canter, ,Reston, Virginia, 1984.

- 15. Van Genuchten, M.T.: A closed form equation for predicting the hydraulic conductivity of unsatured soils, Soil Science Society of America Journal, 44 (1980), 892-898.

- 16. Troisi, S., Fallico, C., Coscarelli, R. and Caramuscio, P.: Considerazioni sulle misure sperimentali dei parametri idrodispersivi di falde sotterranee, in XXII Convegno di Idraulica e Costruzioni Idrauliche, Firenze, 1992.

METHODS FOR RESTORING AQUIFERS

U. Maione

Technical University of Milan, Milan, Italy

ABSTRACT

The paper presents an overview of the most common methods for the reclamation of polluted aquifers. They can be divided in hydraulic and structural methods: the first ones are based on the use of pumping or recharging wells causing a local variation of the flow field and forcing the pollutant to move towards the extraction point or far from the zones to be protected. The structural methods consists in ground or surface constructions which permanently affect the groundwater flow. Vantages and disadvantages of the two methods are compared. Afterwards the use of the mathematical modelling for the design and simulation of the reclamation measures is discussed, with a particular attention to the recent stochastic theories.

1. INTRODUCTION

The reclamation of aquifers is a highly topical matter interesting most developed countries. In Italy, the problem is met mainly in the industrial plainlands of the north, but there are tangible signs that it will soon involve other regions of the country on account of the current widespread environmental pollution that is gradually affecting the aquifers along with an increasing demand for groundwater. Indeed. the serious impoverishment of surface water resources, not only for drinking- water use, is extending that demand for water which, for the moment at least, is less polluted than surface water.

The need to use groundwater resources, while safeguarding their quality in the short and medium term, highlights the application of aquifer reclamation procedures that have been developed and implemented outside Italy. The technical and management authorities must therefore be encouraged to change their way of thinking: until not so long ago, the detection of pollution in a well only led to immediate closure or a change in the depth at which water was taken. (See, for example, the water supply from the Milan aqueduct.) Such solutions were chosen because they were the most economical in the short term, but the long-term result has been that the quality of groundwaters has worsened to the point where their quality and quantity make them totally unusable. If this situation does not change, we may easily forecast ever higher extraction costs and the need to treat waters drawn from aquifers on a massive scale. Such treatments do not only mean that water resources will cost much more than at present: only to a certain extent they guarantee the quality of the waters in that there is no way of thinking that treating waters, with however sophisticated plants, can eliminate all the pollutants they contain.

So what exactly is required to plan, design, implement and manage aquifer reclamation systems?

The salient feature of this field of action is its overtly interdisciplinary nature, which requires the integration of different skills for the solution even of small-scale problems. The work of the analytical chemist or of the biologist normally gives the alarm that focuses interest on the problem; the hydrogeologist singles out the features of the aquifer which is being polluted; the mathematical modeller prepares a programme that must simulate the behaviour of the aquifer and lead to a study of possible alternatives; the administrator must consider the economic and political conditions which the team must use to define the best type of treatment to use in any given case; and so on. At the present time, these experts are trained in the faculties of chemistry, biology, natural sciences, mathematics, engineering, medicine, and so on, and the dialogue among them, as many of us have seen on a personal level, is not always very easy, in Italy at least, where such forms of co-operation are only just beginning.

Considering the growing interest in the matter, it might not be an exaggeration if we began to think of a technician to be trained in this entire set of disciplines on a university level. That is being done in the United States today.

There are a large number of measures that can be taken to prevent the onset of pollution in an aquifer and to restore it. I have no intention here of making a list of them. It would be incomplete and not very significant. What I would like to do is review some of the most common ones, with reference to certain applications already made in Italy, in the hope of arousing discussion among the experts present here today.

2. Aquifer reclamation systems

Action plans for reclaiming aquifers can be subdivided into two essential types: hydraulic and structural.

Hydraulic action consists in the "manipulation" of the groundwater gradient by means of injecting or extracting water so as to move the water and, indirectly, polluting substances in a given way. This manipulation can be done using well systems with different configurations: with some of them, water is extracted by means of a well point or a deep well; with others (Fig.1, from [1]), water is introduced to create a pressure ridge. Whatever the type of action taken, the wells must be located in a very clearly defined manner with very clearly defined flow rates so as to move the underground water and any eventual pollutants in the most suitable way.

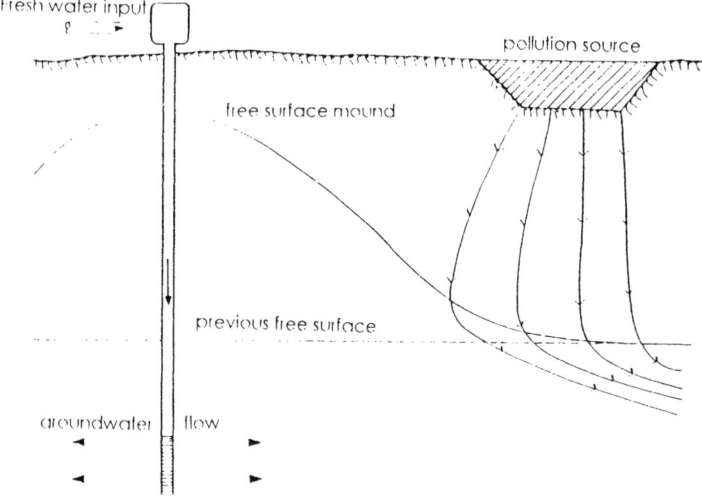

Fig. 1, Principle of the pressure ridge system

Well point systems consist in a number of shallow wells, all close together and connected to a main conduit that is connected in turn to a suction pump. They are used almost exclusively for shallow aquifers on account of the restrictions imposed by the lowering of the static level of the water and by the limits of the pump. These systems have been designed so that the plume of pollutant is completely intercepted.

Conceptually, deep wells are similar to well points. They are used in deep aquifers and are pumped individually.

A pressure ridge is created by injecting non-polluted water into the subsoil by a chain of injection pumps upstream or downstream of the plume of pollutant. Up-gradient injections are made to force the upstream water to flow towards the space occupied by the plume, while the pollutant is collected downstream by a series of extraction pumps. This procedure increases the speed of the water in the plume of pollutant and is used to wash the aquifer. Down-gradient pressure-ridge systems are normally also used for protection against saline intrusion on coastlines.

This method is the surest way of dealing with a groundwater flow of pollutant. The use of recharging and/or extraction wells offers a high level of planning flexibility and definitely means minimal construction costs. The wells can be installed reasonably fast, which is important when a pollution problem has to be solved in a short time-scale. At the present time, use of the well system is the most common method of controlling pollution in aquifers and will probably remain so in the future, too.

It is generally held that, albeit with the advantages we have mentioned, well systems are not the conclusive treatment for polluted aquifers. That is, they are the most immediate and direct means of stopping the migration of a plume of pollutant until a permanent solution is found.

These are some of their specific restrictions [1]:

(i) high management and maintenance costs compared with passive-barriers;

(ii) they can only be used in fine soils;

(iii) some systems may require the use of sophisticated mathematical models to assess their efficiency;

(iv) after installation, the quality of the polluted site has to be monitored continuously.

Hydraulic action involving the use of interception trenches require a trench to be dug below the water level and a conduit to be placed in it. System operation corresponds practically to an indefinite line of extraction wells covering the whole length of the conduit. Usually, a perforated conduit and rough backfill are placed in the trench for greater efficiency in the collection and conveyance of incoming water. Figure 2 shows an interception trench.

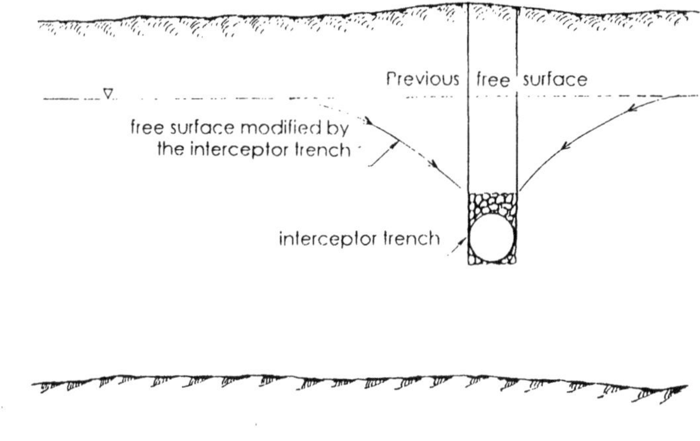

Fig. 2, Hydraulic gradient towards an interceptor trench

There are two precise objectives behind such systems:

(a) prevention: in these cases, people prefer to speak of leachate collection systems;

(b) abatement: true and proper interceptor drains. Their relatively simple construction involves the excavation of a series of trenches in which perforated conduits and gravel-type backfill are placed. There is usually a series of supply pipes converging in a main manifold. The water flows from this manifold towards a sort of sinkhole from which it is pumped up to the surface. If they are used for prevention purposes, their configuration is the comb type. as shown in Figure 3.

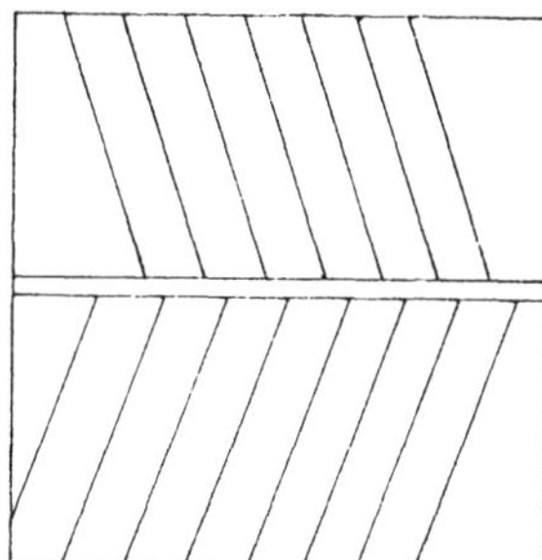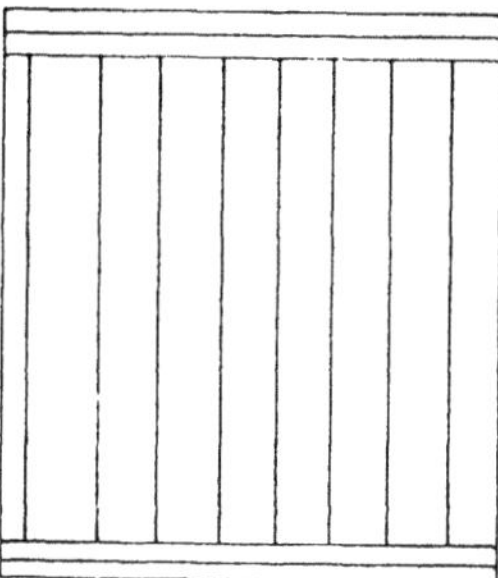

Fig. 3. Examples of disposition of interceptor trenchs.

If on the other hand, the aim is to abate pollution in a given area, there can be relief drains and interceptor drains. The former are installed in areas where the hydraulic gradient is relatively flat. They are generally used to lower the water level beneath a given site or to prevent the pollutant from sinking to lower strata. They can be installed around the entire perimeter of the site.

Interceptor drains collect the water coming from an upstream source so as to prevent it from reaching supply wells located downstream. They are normally installed along the mean direction of the underground flow. Depending on circumstances, one or two interceptor drains may be needed at the base of the aquifer.

The most obvious advantage of these methods lies in their simple construction, while a clear disadvantage is the need for continuous monitoring and assistance. Other advantages include:

(i) low installation costs;

(ii) the leachate can be collected in terrains that are not very permeable;

(iii) even considerable flows can be drained if the perimeter is very large;

(iv) the "captured" pollutant can be controlled;

(v) a lower quantity of fluid to be treated is produced compared with well-point systems.

Typical disadvantages include:

(i) they adapt poorly to terrains that are not very permeable;

(ii) in many situations, their location beneath existing sites is not feasible;

(iii) in the event of dissolved constituents, the water downstream of the line of interceptor drains also has to be monitored.

In addition to hydraulic action, many other solutions can be applied successfully to control latent or effective pollution. They are:

- Impermeable barriers

- Surface controls, liners, impermeable boundaries

Impermeable barriers are made by means of sheets of steel driven into the ground and linked together so as to form a thin waterproof shield. The material used to make the "sheets" also includes wood and concrete. There is some controversy about their being used for treating polluted groundwaters because of problems related with corrosion. The high cost is another main disadvantage of this system. Indeed, the larger the project, the higher are the costs. Then, for satisfactory installation, the soil has to be relatively uniform, loose and devoid of rocky masses.

As regards surface controls, there are three different technologies, normally used together, that are applied especially as a preventive measure. Control of superficial flows does, in fact, minimise the quantity of water flowing towards a very precise site, thereby reducing potential infiltration.

The objective of a surface liner is to hamper the infiltration of surface or rain-water that reaches the site in question.

The adoption of impermeable surface boundaries prevents the downward flow of leachate and/or lessens the pollutant load by means of adsorption processes.

Surface control measures are a relatively low-cost way of reducing possible future pollution of groundwaters, and in this sense, it would be a good thing if they became part of the action plan designed for every type of site. In Italy, these technologies are being applied to a certain extent and many technicians present here today have had a good deal of experience with them.

The design of a site using surface liners, impermeable boundaries or both depends on the hydrological balance of the place in question. Various basic combinations of these techniques can be used.

The first type uses a lower impermeable boundary to maximise the quantity of leachate collected and to minimise the quantity that escapes.

Impermeable surface hoods are prepared to minimise infiltration by endeavouring to increase the tendency towards superficial flowing and evapotranspiration.

The advantages and disadvantages of the method are as follows:

(i) "engineered liners only":

(a) Advantages:

• lower hydrogeological impact.

(b) Disadvantages:

• higher construction costs

(ii) "engineered covers only":

(a) Advantages:

• lower construction costs than liners.

(b) Disadvantages:

• long-term monitoring is required

(i) "engineered covers and liners":

(a) Advantages:

• lower environmental impact

• minimal quantity of leachate collected

• politically and socially acceptable.

(b) Disadvantages:

• high design and construction costs

• high-quality clay or synthetic material is required

• stabilisation times are long.

The so-called grouting system is the process by which liquid, slurry or an emulsion is injected into the ground under pressure. The injected fluid will migrate from its place of origin to occupy a more or less vast area, filling the pores in the ground. As time passes, it tends to solidify, decreasing the previous permeability of the soil and increasing its retention capacity. The "mortars" used are *particulate* or *chemical*. The former are made of water mixed with particulate material that solidifies inside the solid matrix. The latter consist of two or more liquids that take on the form of a gel when they come into contact.

Grouting is a technology that was only recently used as part of groundwater quality-control plans. The success of their position depends essentially on the type of preliminary on-site investigations. In addition, these materials can adapt to the structure of many kinds of ground. On the other hand, we must consider that there is no certainty that the pores in the ground will be completely filled by the injected mortar, so special care has to be taken over the installation procedure.

Slurry walls are one method for encapsulating an area so as to prevent the intrusion of pollutant particles or to prevent them from escaping. A perimeter trench is dug and filled with waterproofing material. Walls of this type can also be positioned upstream of a potentially polluting area (waste disposal, and so on) to block the flow of water (Fig. 4, [1]). They can also be placed around a polluted area to contain the pollutant in it (Fig. 5, [1]). Usually they are efficient when used in combination with superficial liners and drainage wells.

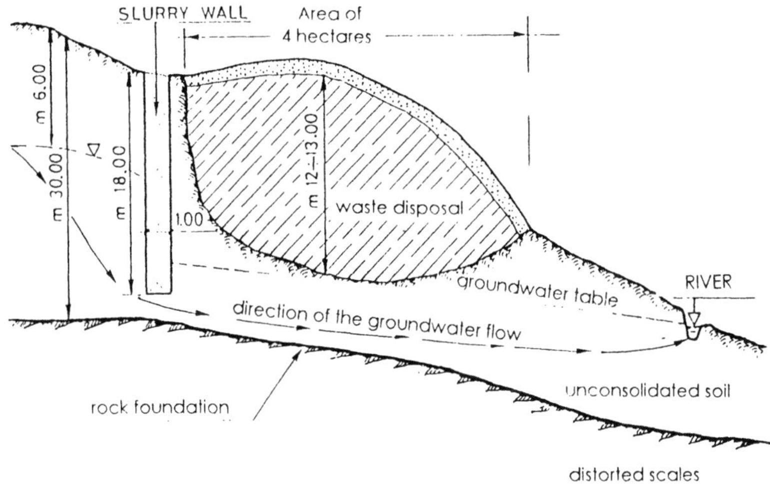

Fig. 4, Scheme of a realisation with a slurry wall.

The decision to adopt a given type of action may only be made on the basis of full-scale, detailed knowledge of the characteristics, origin and extension of the pollution. This cognitive phase is of primary importance and can be decisive for the satisfactory outcome of action taken.

As an example of restoring system in Italy realised with slurry walls there is the municipal waste disposal of the city of Ravenna[2]; the location of the interested area and a sketch of the subsoil layers together with the realised slurry walls are shown in the Figures 6 and 7. The special geology of the area made possible the isolation of the waste disposal by means of a perimetrical slurry wall closed in the silt layer. The pumping plant located into the closed zone assures that the percolation waters don't come out and are conveyed to the treatment plant. The slurry wall is 0.60 m thick and is made by mixture of water cement and bentonite.

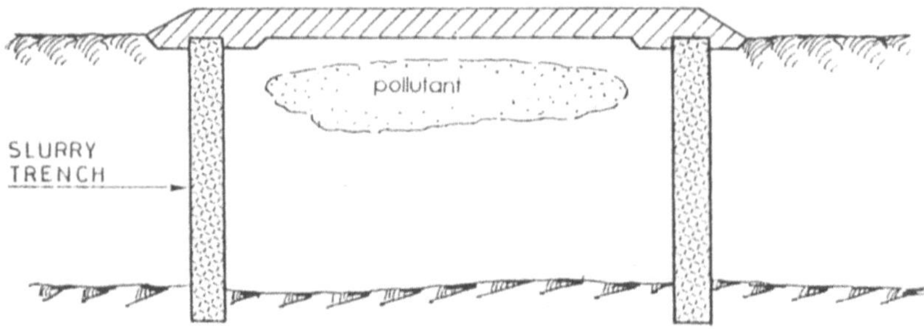

Fig. 5, Isolation of a polluted site [1].

Another slurry wall was realised for the isolation of the Acna Chimica Organica plant near Cengio, in Northern Italy (Figures 8 and 9); in this case the diaphragm is closed in the marl layer. The slurry wall is made with a mixture of water, cement and bentonite.

Fig. 6, Map of the waste disposal of Ravenna [2].

Methods for Restoring Aquifers

3. The use of mathematical models

A tool of crucial importance in this sector (groundwater hydrology) is the mathematical model. Designing the initial stage of reclamation works can also be done by simple mathematical relations [deduced by posing the hypothesis of a simplified operational geometry [1]], but this scaling stage must be followed by a mathematical modelling of the aquifer in the configuration modified by the action taken. The two stages of scaling and verifying often overlap and interact continuously. And the choice of the most appropriate action can only be made after different alternatives have been simulated mathematically.

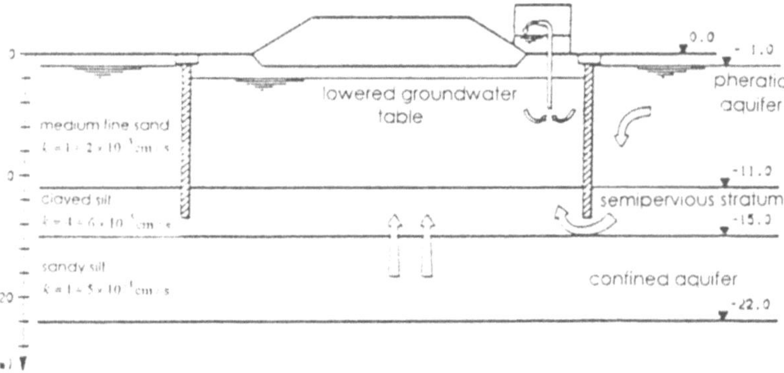

Fig. 7. Scheme of the restoration action at the waste disposal of Ravenna [2].

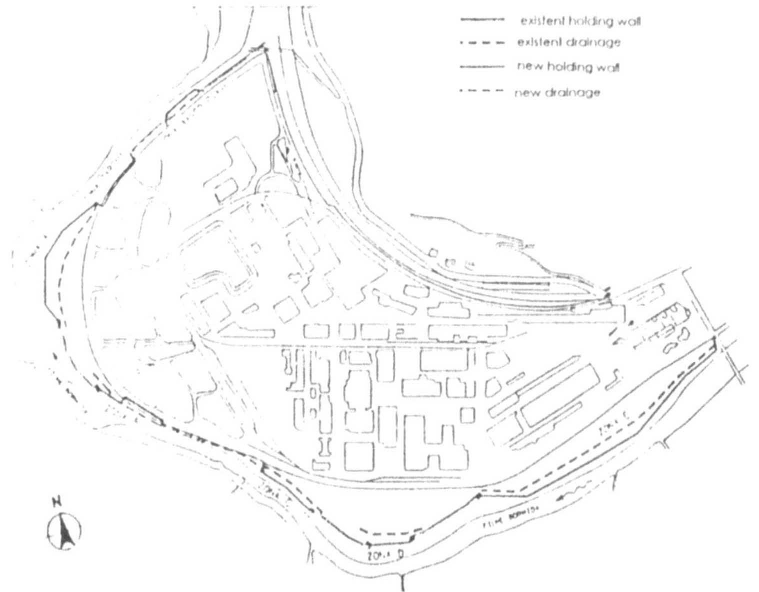

Fig. 8. Map of the plant of Acna Chimica Organica, at Cengio [2].

The mathematical modelling of groundwaters is a branch of the sciences that has achieved results which I can safely call exceptional. In the field of quantitative models of aquifers (management models, without quality components), the literature and the market offer computing programmes suitable for most of the problems the technician has to face ([3], [4] and [5]). Extremely refined mathematical models have also been designed for studying three-dimensional filtration phenomena [6] with computing times optimised by means of special numerical calculation algorithms [7]. Calibration procedures are currently used [8] to make simulations of the real aquifer more reliable.

In this connection, I would like to point out a dichotomy between the mathematical model-maker and the person who uses the model. Normally, the mathematical model is worked out by a specialist who performs some of its applications that are requested when the model is commissioned. Those applications are often the only ones performed by that model because it is so complicated to use that any albeit worthy technician (unless he is the one who wrote the programme or is a specialist in the sector) would be discouraged when it came to trying other applications. In the long run, the mathematical model is not a tool but a self-contained study, valuable as it may be.

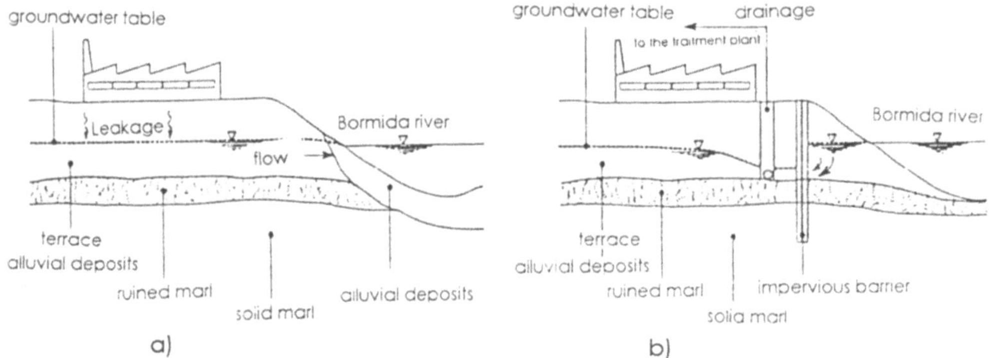

Fig. 9, Sketch of the groundwater situation under the plant, before (a) and after (b) the remediation action[2].

I do not think that the path towards obviating this problem--which, in the end, means a formidable waste of intellectual and financial resources--is to simplify models so that they become easier to use. That may be possible in certain cases, when action is being planned, but when it comes to simulating the behaviour of an aquifer, we have to remember that movement in groundwater aquifers is an extremely complicated matter and it is not possible, therefore, to think of tackling it using uncomplicated mathematical models. We must also consider that the end-user does not want convenience to jeopardise the highest possible precision and reliability of the results.

In my view, the solution lies in the help given by today's information technology, which can now provide all the "filters" that enable the user to apply a mathematical model developed by a specialist for a specific problem to other cases. When I say "filters", I obviously do not mean only those computer programmes that can display the results with very expressive, very realistic graphics, but also those programmes that guide data input procedures in such a way as to minimise errors and which warn the user about possible incongruences that may emerge when the programme is run and thereby generate senseless results or even stop the programme because of serious mathematical errors.

In the modelling of pollutant transport phenomena, we have reached a critical moment. Until very few years ago, the fundamental presupposition was that the dispersive flow could be represented by an equation similar to Fick's law for the molecular diffusion of a solute. The theory that was developed from it, perfected in the fifties, was in fact called Fick's theory and presupposed that the dispersive process was conditioned by dispersion coefficients. For a homogeneous, isotropic aquifer, such dispersion coefficients (dimensionally lengths) were a feature of the aquifer and remained constant for the whole extension of the field of movement. Following the outlines of this classic theory, mathematical models of various types (random walk, finite differences, finite elements) were developed (see [4] for example) and were used widely in technical practice.

But the spread of Fick's theory applications to real aquifers has revealed a series of incongruences. In the first place, the dispersion coefficients (the longitudinal dispersion coefficient in particular) worked out by laboratory measurements are of a lower magnitude that those worked out on the field (the magnitude of the grains making up the porous medium in the laboratory, metres in the field). Subsequently, it was noted that the dispersion coefficients singled out in mathematical models for calibrating pollutant migration could be even higher and all the higher the more the movement of the pollutant affected larger areas. Dispersion coefficients of hundreds of metres were even obtained which suggests that the calibration value coheres little with the magnitude. Finally a few sample experiments conducted with an enormous outlay of means and with extreme thoroughness revealed the shortcoming of Fick's theory when applied to heterogeneous aquifers, like natural ones. That is, it was seen that in order to follow the experimental phenomenon, it would be necessary to use dispersion coefficients that varied (grew) with the distance covered by the pollutant.

The explanation of this behaviour lies in the stochastic theory of transport in heterogeneous aquifers. I do not want to go into this at any great length here today, but I do feel that it would be interesting to provide some food for thought. Briefly, after more accurate surveys, real aquifers which we normally schematise on a local scale ($\approx$100 metres) as constant permeability aquifers (homogeneous aquifers), turn out to be composed of heterogeneous masses with a permeability that varies, sometimes greatly, from one point to another. The statistical distribution of permeability values obtained on a group (not very large) of aquifers is the log-normal with oscillations of the $Y=lnK$ even up to 3-4 [9]. Permeability variations are not, however, casual but follow the spatial trends described by covariance functions. For this kind of problem, a distribution that cohered well with the experimental data was the simple exponential $C_y = \sigma_y^2 \cdot e^{-\frac{r}{\lambda}}$, where λ is the correlation length, the value of which is around 3-5 metres.

The dispersion of pollutants in these porous media is governed both by the dispersive phenomenon on a laboratory scale (which, until a short time ago, was considered to have a Fick type of behaviour) and by the more important phenomenon of differentiated advection. That is, the pollutant is dispersed in the medium and is distributed on different areas from the mean trajectory because of differences in speed encountered in the field of movement from one point to another. An observer who is not aware of the differences in the permeability of the aquifer sees this effect as an "excessive" dispersion of inexplicable origin.

Stochastic theory, on the other hand, fully analyses these variations of permeability and, using various important hypotheses, can provide the solution to the problem. Obviously, since stochastic theory analyses variables as stochastic and not deterministic variables. it gives a solution that represents the "mean" behaviour of all the stochastic population it is handling.

In any case, in the hypothesis of non-reactive pollutant. small variance values (σ^2_y) much lower than 1 and a uniform flow on average, one can theoretically (linear theory [10]) define the trend of the apparent dispersion coefficient that grows with the distance covered by the pollutant and tends towards an a-symptotic value that is a function of the variance value σ^2_y. It is interesting to note that the apparent dispersion coefficient thus defined depends on the statistical characteristics of permeability distribution and must not, therefore, be worked out on the basis of a special measuring survey.

If we wish to apply this result to a real case of pollution, all we have to do is make sure that the working hypotheses underlying the linear theory are respected and then apply a "classic" type of pollutant transport model which operates on an aquifer schematised as homogeneous with dispersion increasing with distance .according to the equation derived from the stochastic theory. If, on the other hand, the phenomenon to be modelled does not for some reason respect the fundamental hypotheses and the behaviour cannot be assimilated with the "mean" behaviour of all the hypothetical aquifers with given statistical parameters, it is not possible to apply the results of the linear theory so simply.

The constraint $\sigma^2_y < 1$ can be overcome. Rinaldo and his collaborators [11] have shown that this limit can be overcome quite well and the results of the theory still cohere with the behaviour of these conspicuously heterogeneous aquifers.

In real cases, the fact that the mean flow is not uniform (action with wells, pumping. and so on) makes the results of the linear theory difficult to apply. In this connection. a numerical research has being carried on at the Milan Polytechnic.

Together with Rinaldo, we have investigated the condition that the particular phenomenon in question should evolve according to the overall behaviour of the population of aquifers with given statistical parameters and we have established that the plume of pollutant in the transverse direction must have an extension greater than 25λ([12] and [13]). Should this extension be lower, the individual case will evolve in a different manner for the mean behaviour.

At this point, it becomes necessary to increase our knowledge of the characteristics of the aquifer with accurate permeability measurements. Studies are in progress at the Milan Polytechnic on this topic, and they indicate that the on-site survey has to be made in great detail [14]. If the results, which are not yet final, are confirmed by numerical proof, they show that we have to measure permeability on a spatial grid with side $\approx 3\lambda$ or measure permeability and hydraulic heads on grids with sides 6λ and 3λ. Considering that the values of λ on a local scale are about 3-5 metres and about some hundreds of metres for problems on a regional scale, we may deduce that such a detailed knowledge of the aquifer presents a real challenge. At the present time, the problem seems difficult to solve. Some help may come from geophysical surveys, as long as the indications emerging from such surveys can be improved.

I would also like to point out that, in this delicate matter, apart from all of the most sophisticated theories, the fundamental role is played by experimental observation. It was true experience that shed doubt on the classic theory and it is by increasing experimental observations that we may hope to gain substantial explanations of the matter.

Experimental work, which demands ever greater sacrifices in this field (compared with numerical processing) really is a burden. In the first place, it has to be done on the spot.

The findings that can be made in the laboratory can be useful (and people are having second thoughts about them) but they can rarely be decisive. On-site experimentation must be conducted with great precision: the locations must be chosen with great care; and the costs are extremely high. For instance, a test to determine dispersion values may be heavily conditioned by the penetration level of observation points, by the stratification of the ground by the very type of test, by the kind of tracer used and, finally, by the mathematical model used to interpret the test. Consequently, in-field tests conducted with a certain degree of approximation can provide a number which can easily be quite credible but cannot do very much to explain the theory. For example, I might quote the sample experimentation of the Borden Site carried out in Canada by Freyberg. et al. [15] and financed by the U.S. Environmental Protection Agency and by the National Science Foundation: 9 wells and 340 piezometers were drilled on an area of 9.600 square metres. 846 aquifer samples were taken to determine particle size and 26 permeability tests were done on site. The experimental campaign was conducted for 3 years with samplings made about every 7 days. You have to understand that few experiments like that can be carried out in this world and none can be carried out in Italy unless a whole army of experts is mobilised and unless a lot of money is allotted for the purpose.

REFERENCES

1. Canter L.W. and Knox R.C.: Groundwater Pollution Control, Lewis Publ. Inc., 1985.
2. De Paoli B. and Marcellino P.: Esperienze di isolamento mediante diaframmi, in: Proc. of the Course on Traitment and restoring of polluted terrains (Trattamento e recupero dei terreni contaminati), Dip. of Hydraulic, Environmental and Surveying Eng., Politecnico di Milano. 1992.
3. Hunt B., Mathematical analysis of groundwater resources, Butterworths, 1983.
4. Bear J. and Verrujt A., Modeling Groundwater Flow and Pollution, D. Reidel Publ. Com., Dordrecht-Boston-Lancaster-Tokyo, 1987.
5. IGWMC - International Groundwater Modeling Center, Price List of Publications and Services Available from IGWMC, Holcomb Research Institute, Indianapolis 1989.
6. Gambolati G., Pini G. and Tucciarelli T., A 3-D finite element conjugate gradient model of subsurface flow with automatic mesh generation, Adv. in Wat. Res., Vol. 1, 1986.
7. Dougherty D.E., PCG solutions of flow problems in random porous media using mixed finite elements, Adv. in Wat. Res., Vol. 13, n. 1, 1990.
8. Giura R. and De Wrachien D., A mathematical model simulating the multilayer system of aquifers of the alluvial plain bounded by the rivers Sesia, Ticino and Po, Proc. of the Int. Conf. on Modern Approach to Groundwater Resources Management, Capri 1982.
9. Delhomme J.P.: Spatial variability and uncertainty in groundwater flow parameters: a geostatistical approach, Water Resources Research, 15(2), 1979, 269-280.
10. Dagan G.: Statistical theory of groundwater flow and transport: pore to laboratory, laboratory to formation, and formation to regional scale, Water Resources Research, 22(9), 1986, 120S-134S.
11. Salandin P. and Rinaldo A.:Numerical experiments on dispersion in heterogeneous porous media, in Computational Methods in Subsurface Hydrology; ed. G. Gambolati, A. Rinaldo, C.A. Brebbia, W.C. Gray & G.F. Pinder, Computational Mechanics Publications and Springer, 1990.

12.Salandin P. and Fiorotto V.: Numerical simulation of non ergodic transport in natural formations, Proc. of the XXV IAHR Congress. Tokyo, Sept. 1993.
13.Guadagnini A., Maione U., Martinoli A. and Tanda M.G.: Sull'applicazione della teoria stocastica del trasporto di inquinanti a singole realizzazioni di mezzi porosi eterogenei. Proc. of the XXIII Convegno di Idraulica e Costruzioni Idrauliche. Firenze. Sept. 1993.
14.Butera I., Grella E., Maione U. and Tanda M.G.: Processi di trasporto in campi eterogenei bidimensionali con condizionamento in trasmissività e carichi. Parte I e II. Proc. of the XXIV Convegno di Idraulica e Costruzioni Idrauliche. Napoli Sept. 1994.
15.Mackay D.M., Freyberg D.L. Roberts I.V. and Cherry J.A. A natural gradient experiment on solute transport in a sand aquifer. Wat. Res.. Vol. 22 n. 13 1986.

HYDRAULIC ISOLATION OF UNCONTROLLED WASTE DISPOSAL SITES

M. Nawalany

Warsaw University of Technology, Warsaw, Poland

ABSTRACT

A notion of hydraulic isolation of uncontrolled waste disposal sites is advocated for the cases when groundwater needs to be protected and all other isolating methods are not possible. Three types of hydraulic isolation that use wells, ditches or drains, respectivelly, are shortly described in the paper. Also the novel idea of geochemical enhancement of the hydraulic isolation is outlined.

1. INTRODUCTION

An occurrence of pollution in an aquifer is a kind of the real life fact which sooner or later calls for the remediation or control action. Though such problem situations are not restricted exclusively to groundwater systems in East Europe, it seems to be a trend there nowadays to clean subsoil that have been polluted by unrestricted industrial activities in the past. As the consequence of this thousands of old uncontrolled/nonisolated dumps are reported to exist in East Europe (though the record of West Europe is not considerably better). Poorly isolated or not isolated at all they endanger groundwater through their physical contact with the underlying aquifers. Transport of pollution that follows does effect detrimentally a quality of groundwater on the local and sometimes on the regional scale. Number of reasons can be listed to explain why despite of difficult economic situation in countries like Poland decision makers attempt to enforce cleaning polluted aquifers or at least introduce some control over pollution. Natural and direct reason is that many municipal intakes or individual wells have been endangered by the pollution being transported by groundwater and therefore this convenient source of potable water may be lost soon. Surprisingly, there is also a considerable psychological factor in the aquifer cleaning trend. Decision makers responsible for environmental protection are more inclined at present to "make some order in the environment" to facilitate and improve the future environmental management. Rapid increase of environmental information in terms of technical means and methods for pollution identification and detection as well as of accessibility to the new technologies that allow to cure (or at least to control) the worse cases makes the decision makers more confident that they can contribute to sustainable development of their own societies. Rapid and considerable changes in environmental law, introduction of modern regulations and procedures make the environmental management more effective. Also strong political position of local communities imposes a pressure for positive changes in environment. Despite of "determination" and "open-minded thinking" that prevails nowadays in the environmental management the economy has its own rights and the lack of money does prevent in many cases to apply most efficient (so expensive) techniques to literally clean pollutes aquifers. What remains are temporal solutions which may stop the most urgent cases from getting worse. One of such solutions is the **active isolation of uncontrolled waste disposal sites**.

When pollutants do percolate from a given waste disposal site (WDS) into groundwater some action is needed that can either annihilate the source of pollution, or isolate the source from the underlying aquifer. Accordingly, the protective actions (measures) can be categorized into two classes :

(a) **Annihilation the source.** Within this class two subclasses are possible

> (a1)   removing the source physically, e.g. by relocating, recycling or utilizing the disposed waste
>
> (a2)   "freezing" the emission, e.g. by changing harmful substances into unharmful ones, by immobilizing the pollutant within the WDS or cutting-off the recharge (roofs, impervious covers etc)

(b)   Isolation of the source. It is assumed within this category that the source is emitting into environment all the time however the flux of pollution does not reach the aquifer. Examples in this category are: drainage systems. encapsulation of the WDS, impervious liners or crack-selfsilting liners and hydraulic isolation.

The latter example does belong to category of active isolation (A.I.). It is based on the principle of forming an artificial local groundwater flow system underneath the WDS. This kind of isolation does allow to control a pollution transport within the underlying aquifer by keeping the pollution confined within some predefined (possibly small) flow domain. Three types of active (hydraulic) isolation is discussed:

- active isolation using (abstraction/recharging) wells

- active isolation using ditches

- active isolation using drains.

An obvious difference between the conventional passive type of isolation and the active isolation of WDS lies in executing technical operations only once (passive) or continuously in time (active). As the consequence **the cost/benefit calculations** and **the assessment of the risk** associated with a use either of the two types od isolation are clearly distinct. Hydraulic isolation can also be enhanced by forming an artificial simpervious layer underneath the WDS. For this geochemical technologies can be applied. Chemical extension of hydraulic isolation is also shortly discussed in the paper.

2. ACTIVE ISOLATION TECHNOLOGY - WELLS

Abstraction wells are commonly used for active isolation of waste disposal sites. Wells are usually located in the close vicinity of the WDS where by abstracting groundwater they withdraw also the WDS-born soluble pollution. However, when a pollution plume is already well-developed in the underlying aquifer abstraction wells are being located far away from the WDS and then are used either for remediation purposes or for deflecting the plume from the main direction of the regional flow thus protecting groundwater intakes or surface waters. Until recently mainly abstraction wells have been used. Their operation, however, often caused undesired side effects like depleting groundwater needed for agriculture, causing land subsidence etc. In the last five years more attention has been turned on making the water balance of the well installations close to zero by using both the abstraction and the recharging wells at the same time.

Wells that are used as a part of the WDS isolating installations should fulfil the following criteria:

-       they ahould be located close to the WDS;

-       they should at least penetrate an aquifer to a depth which is sufficient to capture polluted water from the WDS;

-      by having their water balance equal to zero they should not distort a regional
       groundwater flow;

-      polluted water abstracted from beneath the WDS should be circulated through
       some purification station and after being cleaned it should be returned to the
       aquifer (see Figure 1).

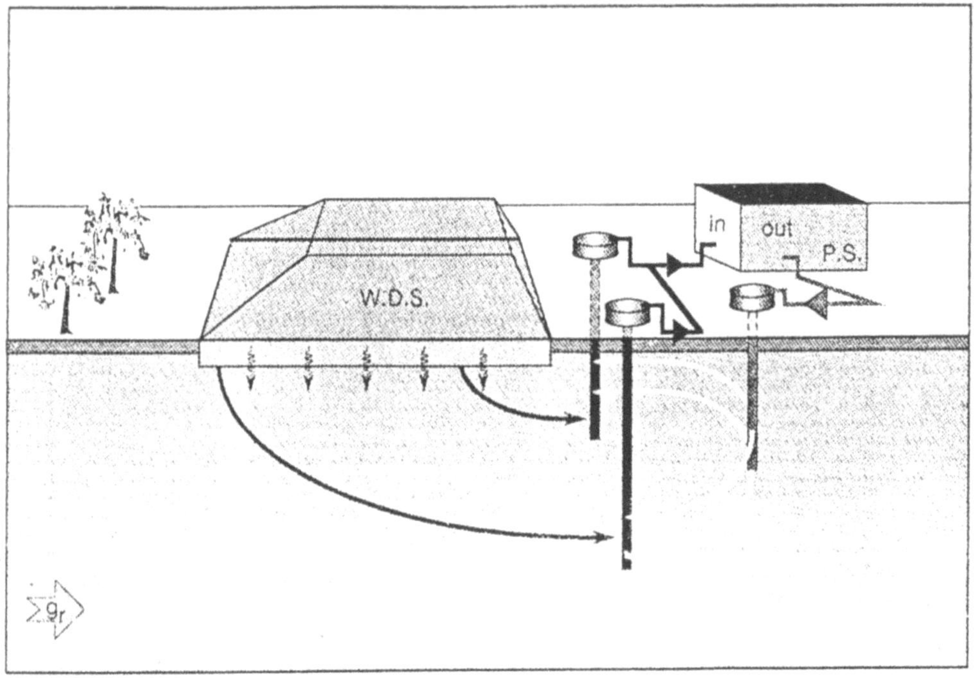

Figure 1. Configuration of the A.I. - Wells.

Hydraulic isolation technology consisting of pumping-recharging wells are usually
analyzed with the use of the classical horizontal two-dimensional model of steady-state
groundwater flow in semi-confined aquifer. Simple geometry, homogeneous parameters
and simple boundary conditions are assumed to simulate a performance of the installation
in terms of capturing the WDS-born trajectories. The same natural conditions for
groundwater flow are retained also for other hydraulic isolation installations discussed in
next chapters. This allows to compare the performances of different technologies.

3. ACTIVE ISOLATION TECHNOLOGY - DITCHES

In case of the uncontrolled or leaking WDS which penetrate a saturation zone of underlying aquifers the application of ditches seems to be the most simple method of hydraulic isolation. Creation of the local groundwater flow system with the A.I. - Ditches does not require an invasion of the subsoil beneath the WDS nor extensive drilling nor complicated pumping installation. Only digging and maintaining ditches is needed and a pump to keep a difference in water heads between the ditches. The whole idea is clearly seen from Figure 2. Naturally, some criteria must be fulfilled when designing the A.I. - Ditches. Below major criteria are listed:

- the ditches should be located along the WDS in such a way that they counteract the regional groundwater flow;
- they should penetrate the aquifer to a depth which is sufficient to create a closed local flow system;
- by having their water balance equal to zero they should not distort a groundwater flow on the regional scale;
- polluted water collected from beneath the WDS by one of the ditches should be circulated through some purification station and after being cleaned it should return to the aquifer through the other ditch.

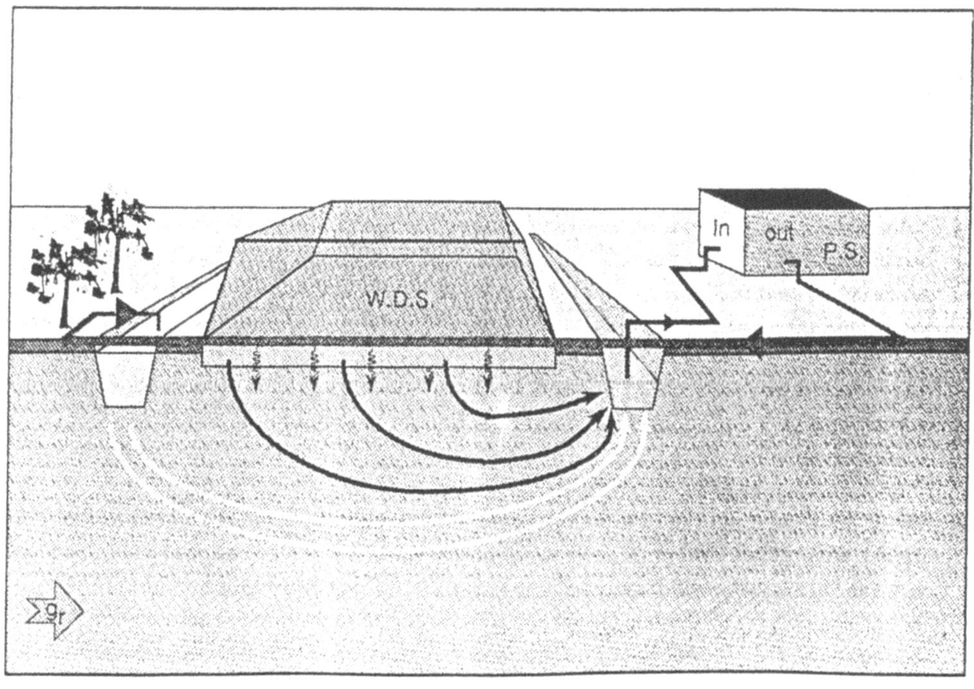

Figure 2. Configuration of the A.I - Ditches.

Hydraulics of ditches is well-krown and numerical calculations for this kind of active isolation is presented in many papers. The advantage of the A.I. - Ditches lies in the simplicity of the installation. The major disadvantages of the ditches is their silting hence lowering their effectiveness in creating a local groundwater system. Silting can be caused by chemical precipitation as well as by biological processes.

4. ACTIVE ISOLATION TECHNOLOGY - DRAINS

Five major premises have made the idea of the A.I. - Drains (see Figure 3) isolating technology feasible, innovative and attractive:

-       existing experience in subterrain irrigation systems used for agricultural purposes for some time (e.g. [1],[2],[3] and [4]);
-       practical and economic feasibility of so-called "trenchless technologies" which can be possibly used to penetrate a groundwater flow domain be neath the WDS without moving the whole mass of collected wastes;
-       simple realization of the zero water mass balance requirement;
-       high controllability of the installation performance as well as of the subsoil environment state with simple measurements;
        potential of combining the technology with other isolating methods.

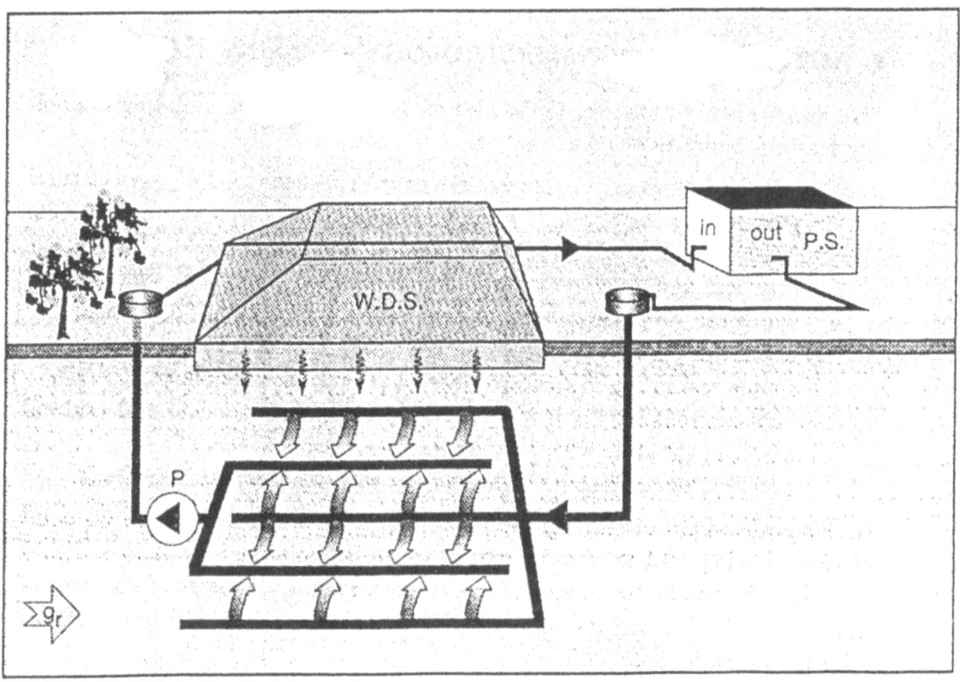

Figure 3. Configuration of the A.I. - Drains.

The A.I - Drains installations should fulfil the following criteria:

- they should be located beneath close to the bottom of the WDS;

- their range should cover the planar projection of the WDS;

- by having their water balance equal to zero they should not distort a regional groundwater flow;

- polluted water abstracted from beneath the WDS should circulate through some purification station and after being cleaned it should be returned to the aquifer.

A performance of a hydraulic isolation technology consisting of abstracting-recharging drains can be analyzed with the use of the analytical (complex variable) vertical two-dimensional model of steady-state groundwater flow in semi-confined aquifer in a presence of drains. Simple geometry, homogeneous parameters and simple boundary conditions are usually assumed for underlying aquifer to simulate the isolation performance in terms of the installation effectiveness in capturing of the WDS-born trajectories. In general case high resolution numerical models are used for purpose [5].

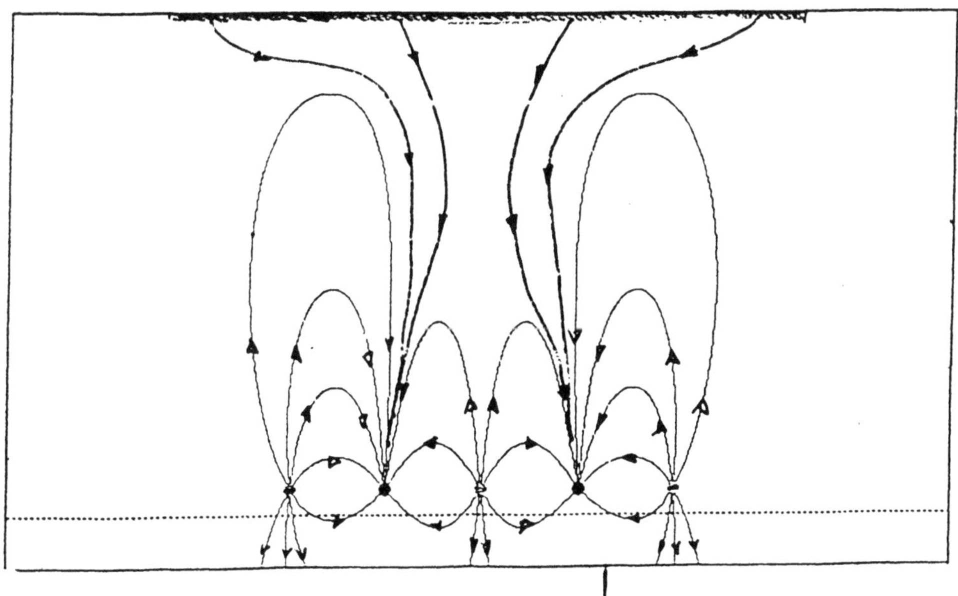

Fig. 4.Flow cells of the local system generated by the A.I.- Drains instalation.

5. CHEMICAL EXTENSION OF HYDRAULIC ISOLATION OF WASTE DISPOSAL SITES.

A combination of the Hydraulic Isolation (HI) of Waste Disposal Sites using active horizontal drains and geochemical isolation technique seems to be very promising method of protecting groundwater. The technique is aimed towards creating an "impervious barrier" using different kinds of clogging reactions that could effectively block pores of the soil along some horizontal zone beneath the WDS. The clogging reactions can be for instance: precipitation of inorganic insoluble salts, polymerization processes or microbial clogging. In any case the agents for the chemical or biochemical reactions need to be delivered to the zone that is designed to be made "impervious". For this the HI - Drains techniques may be used. Mixed geomechanical isolation is going to be realized in two phases: Phase I - chemical agent (A) is pumped into the subsoil through the drains in order to saturate a porous medium in the close vicinity of the drains. Phase II - the second chemical agent (B) is sent to the zone saturated already with the first agent causing an intensive precipitation of product C of the chemical reaction $A + B \rightarrow C \downarrow + D$ thus clogging the soil pores. Consequently, the "impervious" almost horizontal layer (a barrier) is formed beneath the WDS. Forming of the geochemical barrier creates two serious problems: - the problem of controlling a chemical environment within the clogging zone in order to make the clogging reaction possible as well as stable and - the problem of coupling a spatial extent of the clogging reaction with a local flow in porous medium. Flow that transports (delivers) the agents for the clogging reaction(s) is at the same time dependent on the effects of the reaction (on the fraction of the soil pores which are already blocked). Mathematically the problem can be classified as the moving boundary problem. A number of theoretical and practical problems can be envisaged in relation to the chemical extension of hydraulic isolation. Most difficult process to desirable and predict will be an evolution of the time and space extent of the precipitate C. Moving front of the precipitate will cause the reduction of the hydraulic conductivity K around the drains. The reduced hydraulic conductivity caused by clogging pores by the precipite is expected to be lowest along the perimeters of the drains and increasing as the distance from the drain increases. Consequently the "weakest point" of the chemically enhanced hydraulic isolation is located close to the midpoint between two parallel drains. There is a question of how distant each pair of drains should be from each other to guarantee that a reduced hydraulic conductivity in the midpoints will not exceed specified safe value. Also a practical problem will arise how to keep a substance B flowing through the part of the aquifer saturated with a substance A without clogging drains at the very beginning of the process. These problems an at present under investigation in the Institute of Environmental Engineering Systems of Warsaw University of Technology.

6. REMEDIATION

All the hydraulically based isolation techniques discussed until now can be readily used as the remediation techniques. In particular, they can be used for cleaning the aquifer from the pollution plumes having small spatial extent. The latter limitation is the result of the tacit assumption that for creating the local flow system (crucial element of these technologies) only limited (manageable) amount of energy, so money, is needed. If the pollution problem is becoming of regional extent only the A.I. - Wells can be considered feasible though the A.I. - Ditches can be in some cases considered as a candidate.

If the source of groundwater pollution (WDS) is stopped the following three processes (characteristic for all hydraulic isolation techniques) can lead ultimately to cleaning the polluted part of the aquifer:

- removal of polluted water from the soil some hydraulic installation;
- cleaning the water in the purification station;
- reversing the cleaned water to the aquifer.

7. FINAL REMARKS

(i)   The installations described in this paper were compared in terms of their total costs [5]. Although for every combination of soil and WDS parameters A.I. - Wells (with the zero water balance!) are more effective than the A.I. - Drains in terms of costs there is no definite conclusion possible at the moment to indicate them as an always better solution. Further theoretical and experimental investigations of the A.I. - Drains should be undertaken. In particular, the A.I. - Drains installations should be modelled for the unconfined groundwater flow. Especially when the arrangements of the drains different than the parallel one need to be analysed a very refined numerical model needs to be constructed and costly experiments performed.

ii)   As far as the A.I. - Ditches isolation is concerned modeling results indicate that a local groundwater flow system (necessary for isolating the WDS) is created provided a sufficient difference in water heads between the ditches is imposed .

iii)   It can be concluded from the simulations, [5], that for both the A.I. - Wells and the A.I. - Drains relatively small number of hydraulic elements (wells or drains) is needed to control the flow of polluted water beneath the WDS. This statement is again a subject of dependence on the local hydrogeological situation.

iv)   Taking into account sources of uncertainty in the hydraulic isolation of the WDS all the modelling and designing should be subject of robustness/sensitivity evaluation.

v)   A.I. - Drains seems to be atractive alternative for its ability to control flow and transport underneath of the WDS. Also a remediation of the locally polluted equifer can be realized using this technology.

vi) Introduction of artificially created geochemical barriers is also possible using the A.I. - Drains technology.

vii) Modelling the hydraulic isolation of the WDS requires refined numerical methods if analytical model does not exist. High geometrical aspect ratios of the integrated modelling (i.e. when both groundwater flow/transport model must be combined with a model of the installation) cause considerable difficulties in space discretization.

REFERENCES

1. Skaggs, R.W., Water movement factors important to the design and operation of subirrigation systems, Transactions of the ASAE, pp. 1553 - 1561, 1981.

2. Van Bakel, P.I.T., Using drainage system for supplementary irrigation, Irrigation and Drainage Systems, Vol. 2, pp. 125-137, 1988.

3. Walczak, R.T., Van der Ploeg, R.R., and Kirkham, D., An algorithm for the calculation of drain spacing for layered soils, Soil Sci. Soc. Am. J., Vol. 52, pp. 336-340, 1988.

4. Kirkham, D. and Horton, R., The stream function of potential theory for dualpipe subirrigation system, WRR, Vol. 28, No. 2, pp. 373-387, February 1992.

5. Nawalany, M., Loch, J., Sinicyn, S., Active Isolation of Waste Disposal Sites by Hydraulic Means, part II, Models, Report TNO Institute of Applied Geoscience OS 91-42-C, Delft, 1992.

MIX
Papier aus verantwortungsvollen Quellen
Paper from responsible sources
FSC® C105338

If you have any concerns about our products,
you can contact us on
ProductSafety@springernature.com

In case Publisher is established outside the EU,
the EU authorized representative is:
Springer Nature Customer Service Center GmbH
Europaplatz 3, 69115 Heidelberg, Germany

Printed by Libri Plureos GmbH
in Hamburg, Germany